IGNORING NATURE NO MORE

The Case for

COMPASSIONATE CONSERVATION

EDITED BY MARC BEKOFF

THE UNIVERSITY OF CHICAGO PRESS   Chicago & London

MARC BEKOFF is professor emeritus of ecology and evolutionary biology at the University of Colorado Boulder. His numerous books include *The Emotional Lives of Animals*, *The Animal Manifesto*, and *Wild Justice: The Moral Lives of Animals*, the last also published by the University of Chicago Press.

THE UNIVERSITY OF CHICAGO PRESS, *Chicago 60637*

The University of Chicago Press, Ltd., London
© 2013 by The University of Chicago
All rights reserved. Published 2013.
Printed in the United States of America
22 21 20 19 18 17 16 15 14 13   1 2 3 4 5
ISBN-13: 978-0-226-92533-2 (cloth)
ISBN-13: 978-0-226-92535-6 (paper)
ISBN-13: 978-0-226-92536-3 (e-book)

Library of Congress CATALOGING-IN-PUBLICATION DATA

Ignoring nature no more : the case for compassionate conservation / edited by Marc Bekoff.

    pages. cm.

  Includes bibliographical references and index.

   ISBN 978-0-226-92533-2 (cloth : alk. paper) — ISBN 978-0-226-92535-6 (pbk. : alk. paper) — ISBN 978-0-226-92536-3 (e-book)

1. Wildlife conservation—Moral and ethical aspects. I. Bekoff, Marc.

  QL82.I393 2013
  333.95'416—dc23

2013005621

♾ This paper meets the requirements of ANSI/NISO Z39.48-1992 (Permanence of Paper).

I DEDICATE THIS BOOK

to my colleague and friend **JANE GOODALL,**

WHO NEVER IGNORED NATURE

in her groundbreaking and long-term research on chimpanzees

AND IN HER TIRELESS EFFORTS

to make the world a more compassionate and peaceful place

FOR ALL BEINGS

Contents

Foreword

THE TWENTY-SIX COMPELLING ESSAYS in this book differ widely in their descriptions of the nature of the "Nature" that is being ignored (some use a capital N, and some assign a female gender). Indeed, there seems to be more than one "Nature."

One is the product of all of the microbes, plants, animals, chemical elements, lands, waters, processes, and interactions we can observe as inhabitants of the planet Earth (although iconic photos of this Nature were taken from space, none of our authors mentions any other parts of our solar system as Nature). Ignoring this Nature takes the form of seven billion people (going on eleven billion), using and transforming more and more of these natural resources and interfering in more and more of these natural processes and interactions, at faster and faster rates, such that Nature is being irreversibly transformed into simpler and less beautiful forms that eventually may not support life as we know it. "Conservation" is attending to this "Nature."

Another Nature of these essays is the unity of human animals and all other animals. We are said to be so much alike in so many ways that we should not be seen as dualistically separate, and humans should not be seen as privileged or morally superior. Ignoring this Nature takes the form of needless exploitation of and even cruelty toward other animals by humans. There is no consensus on exactly what is "needless" exploitation. "Protectionism" or "Compassion" is attending to this "Nature."

Marc Bekoff has been exploring the intersections of conservation and protectionism for decades, and here has mounted an expedition of thinkers. Some argue that attending to protectionism facilitates attending to conservation (hence the double-barreled "compassionate conservation"). Not all agree, but the discussion reminded me that it would be productive for us all to focus on the many issues on which conservationists and protectionists agree, for example, elimination of unsustainable commercial hunting of wild animals and of subsidized domestic animal confinement husbandry in affluent countries.

The essayists review, invent (sometimes reinvent), and justify dozens of ways to get us to attend to (stop ignoring) these Natures. Readers can sample ethical, educational, religious, scientific, regulatory, nonviolent revolutionary, and economic interventions. Unfortunately this book and its recommendations will never reach most of the billions of hard-working, hungry, aspirational, short-sighted, politically fearful, bellicose, denying, or arrogant Earthlings who are ignoring these Natures. But one of these days, Mother Nature, pushed beyond all sustainable limits, will suddenly and cataclysmically grab our attention (see Paul Gilding, *The Great Disruption* [New York: Bloomsbury Press, 2011]). We may survive as a species, as might a large part of Nature, but life will change as we know it. We will all wish then that we had read and heeded *Ignoring Nature No More.*

BENJAMIN B. BECK

Preface

Who Lives, Who Dies, and Why

It Shouldn't Be All about Us

If you don't know how to fix it, please stop breaking it. . . . You are what you do, not what you say. What you do makes me cry at night. . . . Please make your actions reflect your words. [Severn Suzuki 1992]

THESE WONDERFUL WORDS were spoken by twelve-year-old Severn Suzuki who, along with two other youngsters, all members of the Environmental Children's Organization (ECO), paid their own way to the United Nations Earth Summit in Rio de Janeiro in 1992. Severn started ECO when she was nine. Whenever I watch this speech, given at the close of a plenary session, tears of joy and deep concern come to my eyes. Severn Suzuki is right. We act as if there will be no future generations who will be inheriting the innumerable messes we leave behind.

Ignoring Nature No More raises numerous questions about who we are and what we have done to other animals and ecosystems. It is very much about the question "Who lives, who dies, and why?" We live in a troubled and wounded world that is in dire need of healing. We all should be deeply concerned and perhaps terrified by what we have done and continue to do. We humans are an arrogant lot and we have made huge and horrific global messes that need to be repaired now. Losses of biodiversity are stinging and many are irreversible. The overriding sense of turmoil is apparent to anyone who takes the time to pay attention. Researchers and nonresearchers alike are extremely concerned about unprecedented global losses of biodiversity and how we humans suffer because of our destructive ways. We are animals and we should be proud and aware of our membership in the animal kingdom. Our unique contribution to the wanton decimation of the planet and its many life forms is an insult to other animal beings and also de-means us.

We're Running Out of World: What Must We Do

Humans are a force in nature. "Tell me something I don't know," I hear you say. We're all over the place, big-brained, big-footed, arrogant, invasive, menacing, and marauding mammals. No need to look for mythical Bigfoot: we're here! We leave huge footprints all over the place and have been rather unsuccessful at solving urgent problems. We're simply "running out of world" (Berry 2003, 131). Perhaps we've already run out of world, wild, wildness, and wilderness. Some go so far as to argue we've created a world that's so technologically and socially complex we can't control it (Allenby and Sarewitz 2011).

Our omnipresence and power call for more humility and responsibility, not license to do anything we want. We need to be small footed and big hearted and let each step on earth nourish us. Simply put, there are too many of us and we overconsume in the most selfish and unjust ways, influencing both nonhumans and humans (Mazur 2009; McKibben 2010). Indeed, the major cause of mortality in North American populations of large- and medium-sized mammals is humans (51.8 percent; Collins and Kays 2011). We cause the most deaths in larger North American mammals, including those living in *protected* areas (34.6 percent). To date there has only been one observation of a nonhuman overhunting (Lwanga et al. 2011). Chimpanzees in Uganda's Kibale National Park work together to catch prey, typically red colobus monkeys. Between 1975 and 2007 there was an 89 percent decline in the red colobus population while the chimpanzee population rose about 83 percent.

We're also extremely skilled at denying what is happening right in front of our eyes (Specter 2009; Norgaard 2011), including what we know about the amazing cognitive, emotional, and moral capacities of animals and the pain and suffering they endure (Wicks 2011). As *Homo denialus* we readily "see no evil, hear no evil, or smell no evil," although if we open our senses even a little bit to our surroundings it's impossible to ignore or deny what's happening and the dire consequences of our actions. We can ignore or deny the noise, but not the silence. We ignore and redecorate nature (Bekoff 2006) in incredibly self-serving ways, as if we're the only species who matters. It should not be all about us, but it often is. Nonetheless, I truly believe there is hope (Bekoff 2007a, 2010a) and that the essays included here will help us along. The animals are constantly telling us what they want and need. Their manifesto, simply put, is treat us better or leave us alone (Bekoff 2010a). As we save other animals, we'll be saving ourselves.

Why Do We Ignore Nature and What Can Be Done about It? Ecocide Is a Form of Suicide

When did we begin ignoring nature (Bekoff and Bexell 2010)? Why did we start ignoring our need for untainted and healthy food, clean water, clean air, and reasonable shelter? How did we become so disconnected from nature and an understanding of basic ecological processes? What allows us to tolerate human-induced losses in biodiversity? What can we do about the distance and alienation from nature and other animals that allows us to be so destructive? It's difficult to isolate one or even a few factors as we ponder a number of questions including why we ignore nature, who we are in the grand scheme of things, and what we're doing to animals, plants, entire ecosystems, and precious, unimaginably beautiful, and fragile webs of nature of which we are a major component although we often act as if this isn't the case.

Clearly, in the current hustle-and-bustle-world, it's easy to ignore nature as we run here and there not even knowing why we're doing what we're doing. Being removed from our own daily lives makes it highly unlikely we'll have any meaningful appreciation of our destructive ways and the wide-ranging consequences of them. In addition to ignoring nature because it's convenient to do so, we also live in oblivion and removed from directly knowing and feeling what we're doing because we are so detached from other animals and other nature.

People also are influenced by their failure to recognize that what they do *does* make a difference. Doomsday thinking that there's nothing we can do to turn the tide; a lack of long-term goals, and visions that fail to recognize that future generations really do rely on our goodwill and efforts that we ourselves might not live to see; the need for immediate gratification; outright laziness; hierarchical speciesistic thinking that we're "higher, better, or more valuable" than other animals—a psychological barrier justifying human superiority and exceptionalism, religious proclivities, economic contingencies (it's too expensive to care and make changes), political alignments, and dogma—all factor into negativity. Humans are not "better" or "higher" than individuals of other species. We share many similarities, but we're also different, but different doesn't mean better.

Regardless of why, much of our abuse comes down to the blatant fact that we do indeed ignore nature as if we are immune to both the incredible destruction for which we're responsible and the good things that result from our concern about our destructive ways. Much of it also comes down to rampant anthropocentrism. No matter what we otherwise claim, most of the time "it's all about us." Frankly, it's enough to make one sick, sad, depressed, and unmotivated to do anything. But this is the wrong response, for we *must* do something and must do it now, not when it's convenient, for that time rarely if ever comes. As Zoe Weil (2011) says, we must be "solutionaries." We need to love what we're doing and love the planet for which we're working. We must get out and interact with nature and embrace her beauty and magnificence.

It's easy to deny what's happening or to let others do something about it because "I'm too busy or I'm not at fault." Passing the buck is easy, and scientists often do it so they can avoid thinking about the ethics of their actions or inactions (Wolpe 2006). The fact is each and every one of us is at fault. We are individually and collectively responsible for destroying our one and only Earth, and most people don't seem to give a damn. Or we say we do, but don't do much if anything to change our destructive ways.

What This Book Is All About

This eclectic and forward-looking book was conceived in the spirit of broad and open inquiry. The contributors are experts in their diverse and respective fields and in their essays provide refreshingly new, and in some cases controversial, light. I chose to be an editor, not a censor, in that we need bold and forward-looking ideas and action as we move ahead to heal our wounded world. Each essay deals with a different aspect of the innumerable areas concerned with what's called the human dimension (Evernden 1999; Manfredo et al. 2009; Wilson 2012) and human-animal connections (anthrozoology: http://www.anthrozoology.org/; Bekoff 2007b), and each could stand on its own. Each essay also shows clearly how we need to get back in touch with who we are and the complexity of nature as a whole so we can change our wounding ways. The reference sections alone are invaluable resources not only for information about past research but also for future endeavors. The diverse field of human-animal-environmental relationships is burgeoning and attracts scholars from numerous disciplines. It seems as if there are new articles or books appearing almost daily (see for example Ehrlich 1997; Hilty, Lidicker, and Merenlender 2006; Hawken 2007; Meyer 2007; Sobel 2008; Anderson 2009; Clayton and Myers 2009; Crompton and Kasser 2009; Ehrlich and Ehrlich 2009; Ehrlich and Ornstein 2010; Blumstein and Fernández-Juricic 2010; Cole and Yung 2010; Herzog 2010; Hribal 2010; Koger and Winter 2010; Cooney 2011; Mc-Cardle 2011a, b; Ryan 2011; Wilson 2012).

Some of the basic ideas behind this volume were outlined in an essay that called attention to the need for interdisciplinary input to allay the incredible and unprecedented losses of biodiversity that we're experiencing (Bekoff and Bexell 2010) in what's called the "Anthropocene" (the "age of man"), the current geological period of time when our actions have had enormous and unprecedented global consequences (http://en.wikipedia.org/wiki/Anthropocene). It's rarely a lack of knowledge and concrete data that results in missed opportunities to preserve biodiversity or that informs decisions about who lives and who dies, including animals and ecosystems (Stolzenburg 2008, 2011; Blumstein and Fernández-Juricic 2010; Saylan and Blumstein 2011; Bekoff in press). Rather, losses are typically due to problems of human psychology and social and cultural factors that result in the inadequate protection of animals and their habitats. Science alone doesn't hold the answers to the current crisis, nor does it get people to act. As historian Lynn White (1967) wrote in his classic essay "The Historical Roots of Our Ecological Crisis," "More science and more technology are not going to get us out of the present ecological crisis until we find a new religion, or rethink our old one." More than four decades later this claim still holds: we don't need more science to know we need a new mindset and social movement that center on empathy, compassion, and being proactive. The "putting out the fires" mentality hasn't and can't work, and is not sustainable. Along these lines, Michael Shermer writes, "The majority of our most deeply held beliefs are immune to attack by direct educational tools, especially for those who are not ready to hear contradictory evidence. Belief change comes from a combination of personal psychological readiness and [a] deeper social and cultural shift in the underlying zeitgeist, which is affected in part by education but is more the product of larger and harder-to-define political, economic, religious, and social changes" (Shermer 2011, 4; see also Clayton and Myers 2009; Cooney 2011; Bekoff in press). Joe Zammit-Lucia (2011) says it poignantly, "Conservation is all about people" (see also Schulz 2011).

Of course, science and the results of empirical inquiries are important, but despite all we know, environmental education has largely failed (Saylan and Blumstein 2011). Workable solutions to current and future problems depend on people from different disciplines first knowing about their common interests and then talking *with* one another, not *at* one another, and respecting one another's contributions. All the science in the world won't help us along if we don't get the results out to the masses and we don't learn about how human attitudes are formed, maintained, and changed for the better; how we can get humans to do something to right the wrongs; and how we can have a better understanding of what we can and *must* do. Biologists of many different stripes, psychologists, sociologists, social workers, economists, political scientists, religious scholars, people working in journalism and media, and philosophers have much to say on the notion of ignoring nature and what we can and must do to remedy this practice. Cultural differences also need to be factored in. So it was in this spirit of wide-ranging inquiry that this book was envisioned and realized.

As I was writing these introductory comments an issue of *New Scientist* arrived in which there was an excellent essay called "Unnatural Selection," informing readers "humans have become the biggest force in evolution." I wonder, though, if it's really *unnatural* selection because we're an integral part of nature and it doesn't look as if we're going to disappear any time soon. Michael Le Page writes: "The Zoque people of Mexico hold a ceremony every year during which they grind up a poisonous plant and pour the mixture into a river running through a cave. . . . Any of the river's molly fish that float to the surface are seen as a gift from the gods. The gods seem to be on the side of the fish, though—the fish in the poisoned parts of the river are becoming resistant to the plant's active ingredient, rotenone. If fish can evolve in response to a small religious ceremony, just imagine the effects of all the other changes we are making to the planet." He goes on to note, "We are turning grassland and forests into fields and cities, while polluting the air and water. We are hunting species to the brink of extinction and beyond, as well as introducing new pests and diseases to just about every part of the world. And that's not to mention drastically altering the climate of the entire planet. It is no secret that many—perhaps even most—species living today are likely to be wiped out over the next century or two as a result of this multiple onslaught. What is now becoming clear is that few of the species that survive will live on unchanged" (Le Page 2011, 32–37, http://www.newscientist.com/special/unnatural_selection). We're unrelentingly reminded daily of our destructive ways, both in popular media and scientific literature.

To make his case crystal clear Le Page considers topics including predation by humans and how we go for the biggest and the best animals, climate change, pollution and pollution resistance that crosses species, disease, pests becoming immune to poisons, and invasive species. Contributors to the collection of essays herein also consider some of these topics, and considerably more. We're not only making a difference in places we live. Climate change seems to be the cause for the loss of populations of Adélie and chinstrap penguins across the west Antarctic peninsula (Strain 2011). Our effects are wide ranging and far reaching and we most likely understate the extent of our destructive ways. As some of the essays here point out, sometimes we really don't know what we've done or how to fix what we think we did.

Leaving Our Comfort Zone:
Does Caring and Compassion Mean Letting Go?

The sum total of human behavior translates into nothing less than an epoch-making disaster for global biodiversity, habitat, human rights, the empowerment of women and children, and the full potential of any democracy. That's why rich nations like the

U.S. need to lead the way in showing compassion through human (particularly wom-
ens' [*sic*] rights), through the smart and generous use of green technology, like that
employed in contraceptive options and green legislation. [Tobias 2011]

We all agree our troubled and wounded world needs a lot of compassionate
healing, right now, not when it's convenient. There is a compelling sense of
urgency. We live in a messy, complicated, frustrating, demanding world, and
it's impossible to do the "right" thing all of the time. Compassion is the glue
that holds ecosystems and webs of nature together. We're an integral part of
many beautiful awe-inspiring webs of nature and webs of consumption, and we
all suffer when these complex interrelationships are compromised. We should
work for the planet because we belong to it despite our imagining and acting
as if we stand apart and above nature as natural aliens (Evernden 1999). Pat
Shipman (2011b; see also Shipman 2011a) asks poignantly, "If our species was
born of a world rich with animals, can we flourish in one where biodiversity
is decimated?"

We're all over the place and it's arrogant to think we can pick and choose
where we have impact, for we have impact everywhere. Many different animals
display compassion for others, so it's natural to call upon compassion to alleviate
the suffering of others. It is natural and "animal" to be good, compassionate,
empathic, and moral (Bekoff and Pierce 2009; Keltner 2009; Rifkin 2009;
Bekoff 2010a, 2012; de Waal 2010; Corning 2011) and "more science" will not
necessarily make us better or compassionate (Bekoff and Bexell 2010; Shermer
2011; http://motherjones.com/politics/2011/03/denial-science-chris-mooney).
It's who we are. Wayne Pacelle (2011), president and CEO of the Humane Society of the United States, argues that we need a "humane economy," one that
makes compassion and caring profitable. This will surely make coexistence
more likely. So be it.

Redecorating Nature, Rewilding Our Hearts, and Expanding Our Compassion Footprint

We are continually and inevitably redecorating nature either intentionally or
unintentionally by building homes, shopping centers, and parking lots and also
by decimating species or moving them from one place to another. Redecorating
nature refers to our global tendency, almost a human obsession, to move into
the living rooms of other animals with little or no regard for what we're doing
to them, their friends, and their families. We unrelentingly intrude because
there are too many of us and because it's so easy. The year 2011 saw the world's
seven billionth person and there will be around nine billion by 2050. Because
our population now is still growing and there are more mouths to fill and people

to home, we need to do something with deep internal effects, within ourselves, to prevent even more from being born.

Our big brains should caution us that we can't go on living as we have, but something doesn't seem to click because we're so resistant to changing our ways on a scale that would really make a difference. People are often shy about talking about having fewer babies, but let's face it, there are far too many of us and we don't take lessons from other animals: we need to live within our means. Part of the rewilding strategy (Bekoff in press) is for us to stop making so many of us. An insane and unsustainable reproductive strategy clearly spells doom, and the rampant collateral damage and ripple effects of more and more of us puts us right in the center of blame for what's happening in what's rightfully called the Anthropocene, the period characterized by widespread and egregious human decimation of our one and only planet. Melanie Challenger rightly points out in her book *On Extinction: How We Became Estranged from Nature* that we're not only losing species, but languages, cultures, and ways of life are disappearing as well. As I conduct children's events all around the world I often wonder what the world will look like in the future for someone who's ten years old now? The sobering thought keeps me up at night.

It's difficult to live in the demanding world and not occasionally do unintentional harm. It's the intentional redecorating that needs to be confronted head on before we do so much irreversible harm that nothing will really work to make the world a better place for all beings, human and other-than-human. Redecorating is often a lack of compassion that includes our many serious impacts on diverse ecosystems—building homes in the living rooms of other animals, building highways where they travel, or decimating habitats and not only killing individuals but breaking up closely knit families. If we redecorate nature, and we surely will, we must do it with compassion and caring for the animals we are displacing.

We need to add more compassion to the world and expand our compassion footprint to include individuals of all species, including ourselves (Bekoff 2010a, in press). There will always be trade-offs in deciding who and what to save (Aitken 2004; Macdonald and Service 2007; Bekoff 2010b; Duffy 2010; Leader-Williams, Adams, and Smith 2010; Moore and Nelson 2010; Russell 2010; Brakes and Simmonds 2011; Stolzenburg 2011), and the decisions that need to be made are incredibly difficult and frustrating and require each of us to think deeply so we do the best we can.

We need to ask some hard questions, whether we like it or not. These include: Should we kill in the name of conservation or research? Should individuals be traded off for the good of their own or other species or for ecosystem integrity? Can we really put a price on animals to reflect their value (Vucetich and Nelson 2007; Butler 2010)? Are we trying to do too much? Should we focus our efforts on what seem to be the more soluble problems and realize that we can't do everything? When do we pull the plug and admit defeat? Is it more compassionate to let some species go rather than compromise our efforts because we're spread too thin? How can we rewild our hearts and bring animals back into the picture (Bekoff in press)? How can we build and maintain clear and unobstructed corridors of compassion and coexistence? How can we cross disciplines and work together? How can we walk the talk so that something actually gets done? Would the world be better without us (Weisman 2007)?

So the real question is, what are we going to do with more heartfelt healing concern? Where should we focus heartfelt concern as we try to rewild our hearts (Bekoff in press), rewild urban areas (Beatley 2010; Stefanovic and Scharper 2012), rewild the wild (Foreman 2004; Fraser 2009), and build corridors of coexistence and compassion? Rewilding on any scale, ranging from wide-ranging landscapes to the deeply personal, will mean that we must never ignore nature.

First Do No Harm:
The Case for Compassionate Conservation and Justice for All

If all mankind were to disappear, the world would regenerate back to the rich state of equilibrium that existed ten thousand years ago. If insects were to vanish, the environment would collapse into chaos. [E. O. Wilson n.d.]

Where to begin is a challenge, but perhaps a moral imperative that can ground our collective efforts is "First do no harm" (Bekoff 2010b). In his book *Respect for Nature: A Theory of Environmental Ethics* (1986), one that should be required reading for everyone working in the conservation sciences, philosopher Paul Taylor proposes four broad rules for how we should interact with animals and other elements of nature (172–92). These duties are the rule of nonmaleficence, the rule of noninterference, the rule of fidelity, and the rule of restitutive justice. Taken together Taylor presents compelling arguments that we must take seriously the lives of individual animals. We must be committed to looking for the most humane solutions to the problems at hand, and we have to make compelling and forceful arguments to override these strictures. The interests of all stakeholders must be taken into account, and a "hands off" should always be considered in our deliberations of what is the "best" thing to do.

My colleague Jessica Pierce stresses we need "ethics with teeth." In this vein, some trade-offs, for example, breeding and raising golden hamsters solely to be practice prey for endangered black-footed ferrets (Bekoff 2010b), are not permissible. Until different parties agree, a project might have to be put on hold. We need to decide on what can be compromised, if anything. What values can be tweaked for the common good? The real world demands that we come to terms with incredibly difficult and challenging situations, most of which we've created.

Compassionate conservation has become a buzz phrase, and there are global efforts to support this field (Bekoff 2010b). I had the good fortune to be at the first two international meetings devoted to this topic, the first at the University of Oxford in early September 2010 (Compassionate Conservation 2010) and the second, a special session at the 2011 meeting of Asia for Animals, held in Chengdu, China. Compassionate conservation is a new mindset and social movement that translates discussions and concerns about the well-being of individuals into action. For some it's not strong enough on animal protection but it is a good beginning for getting individual well-being into conversations about a wide variety of conservation projects. Indeed, it's impossible to argue against being compassionate, as a social value, as we try to right many of the wrongs for which we're responsible. Perhaps some of what we take to be gospel needs to be reexamined. Perhaps the answers we come up with are counterintuitive even if we think we have animals' best interests in mind and heart. Thus we need to reconsider big questions such as: Should we try to re-create or restore ecosystems when we really can't do this? Should we be concerned with losses in biodiversity? Is "more" really "better"? Who are we really trying to benefit? Is a hands-off strategy more compassionate than trying to fix all that's wrong? If we accept that we're an integral part of nature, "just one of the group" as ecopsychologist David Abram (2010) says, doing very good and very bad things, should we just let ecosystems and their animal inhabitants evolve however evolution goes?

We also need to ask if we should try to stop possible extinctions when little is done to protect individual animals or to preserve their habitat because of excessive costs. And we also need to consider the species we *choose* to save, most often highly charismatic species, often with little remaining ecological value and sometimes too far gone to have a chance of rebounding to viable population numbers without massive ongoing human efforts. But other less-notables also are dwindling but are no less important to the integrity of ecosystems. Consider honeybees. These amazing animals are responsible for pollinating seven of the world's ten most important crops, but a single bee can contain twenty-five different agrochemicals (Eyres 2011). Should we let wolves and other charismatic species go and save ants? After all is said and done, we're not even very good at protecting charismatic species (Bennett 2011).

A lot of time, energy, and brain power go into trying to right wrongs, so maybe we need to be more selective in what we choose to fix and let other things just go as they may. We might be more successful if we put all our effort into situations that are fixable rather than doing less by trying to do more. Maybe we just cannot do it all and by trying to do everything we're ignoring nature. Like natural selection, some make it and some don't. More and more researchers are taking the view that we can't fix everything and that we need to make choices about who to save and who to let go. In an Internet survey of 583 conservation scientists questioned, 60 percent agreed that criteria should be established for deciding which species to abandon in order to focus on saving others (Rudd 2011). Ninety-nine percent agreed that a serious loss of biodiversity is "likely, very likely, or virtually certain" (ibid.). I wish this weren't so, and I'm sure everyone would agree, but with limited time, money, and person-power it is. Current strategies, sadly, were developed at a time when they may have been more effective if our population had started tailing off and if our levels of global consumption hadn't skyrocketed.

Whether we agree or disagree with some of the essays in this volume or the arguments of others isn't the issue at hand. What is important is that we consider the arguments, accept or reject them, do something about our decisions, and then move out of our comfort zone, think out of the box, and realize that there really is true urgency in this time of crisis. If we allow the proverbial canaries in the coal mine to tell us something's wrong, I'm afraid it'll be too late to do much. We will already have been lamenting "where have all the animals gone," so the early warning systems to which we paid little or no attention will have disappeared right before our eyes and ears. Silence will be stifling. We should have listened to the canary who was warning us that we waited too long to make any meaningful difference. We've been living in an ethical cesspool and drowning in our own filth.

Please read on. Many of the essays could have found a home in a number of different parts, and it was a challenge to figure out where some should be placed. I hope the best-fit approach I used serves them and the topics well. My brief introductions provide a roadmap for each section. The overlap among people writing from different perspectives is indeed exciting and calls for a strong community effort to work on the problems at hand. Taken together they make a strong case for why we must ignore nature no more, move out of our mindset of domination and exploitation, and make compassionate conservation the only acceptable road on which to travel into the future. No one can be against adding more compassion to our troubled world.

References

Abram, David. 2010. *Becoming Animal: An Earthly Cosmology*. New York: Pantheon.
Aitken, Gill. 2004. *A New Approach to Conservation: The Importance of the Individual through Wildlife Rehabilitation*. Burlington, VT: Ashgate.

Allenby, Braden R., and Daniel Sarewicz. 2011. "Out of Control: How to Live in an Unfathomable World." *New Scientist* (May 17): 28–29. http://www.newscientist.com/article/mg21028127.100 -out-of-control-how-to-live-in-an-unfathomable-world.html.

Anderson, Alun, 2009. *After the Ice: Life, Death, and Geopolitics in the New Arctic.* New York: HarperCollins.

Beatley, Timothy. 2010. *Biophilic Cities: Integrating Nature into Urban Design and Planning.* Washington, DC: Island Press.

Bekoff, Marc. 2006. *Animal Passions and Beastly Virtues: Reflection on Redecorating Nature.* Philadelphia: Temple University Press.

———. 2007a. *The Emotional Lives of Animals.* Novato, CA: New World Library.

———, ed. 2007b. *Encyclopedia of Human-Animal Relationships: A Global Exploration of Our Connections with Animals.* Westport, CT: Greenwood Press.

———. 2010a. *The Animal Manifesto: Six Reasons for Expanding Our Compassion Footprint.* Novato, CA: New World Library.

———. 2010b. "Conservation and Compassion: First Do No Harm." *New Scientist* (September 1): 24–25. http://www.newscientist.com/article/mg20727750.100-conservation-and-compassion -first-do-no-harm.html.

———. 2012. "Cooperation and the Evolution of Social Living: Beyond the Constraints and Implications of Misleading Dogmas." In *Cooperation and Altruism,* edited by R. Sussman and C. R. Cloninger, 111–19. New York: Springer.

———. In press. *Rewilding Our Hearts.* Novato, CA: New World Library.

Bekoff, Marc, and Sarah M. Bexell. 2010. "Ignoring Nature: Why We Do It, the Dire Consequences, and the Need for a Paradigm Shift to Save Animals and Habitats and to Redeem Ourselves." *Human Ecology Review* 17: 70–74.

Bekoff, Marc, and Jessica Pierce. 2009. *Wild Justice: The Moral Lives of Animals.* Chicago: University of Chicago Press.

Bennett, Elizabeth L. 2011. "Another Inconvenient Truth: The Failure of Enforcement Systems to Save Charismatic Species." *Oryx* 45: 476–79.

Berry, Robert J. 2003. *God's Book of Works.* London: Continuum.

Blumstein, Daniel T., and Esteban Fernández-Juricic. 2010. *A Primer of Conservation Behavior.* Sunderland, MA: Sinauer Associates.

Brakes, Philippa, and Mark Peter Simmonds, eds. 2011. *Whales and Dolphins: Cognition, Culture, Conservation, and Human Perceptions.* London: Earthscan.

Butler, Rob. 2010. "What Is the Value of the Polar Bear or Any Wildlife?" *Vancouver Sun,* December 29. http://communities.canada.com/vancouversun/blogs/birdwatch/archive/2010/12/20/ what-is-the-value-of-the-polar-bear-or-any-wildlife.aspx.

Challenger, Melanie. 2011. *On Extinction: How We Became Estranged from Nature.* London: Granta.

Clayton, Susan, and Gene Myers. 2009. *Conservation Psychology: Understanding and Promoting Human Care for Nature.* Hoboken, NJ: Wiley-Blackwell.

Cole, David N., and Laurie Yung, eds. 2010. *Beyond Naturalness: Rethinking Park and Wilderness Stewardship in an Era of Rapid Change.* Washington, DC: Island Press.

Collins, Christopher, and Roland Kays. 2011. "Causes of Mortality in North American Populations of Large and Medium-Sized Mammals." *Animal Conservation* 14: 1–10. http://onlinelibrary.wiley .com/doi/10.1111/j.1469-1795.2011.00458.x/full; http://www.bbc.co.uk/blogs/wondermonkey/ 2011/04/unnatural-selection-what-is-ki.shtml.

Compassionate Conservation. 2010. http://www.bornfree.org.uk/comp/compconsymp2010.html.

Cooney, Nick. 2011. *Change of Heart: What Psychology Can Teach Us about Spreading Social Change.* New York: Lantern Books.

Corning, Peter. 2011. *The Fair Society: The Science of Human Behavior and the Pursuit of Social Justice.* Chicago: University of Chicago Press.

Crompton, Tom, and Tim Kasser. 2009. *Meeting Environmental Challenges: The Role of Human Identity.* Surrey, Devon, UK: WWF-UK.

Duffy, Rosaleen. 2010. *Nature Crime: How We're Getting Conservation Wrong*. New Haven, CT: Yale University Press.

Ehrlich, P. 1997. *A World of Wounds: Ecologists and the Human Dilemma*. Oldendorf/Luhe, Germany: Ecology Institute.

Ehrlich, Paul R., and Anne H. Ehrlich. 2009. *The Dominant Animal: Human Evolution and the Environment*. Washington, DC: Island Press.

Ehrlich, Paul R., and Robert Ornstein. 2010. *Humanity on a Tightrope: Thoughts on Empathy, Family, and Big Changes for a Viable Future*. New York: Rowman and Littlefield.

Evernden, Neil. 1999. *The Natural Alien: Humankind and Environment*. 2nd ed. Toronto: University of Toronto Press.

Eyres, Harry. 2011. "The Birds and the Bees." http://www.ft.com/cms/s/2/6534f050-6ba0-11e0-93f8-00144feab49a.html.

Foreman, Dave. 2004. *Rewilding North America: A Vision for Conservation in the 21st Century*. Washington, DC: Island Press.

Fraser, Carolyn. 2009. *Rewilding the World: Dispatches from the Conservation Revolution*. New York: Henry Holt.

Hawken, Paul. 2007. *Blessed Unrest: How the Largest Movement in the World Came into Being and Why No One Saw It Coming*. New York: Viking.

Herzog, Hal. 2010. *Some We Love, Some We Hate, Some We Eat: Why It's So Hard to Think Straight about Animals*. New York: Harper.

Hilty, Jodi A., William Z. Lidicker Jr., and Adina M. Merenlender. 2006. *Corridor Ecology: The Science and Practice of Linking Landscapes for Biodiversity Conservation*. Washington, DC: Island Press.

Hribal, Jason. 2010. *Fear of the Animal Planet: The Hidden History of Animal Resistance*. Oakland, CA: AK Press.

Keltner, Dacher. 2009. *Born to Be Good: The Science of a Meaningful Life*. New York: W. W. Norton.

Koger, Susan M., and Deborah Du Nann Winter, eds. 2010. *The Psychology of Environmental Problems*. 3rd ed. New York: Psychology Press.

Leader-Williams, Nigel, William M. Adams, and Robert J. Smith, eds. 2010. *Trade-Offs in Conservation: Deciding What to Save*. Oxford: Blackwell.

Le Page, Michael. 2011. "Unnatural Selection." *New Scientist* (April 30): 32–37. http://www.newscientist.com/special/unnatural_selection.

Lwanga, Jeremiah S., Thomas T. Struhsaker, P. J. Struhsaker, T. M Butynski, and J. C Mitani. 2011. "Primate Population Dynamics over 32.9 Years at Ngogo, Kibale National Park, Uganda." *American Journal of Primatology* 73: 1–15. DOI: 10.1002/ajp.20965. See also http://www.newscientist.com/article/dn20469-chimps-hunt-monkey-prey-close-to-local-extinction.html.

Macdonald, David, and Katrina Service, eds. 2007. *Key Topics in Conservation Biology*. Malden, MA: Blackwell.

Manfredo, Michael J., Jerry J. Vaske, Perry J. Brown, Daniel J. Decker, and Esther A. Duke, eds. 2009. *Wildlife and Society: The Science of Human Dimensions*. Washington, DC: Island Press.

Mazur, Laurie, ed. 2009. *A Pivotal Moment: Population, Justice, and the Environmental Challenge*. Washington, DC: Island Press.

McCardle, Peggy, Sandra McCune, James A. Griffin, and Valerie Maholmes, eds. 2011a. *How Animals Affect Us: Examining the Influence of Human-Animal Interaction on Child Development and Human Health*. Washington, DC: American Psychological Association

McCardle, Peggy, Sandra McCune, James A. Griffin, Layla Esposito, and Lisa S. Freund, eds. 2011b. *Animals in Our Lives: Human-Animal Interaction in Family, Community, and Therapeutic Settings*. Baltimore: Paul H. Brookes.

McKibben, Bill. 2010. *Earth: Making a Life on a Tough Planet*. New York: Henry Holt.

Meyer, David S. 2007. *The Politics of Protest: Social Movements in America*. New York: Oxford University Press.

Moore, Kathleen Dean, and Michael Nelson, eds. 2010. *Moral Ground: Ethical Action for a Planet in Peril*. San Antonio, TX: Trinity University Press.

Norgaard, Kari Marie. 2011. *Living in Denial: Climate Change, Emotions, and Everyday Life*. Cambridge, MA: MIT Press.

Pacelle, Wayne. 2011. *The Bond: Our Kinship with Animals, Our Call to Defend Them*. New York: William Morrow.

Rifkin, Jeremy. 2009. *The Empathic Generation: The Race to Global Consciousness in a World in Crisis*. New York: Jeremy P. Tarcher/Penguin.

Rudd, Murray A. 2011. "Scientists' Opinions on the Global Status and Management of Biological Diversity." *Conservation Biology* 25: 1165–75. http://www.newscientist.com/article/mg21228383.800-is-it-time-to-let-some-species-go-extinct.html.

Russell, Denise. 2010. *Who Rules the Waves? Piracy, Overfishing, and Mining the Oceans*. New York: Pluto Press.

Ryan, Thomas. 2011. *Animals and Social Work: A Moral Introduction*. New York: Palgrave Macmillan.

Saylan, Charles, and Daniel T. Blumstein. 2011. *The Failure of Environmental Education (and How We Can Fix It)*. Berkeley: University of California Press.

Schultz, P. Wesley. 2011. "Conservation Means Behavior." *Conservation Biology* 25: 1080-–83.

Sharot, Tali. 2011. *The Optimism Bias*. New York: Pantheon Books. See also http://www.time.com/time/health/article/0,8599,2074067,00.html.

Shermer, Michael. 2011. *The Believing Brain*. New York: Times Books.

Shipman, Pat. 2011a. *The Animal Connection: A New Perspective on What Makes Us Human*. New York: W. W. Norton.

———. 2011b. "Creature Contacts." *New Scientist* (2814): 33–36.

Sobel, David. 2008. *Childhood and Nature: Design Principles for Educators*. Portland, ME: Stenhouse Publishers.

Soulé, Michael. 2002. "History's Lesson: Build Another Noah's Ark." *High Country News* 24, no. 9, May 13, 2002. Available at http://www.hcn.org/servlets/hcn.Article?article_id=11219.

Specter, Michael. 2009. *Denialism: How Irrational Thinking Hinders Scientific Progress, Harms the Planet, and Threatens Our Lives*. New York: Penguin.

Stefanovic, Ingrid L., and Stephen B. Scharper, eds. 2012. *The Natural City: Re-Envisioning the Built Environment*. Toronto: University of Toronto Press.

Stolzenburg, William. 2008. *Where the Wild Things Were: Life, Death, and Ecological Wreckage in a Land of Vanishing Predators*. New York: Bloomsbury.

———. 2011. *Rat Island: Predators in Paradise—and the World's Greatest Wildlife Rescue*. New York: Bloomsbury.

Strain, Daniel. 2011. "Penguin Declines May Come Down to Krill." http://www.sciencenews.org/view/generic/id/72595/title/Penguin_declines_may_come_down_to_krill.

Suzuki, Severn. 1992. http://www.youtube.com/watch?v=uZsDliXzyAY.

Taylor, Paul. 1986. *Respect for Nature: A Theory of Environmental Ethics*. Princeton, NJ: Princeton University Press.

Tobias, Michael. 2011. "Pro-Planet Is Pro-Choice." http://blogs.forbes.com/michaeltobias/2011/05/09/pro-planet-is-pro-choice/.

Vucetich, John A., and Michael P. Nelson. 2007. "What Are 60 Warblers Worth? Killing in the Name of Conservation." *Oikos* 116: 1267–78. http://www.isleroyalewolf.org/techpubs/techpubs/JAVpubs_files/Vucetich%20%26%20Nelson%202007.pdf.

de Waal, Frans. 2010. *The Age of Empathy: Nature's Lessons for a Kinder Society*. New York: Three Rivers Press.

Weil, Zoe. 2011. http://zoeweil.com/2011/04/13/rubbing-elbows-with-solutionaries-green-festival-san-francisco/.

Weisman, Alan. 2007. *The World without Us*. New York: St. Martin's Press.

White, Lynn, Jr. 1967. "The Historical Roots of Our Ecologic Crisis." *Science* 155: 1203–7.

Wicks, Deidre. 2011. "Silence and Denial in Everyday Life—The Case of Animal Suffering." *Animals* 1: 186–99. http://www.mdpi.com/2076-2615/1/1/186/pdf.

Wilson, Edward O. n.d. http://www.brainyquote.com/quotes/keywords/vanish.html.

———. 2012. *The Social Conquest of Earth*. New York: Liveright.

Wolpe, Paul R. 2006. "Reasons Scientists Avoid Thinking about Ethics." http://repository.upenn
.edu/cgi/viewcontent.cgi?article=1015&context=neuroethics_pubs&sei-redir=1#search=%22
reasons+scientists+avoid+thining+ethics%22.

Zammit-Lucia, Joe. 2011. "Conservation Is Not about Nature." http://www.iucn.org/involved/
opinion/?8195/Conservation-is-not-about-nature.

Acknowledgments

THIS BOOK consumed me for more than two years. I worked on it in taxis of many different makes and colors; on buses, trains, and planes; and at airports all over the world. Some of the ideas were further developed at various inspirational meetings, including the first international conference on compassionate conservation held in Oxford, England, in 2010; the 2011 Asia for Animals meeting held in Chengdu, China; and at a wonderful meeting on religion and animals held in 2011 in Hawarden, Wales. I thank Celia Deane-Drummond, then at the University of Chester and now at the University of Notre Dame, for inviting me to this wonderful gathering at a most amazing location, The Gladstone (formerly St. Deiniol's) Library. Christie Henry, at the University of Chicago Press, as usual was the consummate "editress." As I've said before, Christie is a gem. She really is and I can't thank her enough for her uncompromising support (and love of red wine)! Abby Collier and Amy Krynak, also at the University of Chicago Press, helped in many ways behind the scenes, and Dawn Hall and Michael Koplow were consummate copyeditors. Many thanks to them all. Also, thanks to Sarah M. Bexell for contributing to some of the ideas that generated this collection of essays and Benjamin Beck for writing the foreword and for always asking hard questions. Three anonymous reviewers made valuable comments on these essays and I'm grateful to them for taking the time to read them. I also thank all of the wonderful people worldwide who are working hard for animals and Earth. Last, but surely not least, I'm deeply grateful to all of my colleagues who took time out of their incredibly busy schedules to write their essays. I well know the commitment it took. Thank you all!

PART ONE

ETHICS, CONSERVATION, AND ANIMAL PROTECTION

TRYING TO MAKE DIFFICULT DECISIONS EASIER

THE ESSAYS in this part are concerned with various topics that fall under the general category of ethics, what are the "right" and "wrong" things to do in a given situation. Of course, given the difficult decisions we have to make involving diverse animals living in different habitats in a wide variety of cultures, there has to be some flexibility. However, without a firm and consistent framework of general guiding principles that would obtain in the best of all possible worlds (which this isn't), it's impossible to make the difficult decisions with which we're faced daily. Modifying or overriding these principles in certain circumstances, for example, first do no harm, would require strong arguments for why this should be done. I often wish I didn't have to make some of the choices that need to be made, but they won't go away if they're ignored. For better or worse we are the surrogate decision makers for other individual animals and for a wide variety of species and ecosystems, and given who we are, we can do just about anything we want. The essays in this section provide the foundation for much of what follows later in this collection.

John Vucetich and Michael Nelson set the stage for much-needed, wide-ranging discussions concerning ethical foundations of conservation. They note that life and nature manifest themselves in different ways, as individual creatures, populations, and ecosystems. Concerns for populations and ecosystems are the focus of conservation, and concerns for individual creatures are the focus of animal welfare ethics. Both are important, but each has a tendency to ignore the other. Their chapter is an attempt to develop an ethic that transcends both perspectives. Further discussion of these issues can be found in Aitken (2004), Bekoff (2006, 2010), and Fraser (2010).

Vucetich and Nelson go on to argue that the ethical foundation of conservation is a shambles and that we can't even provide good answers to the most important unanswered questions in conservation—namely, (1) What is population viability and ecosystem health? (2) How does conservation relate to and sometimes conflict with other legitimate values in life, such as social justice, human liberty, and concern for the welfare of individual nonhuman animals? How should we resolve such conflicts? (3) Do populations and ecosystems have direct moral considerations? These challenging questions have direct on-the-ground consequences for conservation, and the authors argue that we need ethical consensus, not individual decisions, on the matters at hand. These questions are also philosophical or ethical in nature, not purely scientific. Vucetich and Nelson also show the importance of empathy, not only for sentient animals but also with nonsentient beings and ecological collectives. They show how important education is, a theme echoed in many other essays in this collection. Conservation science and humane education need to generate a sense of wonder for nature rather than emphasize prediction and control and must emphasize the ways in which nature is morally relevant.

Paul Waldau also emphasizes the importance of education and empathy and notes that the animal protection and conservation movements are really social movements and that members of each will have to work together in the future if we're to make meaningful progress. He writes, "Conservation and animal protection both call upon fundamental human abilities to recognize realities of other living beings." Waldau draws a powerful and hopeful conclusion: "Conservation insights are also driven by our great need to connect to the *meaning of life*. The two movements discussed in this chapter can, when working *together*, offer a very special hope—namely, that we *now* live in a time in which we can name war, genocide, habitat destruction, countless unnecessary murders of living beings, and global climate change as our heritage and our predicament, but also as *problems* we can choose to face squarely. In a world in which our political systems have slipped into an appalling lack of civility, and religious traditions struggle to gain the spiritual character to promote peace rather than division, the prospect of conservation's protective, constructive, healing insights being linked to the animal movement's power to instill and nurture individuals' ethical character is a soothing one. This can happen if the active citizens in each of these movements will choose to work together. This is one choice *we* can make as a way to celebrate the world we want to live in and leave for our children." Education and empathy can lead to large and significant positive changes in how we treat other animals and the habitats in which they reside. We are obliged to do this for future generations who will have to live with our decisions about who lives and who dies.

Clearly, we do what we do to other animals and ecosystems because we can, and we don't have to answer to individuals of other species who might be wondering what in the world are we doing. Eileen Crist argues that we humans have devastated the natural world because of our sense of entitlement to other species and animals in particular, and our sense that losses of animals and animal suffering do not really matter. She investigates the intellectual and historical roots of the belief in human superiority, or "human supremacy" as she refers to it. Her analysis invites us to scrutinize our hierarchical speciesist assumptions about humans and animals and to recognize the destructive way of life those assumptions support. Our tragic planetary predicament calls for a radical transformation of our understanding of ourselves in relation to our animal kin. We need to pay more attention to Charles Darwin's ideas about evolutionary continuity in which differences among species are recognized as differences in degree, not kind. Crist argues convincingly and powerfully "the long-standing denial or disparagement of animal minds is *causally* implicated in the devastation of the biosphere. Through the portrayal of animals as inferior beings, and eventually even as mechanical entities, the objectification of the natural world and its transformation into a domain of resources was vastly facilitated." Judith Benz-Schwarzburg and Andrew Knight (2011) correctly note that our relationships with other animals show clearly that while we're cognitive relatives we act as if we're moral strangers, alienated from whom other animals truly are.

Dale Peterson and Ben Minteer consider the very complicated bushmeat crisis from different perspectives while echoing some of Vucetich and Nelson's and Waldaus's themes about how concern for individuals comes into conflict with concern for larger entities such as species and ecosystems. Bushmeat, meat from a wide variety of wild terrestrial animals including chimpanzees, gorillas, and bonobos, is harvested unsustainably and is a common source of protein throughout West and Central Africa, where there is an impending food shortage (recent data show that humans also eat almost ninety species of marine mammals; Robards and Reeves 2011). But there is much more here in that cultural differences among of those working on this crisis and the poverty and economic needs of those who benefit from the bushmeat industry means that local interests must be factored into any reasonable or practical solution. There are many stakeholders with different views on the issues at hand. There aren't any quick and simple solutions.

Dale Peterson's essay "Talking about Bushmeat," introduces one of the most urgent and least understood conservation issues, the industrialized killing and selling of wild animals for meat in Central Africa. He begins with a riveting account of what it's like to walk through a bushmeat market. Noting the horrors he also tells how one of his friends rescued a small duiker who had been bound and left to rot in a small cage. The enormous commerce in bushmeat currently removes as much as five million metric tons of wild animal biomass per year from the Congo Basin ecosystem, which is a completely unsustainable take. The bushmeat commerce immediately threatens the well-being and indeed the very existence of the three African great apes (gorillas, chimpanzees, and bonobos), whose numbers are small to begin with, and who are profoundly vulnerable to hunting largely because of their low reproduction rates. We must not ignore this assault on nature, but what can we do about it? As Peterson suggests, we can begin a "cultural conversation" about the problem and nature of bushmeat. He would promote arguments for controlling the bushmeat commerce based on human self-interest (such as protection from some serious public health threats) and human other-interest. The other-interest argument for protecting some species such as the great apes would identify a moral hierarchy based on either an evolutionary closeness to humans or a reasoned calculation of the animal's psychological presence, or both.

Bushmeat appeals to many people in an historical way. Peterson writes, "The big city markets in Central and West Africa have plenty of the boring domestic meats, if you happen to prefer that. But if you want meat that will recall when your family lived out in the village, meat with real flavor from an interesting wild animal, meat taken by skilled hunters from the forest and often cured by a flavorful process of smoking, meat with a real story behind it—*bushmeat*, in other words—you will have to pay a premium, as you would for any other luxury item." Peterson also is sensitive to westerners arrogantly barging into other countries and telling local people what to do. Responding to an accusation posed as a question, "How dare you, a rich westerner, talk of limiting development, which represents the rightful economic advance of impoverished peoples in the Third World?" Peterson agrees that westerners who want to influence conservation elsewhere in the world should be prepared to pay. He also notes that most ordinary Africans do not benefit from the bushmeat trade. Cross-cultural discussions and give and take among the various stakeholders is sorely needed if we're ever to make headway on this challenging crisis.

Also focusing on the bushmeat crisis and discussing various ethical positions, Ben Minteer begins by noting that even if there are disagreements among those people interested in animal protection and those interested in conservation issues such as the bushmeat crisis, they would all agree that the brutal slaughter of dolphins in Taiji, Japan, the wanton killing of mountain gorillas in the war-torn Virunga National Park in the Democratic Republic of Congo, and the slaughter of tigers in the wildlife trade is morally wrong. This "compatibilist" understanding of environmental and animal ethics is much needed in the real world of animal abuse, and it's encouraging that people with different views are talking with one another more and more.

Concerning the bushmeat crisis, Minteer notes, "The intersection of animal rights/welfare and conservation ethics is particularly intriguing in the case of what has become known as the 'bushmeat crisis,' a subject of increasing concern in both the nature conservation and development communities. . . . The bushmeat problem raises an intricate complex of ecological, economic, cultural, and most fundamentally, ethical challenges regarding the survival of species and the welfare of animals, as well as the health and livelihood of some of the poorest and most vulnerable peoples on the planet." And, stressing ethical pragmatism, also needed in the real work of animal abuse, Minteer concludes, "the bushmeat problem appears to be a case in which both animal rights/welfare and a strong nature-centered ethic of conservation would be supported by a strict ban on bushmeat harvest and trade, the establishment of more tightly managed (for biodiversity preservation) protected areas in bushmeat regions, and increased enforcement and interdiction efforts. . . . The upshot is that a feasible, effective, and ethically inclusive policy response to the bushmeat dilemma will require balancing a complex of values and interests, as well as accommodating diverse stakeholders in workable, multilevel partnerships that can reduce human impact on wildlife species and tropical forest systems while improving the food security and livelihood prospects of poor rural people."

The five essays in this section raise many different issues concerning who we are and what we do to other animals and to ecosystems. They highlight, once again, the central role humans play in deciding who lives and who dies. While the problems at hand seem daunting and insoluble, decisions have to be made. The status quo is unacceptable. The real challenge is to come to terms with which, if any, compromises and trade-offs are permissible, and which are not. Flexibility and pluralism are needed in discussions among the human stakeholders. One thing is clear, Western standards will not be easily pushed on to other cultures. Nor should they be, a theme that is taken up in part 5.

References

Aitken, Gill. 2004. *A New Approach to Conservation: The Importance of the Individual through Wildlife Rehabilitation.* Hants, UK: Ashgate.

Bekoff, Marc. 2006. *Animal Emotions and Beastly Virtues: Reflections on Redecorating Nature.* Philadelphia: Temple University Press.

———. 2010. *The Animal Manifesto: Six Reasons for Expanding Our Compassion Footprint.* Novato, CA: New World Library.

Benz-Schwarzburg, Judith, and Andrew Knight. 2011. "Cognitive Relatives Yet Moral Strangers?" *Journal of Applied Ethics* 1: 9–36.

Fraser, David, ed. 2004. "Conservation and Animal Welfare." *Animal Welfare* 19 (2): 121–95.

Robards, Martin, and Randall Reeves. 2011. "The Global Extent and Character of Marine Mammal Consumption by Humans: 1970–2009." *Biological Conservation* 144: 2770–86. http://www.sciencedirect.com/science/article/pii/S0006320711002977.

1

The Infirm Ethical Foundations
of Conservation

John A. Vucetich and Michael P. Nelson

Introduction

THAT CONSERVATION HAS an ethical foundation is widely appreciated. Less appreciated is the shambled condition of that ethical
foundation. This condition is revealed by our inability to answer questions like, What is population viability and ecosystem health? and, Is
conservation motivated only to meet the so-called needs of humans,
or also by respect for nonhuman populations and ecosystems? Some
argue that this ethical uncertainty does not impede the effectiveness
of conservation. We provide examples that suggest otherwise. We
also explain how the source of ethical uncertainty is our mistaken
tendency to think that the morality of our behavior should be judged
more on the consequences of our actions and less on the motivations
that underlie our actions.

Conservation's aim is often thought or said to be to maintain and
restore population viability and ecosystem health. Achieving conservation is difficult, but the framework for conservation's goals seems in
place: Use the best available science and the precautionary principle
as input for a decision-making process that will suggest which actions
will most likely lead to the most desirable outcomes; use politico-legal
force to turn desired actions into law or policy; and include some
environmental education (e.g., media and formal curricula) to build
social support. That education almost always reduces to describing
how humans affect natural systems, as if that will shock or shame us
into supporting conservation.

This framework rests, unfortunately, on an infirm foundation that
casts doubt on whether we really understand the aim of conservation.

The answers to three questions illuminate the inadequacies of the foundation of conservation:

1. What is population viability and ecosystem health?
2. How does conservation relate to and sometimes conflict with other legitimate values in life, such as social justice, human liberty, and concern for the welfare of individuals, nonhuman animals? How should we resolve such conflicts?
3. Do populations and ecosystems deserve direct moral consideration?

These are the most important unanswered questions in conservation. Not having answers that are well defended and widely agreed upon has practical, on-the-ground consequences for conservation. Moreover, none of these questions are purely science questions. They are all philosophical or ethical in nature. This is disturbing because the ethics and philosophy of conservation may well be the most undertreated aspects of conservation. The very nature of conservation is, therefore, up for grabs because its ethical foundation is up for grabs. All the while, few people seem concerned. The need is not for each individual to answer the question in his or her own way; what is needed is the development of ethical consensus, which arises from ethical discourse (Nelson and Vucetich 2011).

An interlocutor might express skepticism: developing ethical consensus where there is none is impossible—not even among conservation professionals. Much evidence, however, speaks to our ability to develop ethical consensus (witness the abolition of slavery, women's suffrage, civil rights). Moreover, if we cannot arrive at a reasonably broad consensus about the three big questions above, then conservation's relationship to society will remain like a nation's tax policy: everyone agrees that tax policy should balance equality and fairness, socialism and libertarianism—but no one agrees on what that means. Instead, we should want conservation's relationship to society to be more like human medicine, which proceeds efficiently because we all agree on the aim (human health) and we all agree, more or less, on what human health means.

Answering "What is the aim of conservation?" is challenging because the question is broad and abstract, while at the same time the particulars of real conservation issues are so varied. It is difficult to identify principles that are general enough to entail most real issues, but not so broad and general as to be vacuous. To say that conservation is about maintaining and restoring population viability and ecosystem health is a bit too vacuous. By answering the three big questions, much that is vacuous will become firm. What follows is an exploration of how to approach the three big questions of conservation and the consequences of failing to take them seriously.

The Three Big Questions of Conservation

1. *What is population viability and ecosystem health?* Conventionally, population viability is assessed by estimating the probability that a population will go extinct over some time frame (Akçakaya, Burgman, and Ginzburg 1999). In principle, it is straightforward to estimate a population's extinction risk and to rank order extinction risk among a set of populations. In practice, both tasks tend to be especially difficult, in large part due to the limited availability of empirical data for most real populations.

Perhaps even more difficult is the task of determining the amount of extinction risk (the probability and time frame) beyond which a population would be considered endangered or not viable. For example, is a 5 percent chance of going extinct in 100 years an acceptably low chance of extinction? Or is a 10 percent chance of going extinct over 200 years more appropriate? No matter how extinction risk might be quantified, why is there so precious little discussion about such a profoundly basic question as, What is an unacceptable risk of extinction?

It seems straightforward to judge ecosystem health in the terms we use to describe ecosystems, that is, by: (i) their species richness and diversity; (ii) the nature of their ecosystem processes (e.g., nitrogen cycling) and ecological processes (e.g., predation or herbivory); (iii) temporal dynamics in these processes; and (iv) the spatial variation of ecosystems across landscapes (e.g., relative frequency of different kinds of ecosystems across landscapes).

One extreme, well-rehearsed perspective considers an ecosystem healthy to the extent that humans have not impacted it. From this perspective humans are a pathogen. Another extreme, well-rehearsed perspective considers an ecosystem healthy to the extent that it can continue providing resources and services that humans need. From this perspective humans are a parasite.

Our attempts to navigate this dichotomous notion of ecosystem health have been inept. For example, as we are increasingly faced with decisions about how to handle conservation-reliant systems (Scott et al. 2010), we find ourselves unable to avoid odd questions like, Is a human-altered ecosystem healthier when humans stop intervening, or when human intervention is used to return it to its prealtered state?

Another circumstance rises from our stumbling through the dichotomous view of ecosystem health. This circumstance, as odd as it is general, is represented by the question: On what portion of the landscape should we protect ecosystem health, and on what portion of the landscape should it be sacrificed for our use? The more familiar forms of this question are: How much wilderness and bioreserve area do we need? and, Should human impact be concentrated (e.g., intensive forestry on a small area) or diluted (e.g., less intensive forestry over a larger area)?

This attitude raises serious ethical questions, such as, On what ethical grounds can we justify respecting some ecosystems, but sacrifice others? This is *Sophie's Choice* manifest in our relationship with nature. The question also represents an ethical tragedy, a situation of our own making that seems to leave us with no acceptable choice. Moreover, this handling of the dichotomy never answers the question, What is a healthy ecosystem?

Despite the well-rehearsed problems with each perspective, each is rooted in a fundamental truth: humans can ruin ecosystems and humans need what ecosystems provide. But these perspectives also require believing that humans are separate from nature and require denying nature's intrinsic value. Both beliefs are unwise. Is it possible to develop a unified notion of ecosystem health that simultaneously recognizes: (i) humans can ruin ecosystems; (ii) humans need what ecosystems provide; (iii) humans are not separate from nature; and (iv) the value of healthy ecosystems for the sake of the ecosystem's interest, not just our own interest? What portion of conservation professionals concern themselves with this problem?

2. *How does conservation relate to and sometimes conflict with other legitimate values in life, such as social justice, human liberty, and concern for the welfare of individuals, nonhuman animals? How should we resolve such conflicts?* One approach to this question is to consider a useful definition of sustainability, which is "meeting human needs in a socially just manner without depriving ecosystems of their health [or populations of their viability]" (Nelson and Vucetich 2009c; Vucetich and Nelson 2010). Received definitions of sustainability suggest our unwillingness to, for example, sacrifice social justice in exchange for conservation and raise more particular questions like: Is it socially unjust to deprive a human community of their mode of living, if their mode of living deprives a nonhuman population of its viability or an ecosystem of its health? This question can be answered, but doing so requires: (i) a better understanding of what ecosystem health is; and (ii) an interest and ability to understand the nature of social justice, an interest and ability that seems well beyond the majority of conservation professionals and outside of the realm of what we normally think of as conservation science.

These questions would be ridiculous for anyone thinking that a particular conservation action was absolutely necessary for the survival or basic welfare of humanity. In that case, one might willingly pay almost any price for the conservation. The circumstance is, however, far more complex. Survival of the human species does not, for example, depend on Kansas having intact grassland eco-systems or the Pacific Ocean having blue whales. We already have a pretty good idea about how humans can survive without these populations or ecosystems.

Still, we cannot ignore the "13th rivet" metaphor, which explains how the loss of any particular species or ecosystem may not be important for the wel-fare of humanity, but the collective loss of many populations and ecosystems is. This raises the problem of *how* we go about deciding how we ought to treat any particular population or ecosystem. For every proposed conservation action, we must know how/whether the benefits of that particular action are worth the ethical costs that action might incur on social justice, or animal welfare, or whatever the costs may be.

The point is, conservation is not the only legitimate value in society. Par-ticular conservation actions sometimes conflict with other values, and no particular conservation action always and automatically trumps every other value. Consequently, knowing conservation's role in society requires knowing how and why populations and ecosystems are valuable. In particular, we need to know how they are valuable beyond their utility to humans.

3. *Do populations and ecosystems deserve direct moral consideration?* This question is critical not only for conservation, and the academic field of environmental ethicists has generated a great deal of insight about how the question might be answered,[1] though it is largely unknown to many conservation professionals.

An important line of reasoning has been that direct moral consideration should be extended to anything possessing a morally relevant trait. Many con-sider sentience and the capacity for reason to be morally relevant traits, and some consider them to be the only morally relevant traits. If so, ecological collectives would not deserve direct moral consideration because they are not sentient or capable of reason. Another school of thought known as *biocentrism* argues that being alive is the morally relevant trait. While some of these scholars argue that ecological collectives are morally relevant because they are living *things*, others argue they do not deserve moral consideration because they are not living *individuals*. Each of these approaches represents a kind of thinking known as *extensionism*.

By contrast, some professional ethicists have argued that ecological collec-tives deserve direct moral consideration because they are the will of some deity. Ironically, some theological consideration suggests that only humans deserve direct moral consideration.

Another more secular approach has been to argue that ecological collec-tives deserve direct moral consideration because they and we are members of a shared biotic community, and all community members deserve moral consideration. This was Aldo Leopold's contribution to environmental ethics. Deep Ecologists approach this question by first recognizing that humans deserve direct moral consideration, and then by recognizing that humans and ecological collectives are indistinguishable, and for these reasons ecological collectives deserve direct moral consideration.

In the process of developing these insights, some environmental ethicists have discovered a more basic challenge, which is, knowing what exactly is meant by the term *direct moral consideration*. First, as a matter of vocabulary, environmental ethicists generally say that a thing deserves direct moral consideration if it has intrinsic value, in contrast to having only instrumental (or use) value. The trouble is, what exactly is meant by intrinsic value.

Intrinsic value could be something that exists within certain things; implying intrinsic value is an objective property that can be discovered. In this case ethicists say to be intrinsically valuable is to be valuable in and of itself. However, intrinsic value may only exist in the mind of the valuer. In this case, intrinsic value would be value in addition to use value. Alternatively, intrinsic value may be more relational, that is, something that emerges from a valuer's relationship with certain things. Uncertainly about the meaning of intrinsic value amplifies the difficulty of answering the question, Do populations and ecosystems deserve direct moral consideration?

Answering this question would solve a great challenge for conservation. However, answering this question in the affirmative creates even more difficult ethical questions for conservation. Specifically, how to weigh and adjudicate among the disparate interests of humans and nonhumans.

Practical Implications

We have made a case that conservation's meaning, purpose, and relationship to the rest of society are inadequately understood. While many people believe that an infirm ethical foundation is no impediment to conservation (Norton 1994), there are many examples to the contrary, for example:

- *US Endangered Species Act.* The definition of endangered species in what is, arguably, the most powerful environmental law in the world is one that is not "in danger of extinction throughout all or a significant portion of its range." So, how much risk is too much risk? In general, judgments about excessive risks depend upon the consequences. For example, what counts as excessive risk for having rain on your picnic differs from what counts as excessive risk of dying due to the failure of your car's brakes. Similarly, what counts as an excessive risk of extinction will depend upon whether we think the American burying beetle is valuable only for human welfare or if it is also intrinsically valuable. In other words, appropriate conservation requires answering the three big questions of conservation. The same conditions arise when considering the meaning of "significant portion of range" (Vucetich, Nelson, and Phillips 2006; Nelson, Phillips, and Vucetich 2007; Waples et al. 2007a, b; Carroll et al. 2010).

- *Conservation-Reliant Species.* There is an increasing awareness of the difficulty of knowing how to manage species that require perpetual human support or that may never be recovered (Scott et al. 2010). Polar bears and caribou are important examples. The great concern here is spending resources that will lead to no benefit. If ethics depend on consequences then this concern is appropriate. However, if ethics depends importantly on motivation, which it does, and if these species are intrinsically valuable, then this concern is moot. Our obligations to other humans, as a parallel, are not reducible to their benefit to us because we believe all humans possess intrinsic value. In other words, appropriate conservation requires us to answer the second and third big questions of conservation. These questions about conservation of reliant species, unrecoverable systems, and other hopeless cases apply to hundreds of species and hundreds of thousands of square miles of the earth's surface.

- *Conservation's Conflict with Animal Welfare.* Many conservation actions, including the control of exotic and invasive species, involve killing individual creatures. Is the cost of killing hundreds of individual barred owls worth the benefit of protecting populations of northern spotted owls (Welch 2009)? A common response is that the needs of conservation (here preserving northern spotted owl populations) trump the welfare of individual animals. However, one of the greatest developments in twentieth-century ethics has been the development of reasons to think that nonhuman animals and ecological collectives deserve direct moral consideration. Society's appreciation for these reasons is increasingly apparent (e.g., Animal Welfare Act and Endangered Species Act). The ethical thing to do is not to deny the validity of one of these moral developments but to work toward an ethic that accommodates this conflict (Vucetich and Nelson in review). In other words, appropriate conservation requires answering the second big question of conservation.

The Ethics of Control and Consequence

We are all familiar with the narrative explaining how our conservation crisis has roots in our historic fetish for controlling nature (White 1967; Holling and Meffe 1996). However, this pathology has roots in a more basic limitation of Western ethics, which is—ironically—its failure to seriously confront the question: What deserves direct moral consideration?[2] Despite this long-standing limitation of Western ethics, its history provides important clues for how we might approach the question.

To capture the salient elements of this history in a few hundred words we paint with the crudest strokes:[3] The ethic of those living prior to Aristotle, represented by characters of Homeric literature, was built on a belief that life had a purpose (what Aristotle referred to as *telos*), and that purpose was to be a good warrior. Moreover, ethics were the vehicle between how-we-are and how-we-ought-to-be, and the engine of this vehicle was virtues that, when manifest, give rise to how-we-ought-to-be. In Homeric days, the virtues included courage and cunningness, which were apropos given the purpose.

As the Homeric period gave way to the rise of Greek city-states, Aristotle codified the purpose of human life as being a good citizen of the city-state, and the accompanying virtues included justice, prudence, temperance, and magnanimity. As Europe fell into the Dark Ages, theologians and monarchs dictated a person's purpose, which was to atone for original sin and get to heaven. The virtues giving rise to this purpose were faith, hope, and humility. Importantly, Aristotle would have considered humility a vice. This basic framework for ethics (i.e., a conjoining of purpose and virtue) remained unchanged for more than two millennia. Today we call this framework *virtue ethics*. The essence of this ancient ethic focuses on how a moral agent should behave without much attention given to understanding who the moral patients are.[4] That is, virtue ethics does not, by itself, try to answer the question, What deserves direct moral consideration?

With the Enlightenment, as the story goes, reason overthrew religious and monarchical tyranny. Another victim of the coup was Aristotle's metaphysics, especially his telos or notion of purpose, and subsequently our interest in virtue ethics. The difficulty had been that a sense of telos conjoined to religious and monarchical tyranny had been the primary motivators for behavior for the past thousand years. An ethical crisis emerged: On what now would ethics and behavior be based?

Ethicists occupied themselves with this question for the next couple of hundred years: Kant suggested that reason alone could be the foundation for determining what is ethical; and Hume suggested that emotion and feeling, along with intellect, should be the foundation for judging what is ethical. These developments in ethics were triumphs of human liberty. They were beautifully anthropocentric and contrasted to prior tendencies for life to be theocentric or monarchicentric.

Despite the insight that Kant and Hume offered, each account had the effect of revealing the inadequacies of the other. Kierkegaard took the failures of Kant and Hume as a basis for suggesting that we are free to decide whether ethics (reason) or aesthetics (feeling) should be the foundation for judging what is right and wrong. Nietzsche took the failures of Kant and Hume as evidence that personal will was the only sensible basis for deciding how to behave. Hitler and Mussolini seem to be manifestations (though perhaps perverse) of his ideas.

Bentham and Mill suggested that the goodness of an action should be judged on the basis of maximizing happiness and minimizing suffering of humans. Their thoughts led to Consequentialism and Utilitarianism,[5] which have been for the past 150 years the dominant framework for thinking about ethics generally and certainly the dominant form of ethical thinking within conservation.

A growing number of scholars are increasingly disturbed by several, well-rehearsed weakness of Consequentialism and Utilitarianism. For example, Utilitarianism is limited by our inability to judge and quantify happiness (Sen 1987; Putnam 2002; Moore 2004), especially in nonhumans. Utilitarianism does not provide a useful way of comparing happiness among, for example, a person, a goat, a population of Puerto Rican parrots, and a mangrove ecosystem. Human psychology also seems predisposed to overemphasize our happiness and underemphasize the happiness of others (Vucetich and Nelson 2007). In the absence of any constraints, Consequentialism reduces to an ethic for which "ends-justify-the-means" is the primary principle for judging morality, leaving open the possibility that any particular behavior (e.g., child slavery) could be justified if the benefits of the behavior outweighed the costs (Rachels and Rachels 2009).

Another limitation of Consequentialism is our inability to reliably predict the consequences of our actions except in the simplest of cases. Again this weakness is accentuated when thinking about our relationship with the environment, where the causal relationships between humans and nature are complex. According to Consequentialism, being unable to reliably predict consequences leaves one unable to know the morality of an action (including conservation actions). The unthinkability of being left in a state of amorality encourages us to exaggerate our ability to predict the consequences of our actions. Our obsession with controlling nature, which others have argued is a root cause of our environmental crisis (Holling and Meffe 1996), rises from our commitments to Consequentialism.

Consequentialism's singular focus on consequences distracts our attention from being concerned with the motivations that underlie our actions. Focus on consequences runs contrary to a basic tenet that ethics is primarily about assessing the motivations for our actions. Lack of concern for motivation explains our inability to appreciate or answer the three big questions of conservation.

The most fundamental limitation of Utilitarianism is the inability to answer the question, What deserves direct moral consideration? The history of Utilitarianism suggests that moral consideration extends to those who can experience happiness and suffering.[6] But Utilitarianism cannot answer the question, What counts as happiness or suffering? nor is it equipped to determine who is capable of experiencing happiness and suffering. Science can answer some of these questions (Chandroo, Duncan, and Moccia 2004; Elwood and Appel 2009). However, other aspects of knowing who suffers and what counts as suffering may be metaphysical.[7] For example, is an ecosystem overrun with exotic species an ecosystem that suffers? Science can describe the consequences of being overrun with exotic species but cannot say whether that counts as suffering, or whether an intact ecosystem counts as a flourishing ecosystem.[8] Knowing whether an ecosystem can or cannot flourish may well require insights from outside the boundaries of science and ethics.

Another form of Consequentialism that is especially influential among conservation professionals is Pragmatism (Norton 1994; Katz and Light 1996; Minteer and Collins 2005; Lockwood and Reiners 2009). The essential tenet of Pragmatism is that truth or meaning ought to be judged by practical consequences. A pragmatic ethic is judged, therefore, by its ability to solve ethical problems, as those problems are perceived. Although Pragmatism may seem commonsensical, it has long been deeply controversial among ethicists. Pragmatism is especially vulnerable to the criticism that the ends do not justify the means. Aside from the previously mentioned problems with thinking that ends justify means, Pragmatism is especially troublesome for conservation to the extent that we have not adequately identified the ends. That is, we still have not adequately answered the question, What is the aim of conservation? This makes conservation motivated by pragmatism like a missile without a guidance system.[9]

The Ethics of Virtue

The depraved morality that Utilitarianism offers has led a growing number of ethicists to think that an environmentally sustainable life requires rediscovering virtue ethics and reinventing it for contemporary life (Cafaro 2001; Sandler and Cafaro 2005; Hursthouse 2007; Sandler 2007). Such reinvigoration may also be necessary for shoring up the foundation of conservation and adequately specifying conservation's purpose. Recall that virtue ethics involves three aspects: (i) identifying the purpose of a person, (ii) identifying virtues necessary for manifesting one's purpose, and (iii) engaging in activities that promote the virtues.

With some reflection, it seems that the purpose of a person living a sustainable life would have to be *to treat others as one would be treated, if one were in their position.* This principle rises from the simple commitment that ethics be rationally consistent and is known as the principle of ethical consistency (PEC) (Gensler 1996). Application of PEC requires empathy, a vivid, knowledge-based imagination about another's circumstance, situation, or perspective. Empathy is not an emotion, but a capacity that depends on objective, empirical knowledge (discovered by science and disseminated by education) about the conditions and capacities of others (to flourish and suffer). Empathy is required because PEC requires one to treat others only as one would consent to be treated if one were in their same situation. Principles of psychology indicate that one's empathy for an object is limited by one's familiarity with the object and the extent to which one observes similarity between one's self and the object (Preston and de Waal 2002).

For PEC to be useful in a conservation context empathy with nonsentient beings and ecological collectives would have to be possible. It is possible because we can observe similarity and have familiarity with nonsentient beings and ecological collectives. Not only is such empathy possible, we also admire those who exemplify such empathy. If the reader is unsure, go back and read Shel Silverstein's *The Giving Tree* and consider Aldo Leopold's capacity to "think like a mountain."[10]

Because PEC rises from principles of consistency, we are obligated to apply PEC consistently—that is, whenever possible. To do otherwise is to apply PEC arbitrarily, which would be unethical. Because we can become familiar and observe similarity with any living thing (including plants and fungus, and possibly unicellular organisms, and ecological collectives, such as species and ecosystems), we ought to do so.[11]

PEC is important not only for being the purpose behind a virtue ethic applied to conservation ethics, but also because it answers the third big question of conservation, What deserves direct moral consideration? Knowing that the viability and health of populations and ecosystems is morally relevant for populations' and ecosystems' sake, and not only because we depend on their viability and health, provides much guidance for answering the first big question of conservation. Moreover, understanding that PEC applies to humans, nonhuman individuals, and ecological collectives leads to critical insight for responding to ethical conflicts that arise when conservation seems to conflict with social justice, human liberty, and concern for the welfare of individual animals. Although it would be valuable to elaborate further, here we only point out that the resolution to such conflict rises from the same principles we use regularly in our everyday lives to resolve other ethical conflicts, such as how do we balance a decision to be a good conservation scientist and our responsibility to be an environmental advocate (Nelson and Vucetich 2009b).

For emphasis, the claim is not merely that PEC is an appropriate rule to live by, but that a sustainable relationship with nature requires that manifesting PEC be the *purpose* of one's life, one's reason for living, and the overarching principle that guides all of one's actions and decisions. PEC would be the dominant narrative thread in one's life. The process of maturation would be defined by the process of continually improving one's ability to manifest PEC.

Given PEC as a purpose, here is a candidate list of virtues that would seem necessary and sufficient for promoting the manifestation of one's purpose:

1. *Constancy of Purpose, or knowing that one has a purpose.* If virtues help one manifest one's purpose, it may seem redundant that the first virtue is merely a reminder that one's life has a purpose. However, the postmodern world is characterized by, among other features, the decline of vocationalism and the rise of professionalism. With this shift, and the residual influences of Social Darwinism, the notion that life might have any purpose other than securing resources and safety is relatively unfamiliar to most. This virtue includes knowing that manifesting PEC is what makes one happy. For these reasons, the simplest and most basic virtue is simply knowing that one has a purpose and that this purpose is manifesting PEC.[12]

2. *Self-empowerment, or knowing that one is living a sustainable life, does not depend on others.* The postmodern world is fixated on "Tragedy of the Commons," the thought that individuals acting independently in their own self-interest ultimately destroy shared resources, in the absence of conditions that are difficult to accommodate. Understanding the Tragedy of the Commons is wise. However, being fixated on Tragedy of the Commons in a world committed to Consequentialism is devastating. Together they strip away all motivation to behave sustainably unless everyone else does the same (Nelson and Vucetich 2009a). However, if a sustainable (ethical) life is defined more by the motivations of our behaviors and less by their consequences, then living an ethical life has nothing to do with others' behavior and depends only on one's self.

3. *Empathy, or working to increase one's capacity to be empathetic with all humans, nonhuman individuals, and ecological collectives.* This needs to be a virtue because we are not a particularly empathetic people, and empathy is central to PEC.

4. *Connectedness, or seeing connectedness among all living things.* This needs to be a virtue because connectedness is the medium used to see that humans and nature are part of the same community, and because we have a long history of denying this connectedness. While many people would already be familiar with the value of this and the previous virtue, what might be less familiar is the role that conservation science plays in developing these virtues (see below).

5. *Sharing with one's community.* PEC raises a concern that ethicists refer to as *ethical overload*, the challenge of how to care for so many moral patients, many of whom have conflicting moral needs. While this is an important challenge, principles that are known to all of us are quite capable of handling this challenge. In particular, the principle of knowing when it is right to share with another. For example, should we share our lunch with a person we had constantly reminded to bring his or her own lunch? Perhaps. Doing so would be generous. But not sharing might also be appropriately just. The decision depends importantly on our motivation. To share or not might also depend on whether there was good reason for the person to have not brought a lunch. This case demonstrates the sophisticated mechanisms we already have for knowing when to share. The radical shift is to see sharing as a primary virtue for relating to nature. For example, knowing whether it is right to hunt wolves only requires knowing whether it is right for hunters to share deer and elk with wolves. One's obligation to grow wise in knowing when to share doesn't conflict with one's obligation to care for (i.e., being interested to share) with everything.

6. *Mourning, or knowing to grieve in the face of tragedy.* When confronted with challenges like whether to kill wolves or allow caribou populations to suffer greater extinction risk (DeCesare et al. 2010), we tend to deny the tragedy of the circumstance by claiming either that individual wolves do not count (if our concern was focused on the conservation of caribou populations) or that caribou populations do not count (if our concern was focused on individual wolves). Our tendency to deny tragedy is a general tendency reflected in our strong preference for Hollywood endings over those of Shakespearian tragedies. Mourning the circumstance of having to decide between wolves and caribou forces us to see that wolves are not the ultimate cause of caribou decline. Instead, mourning the tragedy motivates us to confront the ultimate cause, which is the overexploitation of boreal forests and gas exploration (Wittmer et al. 2007). Mourning in the face of tragedy is a virtue because it focuses attention on ultimate problems and encourages avoiding tragedies in the first place.

If this framework is sensible, then the greatest disappointment in a virtue ethic rooted in PEC is that it all seems too trite. That is, haven't we known since kindergarten that we ought to follow the Golden Rule? This reaction misses the salient point, which is about the underappreciated role of motivation (virtues) and the exaggerated role of consequences in our ethical thought. None of us recall a world that wasn't thoroughly dominated by Consequentialist thought. In this sense, virtue ethics is an unfamiliar mode of thought. However, because virtue ethics dominated the Western mind for nearly two millennia, it is Consequentialist thinking that in the long run has been the unusual mode of thinking. Embracing virtue ethics and living sustainably will require the greatest shift in ethical thought in 400 years.[13]

The Purpose of Conservation Science and Education

Promoting the virtues requires contributions from individuals, communities, and institutions. Conservation science and education have a critical and unique role to play in promoting the virtues. Only science and education can discover and disseminate the objective, empirical knowledge necessary for increasing our capacity to empathize with others (including nonhumans) and see connections in nature. Empathy and seeing connectedness are keys to seeing how humans and nature are part of the same moral community. In this sense, science and education are critical for knowing how nonhumans are morally relevant. One might say that the central purpose of conservation science and education is to generate a sense of wonder for nature (Carson 1965; Moore 2005).

This view is a radical departure from the received view that the purpose of science is to predict and control nature. This view also departs from a commonly held view that the aim of environmental education is to shock or shame us into behaving sustainably by showing us how we damage nature and how we can mitigate our damage (Erhlich 1995). We won't care about the damage we cause or how to mitigate it until science and education foster a sense for how nature is morally relevant.

This new conservation science would differ in tangible ways from the old. We would, for example, be more interested in questions that increase our capacity for empathy than for control. We would also be more interested in communicating to larger audiences of the general public how and why our knowledge of nature leads us to love it. Our best guidance for this new science will likely come from contemplating the lives of heroes like Rachel Carson and Aldo Leopold. It is not so much their tangible accomplishments that impress us, instead the accomplishment that impresses us most is the kind of people they had become and the virtues they manifested.

Acknowledgments

J. A. V. was supported, in part, by the US National Science Foundation (DEB-0424562). The views expressed here do not necessarily represent those of the US National Science Foundation.

Notes

1. For a comprehensive review of these insights see DesJardins (2000) or Jamieson (2008).

2. The treatment of what is morally relevant has been handled very differently in different cultures. Traditional people of North America, for example, take for granted that many nonhuman things deserve direct moral consideration (Callicott and Nelson 2003; Moore et al. 2007).

3. The history of ethics and interpretation of virtue ethics presented here summarize the pioneer-

ing scholarship of A. MacIntyre (1984), whose work has sparked a resurgence of ethicists' interest in virtue ethics that continues growing to this day.

4. A moral agent has the capacity to behave morally, and a moral patient is something that deserves direct moral consideration.

5. Utilitarianism is a more specific form of Consequentialism. The central tenet of Consequentialism is that the rightness of an action is determined by the consequences of an action. Utilitarianism presumes that the consequence of an action should aim to produce the most possible utility, happiness, or pleasure for the most people.

6. Scholars actively debate whether sentience is merely the capacity to suffer or the ability to be conscious of experienced happiness or suffering. There is also debate about the meaning of "consciousness" (Duncan 2006).

7. We mean metaphysical in the sense that some claims about the nature of reality cannot be answered by science or ethics alone.

8. "Happiness" might be inappropriately narrow for the purpose of this conversation. "Flourishing" is likely more appropriate (Cuomo 1998).

9. P. Pister, the fish conservationist, shared this expression with us.

10. Empathizing with nonhumans often raises concerns about anthropomorphism (de Waal 1999; Sober 2005; Bekoff 2006). That concern is appropriately handled by the careful use of language and knowledge about the biology of the nonhuman being spoken of.

11. We are not the first to suggest that the key to conservation ethics is recognizing how PEC is applicable to humans, nonhuman individuals, and ecological collectives (see e.g., Gould 1990).

12. MacIntyre (1984) provides a complete explanation for why constancy of purpose should be a virtue.

13. MacIntyre (1984) believed that we would not make the transition from Utilitarianism to virtue ethics smoothly and believes we are now entering the beginning of what might be called the second Dark Ages.

References

Açakaya, H. R., M. A. Burgman, and L. R. Ginzburg. 1999. *Applied Population Ecology.* 2nd ed. Sunderland, MA: Sinauer Associates.

Bekoff, M. 2006. "Animal Passions and Beastly Virtues: Cognitive Ethology as the Unifying Science for Understanding the Subjective, Emotional, Empathic, and Moral Lives of Animals." *Zygon* 41: 71–104.

Cafaro, P. 2001. "Thoreau, Leopold, and Carson: Toward an Environmental Virtue Ethics." *Environmental Ethics* 22: 3–17.

Callicott, J. B., and M. P. Nelson. 2003. *American Indian Environmental Ethics: An Ojibwa Case Study.* Englewood Cliffs, NJ: Prentice Hall.

Carroll C., J. A. Vucetich, M. P. Nelson, D. J. Rohlf, and M. K. Phillips. 2010. "Geography and Recovery under the U.S. Endangered Species Act." *Conservation Biology* 24: 395–403.

Carson, Rachel L. 1965. *The Sense of Wonder.* New York: Harper and Row.

Chandroo, K. P., I. Duncan, and R. D. Moccia. 2004. "Can Fish Suffer? Perspectives on Sentience, Pain, Fear and Stress." *Applied Animal Behaviour Science* 86: 225–50.

Cuomo, C. J. 1998. *Feminism and Ecological Communities: An Ethic of Flourishing.* New York: Routledge.

DeCesare, N. J., M. Hebblewhite, H. S. Robinson, and M. Musiani. 2010. "Endangered Apparently: The Role of Apparent Competition in Endangered Species Conservation." *Animal Conservation* 13: 353–62.

DesJardins, J. R. 2000. *Environmental Ethics: An Introduction to Environmental Philosophy.* New York: Wadsworth.

Duncan, I. 2006. "The Changing Concept of Animal Sentience." *Applied Animal Behavior Science* 100: 11–19.

Ehrlich, P. R. 1995. "The Scale of Human Enterprise and Biodiversity Loss." In *Extinction Rates,* edited by J. H. Lawton R. M. May, 214–24. Oxford: Oxford University Press.

Elwood, R. W., and M. Appel. 2009. "Pain Experience in Hermit Crabs?" *Animal Behaviour* 77: 1243–46.

Gensler, H. J. 1996. *Formal Ethics*. London: Routledge, chaps. 5 and 6.

Gould, S. J. 1990. "The Golden Rule? A Proper Scale of our Environmental Crisis." *Natural History* 9: 24–30. Reprinted in L. Pojman and P. Pojman, eds. 2007. *Environmental Ethics: Readings in Theory and Application*. 5th ed. New York: Wadsworth.

Holling, C. S., and G. K. Meffe. 1996. "Command and Control and the Pathology of Natural Resource Management." *Conservation Biology* 10: 328–37.

Hursthouse, R. 2007. "Environmental Virtue Ethics." In *Working Virtue: Virtue Ethics and Contemporary Moral Problems*, edited by R. L. Walker and P. J. Ivanhoe, 155–71. New York: Oxford University Press.

Jamieson, D. 2008. *Ethics and the Environment: An Introduction*. Cambridge: Cambridge University Press.

Katz, E., and A. Light, eds. 1996. *Environmental Pragmatism*. London: Routledge.

Lockwood, J., and W. Reiners. 2009. *Philosophical Foundations for the Practices of Ecology*. Cambridge: Cambridge University Press.

MacIntyre, A. 1984. *After Virtue: A Study in Moral Theory*. 2nd ed. Notre Dame, IN: University of Notre Dame Press.

Minteer, B. A., and J. P. Collins. 2005. "Why We Need an Ecological Ethics." *Frontiers in Ecology and the Environment* 3: 332–37.

Moore, K. 2004. *Pine Island Paradox*. Minneapolis: Milkweed.

Moore, K. D. 2005. "The Truth of the Barnacles: Rachel Carson and the Moral Significance of Wonder." *Environmental Ethics* 27: 265–77.

Moore, K. D., K. Peters, T. Jojola, and A. Lacy, eds. 2007. *How It Is: The Native American Philosophy of V. F. Cordova*. Tucson: University of Arizona Press.

Nelson, M. P., M. Phillips, and J. A. Vucetich. 2007. "Normativity and the Meaning of Endangered, a Comment on Waples et al." *Conservation Biology* 21: 1646–48.

Nelson, M. P., and J. A. Vucetich. 2009a. "Abandon Hope." *The Ecologist* (March).

———. 2009b. "On Advocacy by Environmental Scientists: What, Whether, Why, and How." *Conservation Biology* 23: 1090–101.

———. 2009c. "True Sustainability Needs an Ethical Revolution." *The Ecologist* (December 31). www.theecologist.org/blogs_and_comments/commentators/other_comments/383966/true_sustainability_needs_an_ethical_revolution.html.

———. 2011. "Environmental Ethics and Wildlife Management." In *Human Dimensions of Wildlife Management*, edited by W. Siemer, D. Decker, and S. Riley, forthcoming. Baltimore: Johns Hopkins University Press.

Norton, B. 1994. *Toward a Unity among Environmentalists*. Oxford: Oxford University Press.

Preston, S. D., and F. B. M. de Waal. 2002. "Empathy: Its Ultimate and Proximate Bases." *Behavioral and Brain Sciences* 25: 1–7.

Putnam, H. 2002. *The Collapse of the Fact/Value Dichotomy and Other Essays*. Cambridge, MA: Harvard University Press.

Rachels, J., and S. Rachels. 2009. *The Elements of Moral Philosophy*. 6th ed. New York: McGraw-Hill.

Sandler, R. 2007. *Character and Environment: A Virtue-Oriented Approach to Environmental Ethics*. New York: Columbia University Press.

Sandler, R., and P. Cafaro, eds. 2005. *Environmental Virtue Ethics*. New York: Rowman and Littlefield.

Scott, J. M., D. D. Goble, A. M. Haines, J. A. Wiens, and M. C. Nee. 2010. "Conservation-Reliant Species and the Future of Conservation. *Conservation Letters* 3: 91–97.

Sen, A. 1987. *On Ethics and Economics*. New York: Blackwell.

Sober, E. 2005. "Comparative Psychology Meets Evolutionary Biology: Morgan's Canon and Cladistic Parsimony." In *Thinking with Animals*, edited by L. Daston and G. Mitman, 85–99. New York: Columbia University Press.

Vucetich, J. A., and M. P. Nelson. 2007. "What Are 60 Warblers Worth? Killing in the Name of Conservation." *Oikos* 116: 1267–78.

————. 2010. "Sustainability: Virtuous or Vulgar?" *Bioscience* 60: 539–44.

————. In review. "Resolving the Conflict between Conservation and Animal Welfare. *Conservation Biology.*

Vucetich, J. A., M. P. Nelson, and M. K. Phillips. 2006. "The Normative Dimension and Legal Meaning of 'Endangered' and 'Recovery' within the United States' Endangered Species Act." *Conservation Biology* 20: 1383–90.

de Waal, F. 1999. "Anthropomorphism and Anthropodenial: Consistency in Our Thinking about Humans and Other Animals." *Philosophical Topics* 27: 255–80.

Waples R. S., P. B. Adams, J. Bohnsack, and B. L. Taylor. 2007a. "A Biological Framework for Evaluating Whether a Species Is Threatened or Endangered in a 'Significant Portion of Its Range.'" *Conservation Biology* 21: 964–974.

————. 2007b. "Normativity Redux." *Conservation Biology* 21: 1649–50.

Welch, C. 2009. "The Spotted Owl's New Nemesis." *Smithsonian* magazine (January).

White, L. T. 1967. "The Historical Roots of Our Ecological Crisis." *Science,* 155: 1203–7.

Wittmer, H. U., B. N. McLellan, R. Serrouya, and C. D. Apps. 2007. "Changes in Landscape Composition Influence the Decline of a Threatened Woodland Caribou Population." *Journal of Animal Ecology* 76: 568–79.

2

Venturing beyond the Tyranny of Small Differences

The Animal Protection Movement, Conservation, and Environmental Education

Paul Waldau

AS ONE EXAMINES the relationship of the conservation movement to another vibrant, worldwide movement known by various names such as "animal protection," "animal rights," and "animal liberation," one finds *many* reasons these "causes" ought to have an arm-in-arm relationship. Historically, however, the relationship between these two megamovements has often been described as tense—differences have at times prompted advocates of each of these social movements to remain silent about key insights of the other movement and, thereby, *ignore* fundamental features of nature.

To be sure, the differences between the two movements, some of which are described below, are at times real. In this chapter, however, I argue that the differences are, from both longer and deeper perspectives, *relatively minor* because these differences by no stretch of the imagination justify advocates of each social movement undermining advocates of the other movement. Instead, this chapter suggests, the relationship is one of natural allies, even siblings. However one chooses to describe the possible synergies, it is without question that the insights and prospects of each movement offer important lessons and opportunities for the other. It is possible, then, to reaffirm the conservation movement as an animal protection movement just as one can also affirm the animal protection movement as advancing the core insights of conservation.

By the term *animal movement* is meant a startlingly broad and successful social movement found in many countries around the world today. The choice of *animal movement* to name these worldwide animal protection efforts in no way is meant to suggest that *other* major social movements, such as the conservation movement, do not involve important protections, recognition of moral rights, or even "liberation" for living beings beyond our species. In fact, any number of social movements clearly do promote efforts to develop what can be rightfully thought of as "protection" for an astonishing array of living beings outside our own species. But of the greatest relevance to the points made in this chapter, the conservation movement has from its inception had decisively important animal protection features that make it the natural ally of the worldwide animal protection movement.

The relationship of the animal movement to the conservation movement is here framed in terms of three questions. First, why is the animal movement important to the worldwide conservation movement? I answer this opening question by pointing to the rapid growth of the animal movement and its success with certain groups not reached by forms of the conservation movement.

Second, why has the animal protection movement proceeded at a different pace and in different ways than has the conservation movement? The answer to this question, which considers not only historical causes but also contemporary realities, addresses how both of these social movements call upon a set of human skills that in part overlap and yet at critical points are distinctive from one another.

Third, what are the most relevant features of the animal protection movement to the conservation movement? The answer points to important developments in law and legal education, to a field known as "religion and animals," and to the value of animal studies for the development of critical thinking and a deeper sense of the riches and limits of the humanities and our sciences.

Background on the Animal Movement

The animal movement is extremely diverse, a feature related to, first, the fact that humans encounter an astonishing variety of living beings beyond the species line *and*, second, the animal movement's deep, sustaining roots in a variety of ancient cultural traditions. Some people characterize this diversity as a strength because it produces multiple facets in a social movement attempting to address an extraordinary range of problems. Others, however, find the diversity to be a lack of coherence.

What often comes to mind today when people around the world are asked about the animal movement (under any of its many names, such as "animal rights" or "animal protection") are modern animal protection efforts that gained much recognition in the latter part of the nineteenth century and then took off in the 1970s in industrialized countries around the world. This modern movement is attempting in many locales to deal with an astonishing range of human-caused problems, and these efforts are being worked out in unique ways in different cultures. This diversity cannot be easily understood unless local efforts are connected to ancient roots that shape values, language, and political possibilities. As suggested recently,

> Concern for [nonhuman animals] is such an ancient theme that scholars often have identified it with the very beginning of religion. The best-selling scholar Karen Armstrong in her 2006 book *The Great Transformation* observed how care about all living beings became a hallmark of the important time in prehistory called "the axial age." During the period from 900 BCE to 200 BCE, ancient religious sages in China, India, Israel, and Greece taught that "your concern must somehow extend to the entire world. . . . Each tradition developed its own formulation of the Golden Rule: do not do to others what you would not have done to you. As far as the Axial sages were concerned, respect for the sacred rights of *all beings*—not orthodox belief—was religion." (Waldau 2011)

These roots today are alive and well. One not only finds a fascination with other-than-human living beings to be common among human cultures and many religions but also a recurring preoccupation in many individuals' lives even when leading cultural institutions (such as those involved in legal systems or established religious traditions) promote a radical human-centeredness as a justification for dominating all other living beings. Today, then, there are many powerful forms of concern for animal protection, the best-known example of which is the ethic of kindness and anticruelty that is easy to identify across place and time.

The fact that the antecedents of today's animal movement are ancient and widespread has not, to be sure, produced consistent results. The prominent commentator Matthew Scully, who happened also to have functioned as the senior speechwriter of President George W. Bush, observed in a best-selling book,

> No age has ever been more solicitous to animals, more curious and caring. Yet no age has ever inflicted upon animals such massive punishments with such complete disregard, as witness scenes to be found on any given day at any modern industrial farm. (Scully 2002, x)

However one chooses to characterize or caricature the animal movement, it must be said that its modern form has found much success even though advocacy for "the cause" has often had to go forward in highly polarized circumstances. The polarization, which has impacted the movement in important ways, comes from diverse sources. Some has been prompted by people and industries that use nonhuman animals as mere resources—opposition from such quarters might easily be understood as an instance of Upton Sinclair's insight that "it is difficult to get a man to understand something when his salary depends upon his *not* understanding it."

Advocates of conservation and environmental protections have also experienced opposition inspired solely by reluctance to give up an income stream. Similarly, conservation advocates have seen opposition grounded in a diverse array of claims about "tradition," private property rights, and privileges of many other sorts. A common element in much of the opposition to the animal movement is thus familiar to many conservationists—an abiding, almost autistic human-centeredness anchored in a formidable coalition of business, political, and religious rationalizations about why the more-than-human world (Abram 1996) was made for our species.

But some of the opposition to the animal movement stems from the discomfort that advocates of conservation and environmental protection have with the animal movement. Such opposition has been inspired by more than conservationists' well-known frustration with some animal protectionists' insistence that we must protect even invasive species that are harming local flora and fauna. In fact, the tenor and depth of some conservationists' opposition to the animal movement is often more quotidian, as it were, and reflects ways in which conservationists and environmentalists can at times *ignore nature*. Consider two examples, although many more could be listed.

> A faculty member at a major American law school laments that even in the ecology-conscious northwest of the United States there are conservationists known for aggressive challenges to power plant emissions of greenhouse gasses who simply refuse to recognize the impacts of greenhouse gas emissions from factory farming. What is tragic about this, of course, is that there are major reports, such as the 2006 report of the United Nations Food and Agriculture Organization entitled "Livestock's Long Shadow" (FAO 2006) that reveal startlingly high figures for the emission of greenhouse gases from industrialized agriculture—in fact, as this report confirms, industrialized agriculture as a sector *out emits*, as it were, the *entire* transportation sector.

> Couple this with the tremendous pollution and social dislocation problems created by industrialized agriculture as described in a 2008 report published jointly by the Pew Charitable Trusts and the Johns Hopkins School of Public Health (2008) and there are many reasons environmentalists and conservationists have reason to join the animal movement in decrying factory farming.

Ironically, focusing on industrialized agriculture's harms to nonhuman animals is a classic inconvenient question that even Al Gore avoided in his poignant *An Inconvenient Truth* (2006). Such myopia may be driven by the denial and/or self-indulgence of those used to eating the consumer products of factory farming, but it is, nonetheless, a paradigmatic form of ignoring nature. What could possibly justify such myopia and continued silence in the face of a judgment (PEW Commission 2008) that "by most measures, confined animal production systems in common use today fall short of current ethical and societal standards"?

One explanation for the refusal of conservation-minded individuals or organizations to address factory farming may lie in the tyranny of small differences. Heavy focus on a few differences operates as a kind of political correctness and allows conservationists to ignore the natural consequences—deep harm to the nonhuman animals, the workers, the local communities' econiche, watersheds, and the larger community of life's future as impacted by global climate disruption. What else could prevent conservationists, so like animal protectionists in repudiating rabid forms of human-centeredness, from working hand in glove with the animal movement on this issue? Fairness and balance require acknowledgment, to be sure, that there have been equally appalling versions of this phenomenon among some animal movement people when it comes to supporting certain environmental causes (some of these problems are described below).

A second example of some conservationists' caricature of the animal movement helps nuance how the tyranny of small differences can prompt otherwise committed people to ignore nature. A conservationist affiliated with a major midwestern law school observed to me at a conference on animal law at Harvard Law School in February 2011 that his conservation colleagues on campus *oppose* establishment of an animal law course at their university. Since contemporary courses in animal law often give much attention to protections for *individual* nonhuman animals, those who oppose the inauguration of an animal law course, which can now be found in more than two-thirds of accredited American law schools, may be motivated by their fear of a loss of human privilege. If so, they share a fear experienced by fundamentalist Christian communities who opposed any talk of "animal rights" because they deem such talk to hint at "moral equivalency" (this was the explanation given to me at a November 2010 meeting in Washington, DC, of representatives of conservative religious leaders discussing how their communities could foster animal protection). Again, fairness and balance require acknowledgment that there are *many* secular and religious communities and individuals, as well as intrepid conservationists, who absolutely refuse to allow such a fear to banish animal protection concerns from their discussions.

The animal movement has, then, both supporters and opponents in high and low places. But consider the movement's profile in certain public policy discussions.

Public Policy Ignoring Nature?

In some ways the animal movement is a factor in various policy discussion circles around the world—for example, each year thousands of legislative proposals are now being made that would, if enacted, put some additional protection into effect. Yet, in major ways, the animal movement continues to be marginalized in important public policy discussions—in the United States, for example, the major graduate programs where students study "public policy" are dominated completely by a worldview (as evidenced by curriculum, faculty interests, and publications) that advances a humans-first agenda that is the bane not only of animal movement advocates worldwide but also conservationists and environmentalists. As the environmental educator David Orr (1994, 5) has suggested, "The truth is that without significant precautions, education can equip people merely to be more effective vandals of the earth."

Yet despite many kinds of opposition and the often polarized milieu in which the ideologically and geographically diverse animal movement now goes forward, many segments of the movement are moving astonishingly quickly. Other segments are, it is important to note, moving at a slow pace, and still further segments are not moving at all. Such a pattern is not uncommon for social movements, of course, for as one philosopher-filmmaker suggests (Pontecorvo 1966), "It's hard enough to start a revolution, even harder still to sustain it, and hardest of all to win it. But it is only afterwards, once we've won, that the real difficulties begin."

Overall, however, there is reason to agree with one historian (Beers 2006) who concludes that the modern version of the animal movement has moved astonishingly quickly and has had "a far greater impact on society than previously suggested." This historian also suggests that the animal movement has some fundamental features that any other social movement might easily emulate. These include multiplicity of strategies for a multi-issue agenda, diversity of social background and ideological orientation of those within the movement— denominated radicals or moderates or conservatives—powerful membership demographics dominated by both women and youth, and an overlap between leaders and activists in animal protection and other social movements, such as civil rights and women's rights.

With this sketch in mind, we turn to three questions that will help anyone see the animal movement as an important cousin of the conservation movement and general education efforts pertaining to the importance of ecological visions.

First Question—Why Is the Animal Movement Important to the Worldwide Conservation Movement?

The animal movement has had success with certain groups that are crucial to all forms of the conservation and environmental movements. Animal protection is, for example, a key concern of the younger generation. While even the mere mention of the controversial group People for the Ethical Treatment of

Animals puts off both many animal protectionists and conservationists, it is worth mulling over what the following study suggests about the appeal of animal protection to young people. In 2006, Label Networks, Inc., a marketing company that described itself as "the leading global youth culture marketing intelligence + research company authentically measuring the most trendsetting and mainstream subcultures in the world," published its latest "Humanitarian Youth Culture Study." According to the study, "PETA is the #1 overall nonprofit organization that 13–24-year-olds in North America would volunteer for . . . peaking among 13–14-year-olds at 29.1% of this age group." What is significant is that PETA held nearly a two-to-one margin over the runner-up (the Red Cross, which had just received great publicity after the 2005 devastation of Hurricane Katrina).

Similar points could be made about the role of women in the animal protection movement—overall today, entering classes in US veterinary schools are about 85 percent female, with some entering classes *entirely* female. For six straight years (1999–2005) the graduate program I directed at the Center for Animals and Public Policy at Tufts University's Cummings School of Veterinary Medicine had only female students (the class sizes ranged from eight to thirteen), and in some years the pool of thirty to fifty applicants included only one or two males.

Further, the animal movement has fostered important education frontiers— a telling example involving Harvard Law School appears below. Equally revealing developments regularly appear today in an astonishing number of academic fields that can now be meaningfully grouped under umbrella terms like "animal studies," "human-animal studies," and "anthrozoology." When assessing the "why?" of such developments, one obvious answer is that the animal movement has appeal to individuals because it addresses issues that directly involve the day-to-day world of these individuals and thereby brings meaning in existentially significant ways—more on this below.

Admittedly, many people today remain, because of the tyranny of small differences, unconvinced that the animal movement's insights have any commonsense features at all. But it is important to consider further what any "tyranny" of relatively minor differences really means when such differences are allowed to overshadow both important essential similarities and the political advantage that making common cause would create. Indeed, from the vantage point of those *not* intensely involved with either one of these movements, the similarities of the two movements greatly outweigh the dissimilarities. This example brings home an obvious point—*something is amiss* when the mere fact of relatively unimportant differences is allowed to obscure obvious similarities and shared purpose.

Second Question—Why Does the Animal Protection Movement Move at a Different Pace, and in Different Ways, Than the Conservation Movement?

Conservation and animal protection both call upon fundamental human abilities to recognize realities of other living beings. The animal movement calls upon individuals, businesses, and governments to notice other animals and take them seriously as individuals and members of communities that live in habitats that are essential to their well-being. Similarly, conservation-minded active citizens call upon individuals, businesses, and governments to notice and take seriously the other-than-human world, to protect species, ecological niches (including communities), and biodiversity, and to keep these realities high on the list of public policy priorities.

The overlap is impossible to ignore—other-than-human living beings count, although there is some difference as to whether the focal point is on sentient individuals mattering in and of themselves or as members of species functioning in ecosystems. This difference is important, but it does not eclipse the common insight that individuals in their communities must thrive in habitats that are essential to their well-being. Beyond any difference is, in fact, a panoply of shared insights that drive any human individual to value the individual lives and communities beyond the species line. In both conservation and the animal movement, then, nonhuman "others" are included within our moral circle. While the animal movement is sometimes caricatured as including only those "others" and "neighbors" that happen to be dogs and cats in our households or in nearby feral colonies, the animal movement has many segments. For example, local animal protectionists often work diligently on behalf of local or migrating birds, the nonhuman primates that are our close evolutionary cousins, many charismatic megafauna such as elephants and tigers and wolves, and also many tiny mammals, amphibians, and reptilians in our backyards and beyond. Also included are the raccoons, mice, deer, coyotes, and so many more who are "out there" beyond the human community.

As each of these movements calls upon ethics, its advocates must struggle against a powerful, often ignorance-driven set of practices cloaked in human-centered justifications that, frankly, have long harmed both humans and nonhumans. It cannot be denied that "ethics" in conservation contexts sometimes does not mean exactly what "ethics" means in a variety of animal movement contexts. But by and large, both movements rely on versions of ethics that are, to mix metaphors, cousins, natural allies, and kindred spirits. Clearly, those who would protect nonhuman animals are necessarily committed to protecting those living beings' social and ecological worlds. How else can a nonhuman animal thrive if not in a healthy habitat for itself, its offspring, and its larger community? This is true whether the nonhuman lives with a human family as a companion animal or lives beyond human communities.

One of the motivating geniuses of the animal movement is that it puts front and center a version of ethics that is highly relevant and comprehensible to individual human actors—the level of individual-to-individual encounters that humans have with nonhuman neighbors they encounter. Accordingly, a theme in animal protection is emphasis on the choice-to-choice, day-to-day features of not harming *individual* nonhuman others. Thus one paradigm of animal protection centers on saving specific animals from specific harms in specific contexts. Such actions are, from an ethics standpoint, easily understood by human individuals—in other words, preventing harm to individuals is something each of us natively comprehends and can do in many circumstances. This explains why so many animal protectionists start with a concern for animal shelters, feral cats and dogs, and local fauna. Further, from the standpoint of broad theoretical considerations, there are multiple grounds for suggesting that individual-to-individual sensibilities are not only important but also have a pronounced realism and practicality to them—everyday experience for many people confirms that caring about other individuals is a profoundly important experience (this may in part be explained by Viktor Frankl's suggestion below that "self-actualization is possible only as a side-effect of self-transcendence"). There is also a Darwinian account—namely, Robert Trivers's reciprocal altruism analysis (Trivers 2011 is a good summary), that suggests a biological basis for altruistic caring—in other words, individual humans are products of evolutionary forces that created survival advantages for those living beings that had a predisposition for individual-to-individual cooperation. Other ethical theories as different as theistic Christian ethics, nontheistic Buddhist ethics, feminist care ethics, and the simple rational calculations of Peter Singer's utilitarian approach in the tradition of Jeremy Bentham also provide ways of understanding that one-on-one caring of this kind has a central place in human life.

Conservation does not emphasize the foundational importance of individual-to-individual encounters in quite this way, of course, although the ethics-based insights that are the heartbeat of the conservation vision nonetheless can have great personal meaning. True, some animal movement advocates feel that conservation efforts are less existentially relevant to the daily lives of human actors because they are less immediately "do-able" in the sense of having discernible, immediate impacts *on specific living beings*. Why? Because a focus on taking responsibility for one's actions as those actions impact discernible, local individuals cashes out as a focus on choice-to-choice, day-to-day ethics. Focus at the individual-to-individual level provides a very personal, existentially meaningful answer to the root question of all ethics—namely, "who and what are the others in my larger community?"

Paying attention to this question obviously has great potential for sharpening ethical skills since individual-to-individual protections are something each of us as an individual intuitively understands. Further, this skill thus has enormous potential for *remedying any tendency to ignore nature*. This benefit has crucial spin-offs completely relevant to conservation since it can foster recognition of the importance of habitat and group interactions to the lives of the individual nonhuman animals one notices. The protection emphases of the animal movement thus create important personal and educational dynamics for *conservation*-inspired lessons about habitat, wildlife communities, and econiche considerations, all of which are building blocks that conservationists can use for their own insights about the importance of taking full responsibility for living in a shared environment.

Third Question—What Are the *Most* Relevant Features of the Animal Movement to the Conservation Movement?

Consider first this development in law and legal education.

> "Animal law" has become a distinct field in the last decade, and the history of this development is striking. When Harvard Law School adopted its first animal law course in 2000 as a result of petitions signed by scores of students year after year, the American legal education establishment took notice, even though fewer than a dozen American law schools then offered such a course. By the year 2002, more than 40 of the almost 200 law schools accredited by the American Bar Association offered such a course. As of spring 2010, the number of American law schools offering at least one animal law course will exceed 130. This tenfold increase within a decade has been driven by student requests. (Waldau 2011, 86)

When students seek education and invest it with personal commitment, this creates optimal circumstances for teaching important skills like rigor, critical thinking about the role of values, and recognition of the power of diversity and tolerance. Further, looking beyond the species line on the basis of these skills fosters a deeper sense of both the riches of the humanities and, as importantly, their limits as well. Similarly, animal law as a topic fosters a deeper sense of the riches of our sciences and their limits. Sciences can tell us certain things about which animals have which features, but in the words of Erwin Schrödinger, the 1933 Nobel Prize winner in physics, even though science "supplies a lot of factual information, and puts all our experience in magnificently coherent order," it nonetheless "keeps terribly silent about everything close to our hearts, everything that really counts."

Animal law, it turns out, is but one field that has exploded in popularity about this matter "close to our hearts." Other developments include the rapid spread of courses in the newly emerged field of religion and animals, history-based inquiries to our past treatment of other animals, social science–based approaches that illuminate current realities around the world, courses focusing on our deeply moving literature and other arts dealing with nonhuman animals, and cross-cultural animal studies. Today, thus, university-level education is seeing entire degree-granting programs at both the undergraduate and graduate levels emerge in multiple fields. These education-based developments clearly dovetail with the development of awareness that reaches beyond the species line, all of which helps conservation and environmental and ecological awareness develop ever more fully.

Conservation's Lessons about Animal Protection

As suggested in the second paragraph of this chapter, there is a very real "two-way street" feature of humans' extending their attention and care across the species line—the animal movement is impoverished if it fails to take to heart the core insights of the conservation movement. These insights' relevance to animal protection in general is evident in many fields, but particularly easy to see in conservation biology, restoration biology, recuperation biology, and perhaps most obviously in the national and international efforts to reduce trade in wildlife, such as the efforts of the Traffic group (http://traffic.org) and the people who run the Wildlife Enforcement Monitoring System (WEMS) Initiative (http://wems-initiative.org/). Conservation often goes well beyond well-known, even iconic nonhuman animals that draw the attention of many in the animal protection movement—for example, the living beings now dominating the International Union for Conservation of Nature (IUCN) Red List, or the work of the Center for Biodiversity, are the smaller, less well known animals that the animal protection movement rarely focuses on in any detail. The natural connections between conservation and animal protection are also evident in the way that many different kinds of scientists today openly call out their desire to protect the species they study—a recent example appears in the subtitle of Diana Reiss's 2011 book *The Dolphin in the Mirror: Exploring Dolphin Minds and Saving Dolphin Lives*.

It is important to recognize, then, that animal protection can come in truncated, ultimately harmful forms. Doing so makes it easier to see why naming companion animals as representatives of the whole of the more-than-human world may seem appealing but nonetheless involves serious risks. Thus even though the emergence of dogs and cats as humans' favorite domesticated animals surely has opened many a mind and heart to the depths and breadths of caring beyond the species line, dangerous implications are nearby if one fails to see the human-centered features of such valuing. Dogs and cats may have a certain quality of life when brought into human households, but unreflectively making the category "pets" fully representative of *all* nonhumans can foster a mentality whereby lots of different nondogs and noncats are made "pets." Such a companion-animal-centered paradigm drives the incredibly harmful trade in exotic species.

Similarly, an exclusive concern for one's own family dog or cat can cause people to play down the harsh realities suffered by other nonhuman animals used in research by the pet industry. Lots of unowned dogs, cats, and other animals are consumed in the never-ending search for marketable products and medical breakthroughs to be sold to citizens in industrialized countries who own the cats and dogs on which the equivalent of one-hundred-plus billion US dollars is spent annually. Foregrounding the needs and pleasure of dogs and cats also creates pressures on other animals because dogs and cats are often fed products marketed by the food animal industry. Lastly, dangers exist for companion animals by virtue of their very popularity, since overbreeding and inbreeding creates any number of serious medical problems that compromise individual animals' lives.

In summary, making companion animals the definitive representative of all other-than-human animals has risks, especially if someone willingly refuses to see that the companion animal category can create extensions of human power. Such extensions of human dominance can exacerbate the already profound harms and imbalances challenged by both the animal movement and conservationists. Consider how protecting feral cat colonies at all costs (that is, without any acknowledgment of how human support of such animals amounts to humans harming the wildlife the cats impact) is really just another form of human-centeredness. Protecting feral cats from disease and starvation is laudable, but it can be done in ways that are very destructive of the fauna and flora where the feral cat colony exists. If the failure to see these consequences is caused by ignorance or mere favoritism of cats, focusing on individual-level protections for the feral cats alone can be seriously out of balance with both conservation insights and key animal protection insights.

It turns out that even simple, pleasurable acts, like walking with dogs in conservation areas, can thrust human domination into the center of ecological niches that are disturbed and imbalanced by such presences. This is no way to challenge the salient fact that walking with dogs is among the most beautiful experiences in the world, but when such pleasures are indulged with no sense at all of their consequences, the result is merely another mindless *human* privilege that produces a harmful human presence for individual wildlife and the larger ecosystem.

Focusing solely on companion animals as the paradigmatic form of animal protection can, then, prompt a failure to see one of the truly significant paradigms of all life—namely, *free-living* nonhuman animals. As many cultures have attested, it is the presence of lives not under the dominion of the human community that is so important to understanding humans' place in our larger, more-than-human community. Thus, while it is important that companion animals open doors (so evident in the popularity of special legal protections now widely proposed for dogs and cats), this key insight cannot be allowed to obscure the impressive insight that sits at heart of the conservation movement: the paradigmatic status of the free-living individuals in a community that we call "wildlife." Suffice it to say that there are many reflective circles in the animal movement that advance the same insight, but there are also any number of circles where this key insight is ignored. In one sense, then, the animal protection as a whole is clearly a beneficiary in countless ways of the conservation movement's powerful insight that one ignores nature if one fails to consider humans' connectedness to all nonhuman animals because one is concentrating solely on "pets."

Shared Visions and Ethics

There are, in fact, many ways in which such mutual enhancement between the animal movement and conservation takes place, all of which supports yet another conclusion: human individuals need to learn a robust version of ethics, one that prompts individuals to look locally and look more broadly. This is the only way humans can reach a full understanding of how to act in ways that benefit a series of nested communities—the individual himself or herself, the individual's family (including companion animals) and local community, the human species as whole, local fauna, and the larger communities of life like the living, interconnected webs that are one's econiche, ecosystem, and the Earth.

There is another class of benefits as well—the social realities created by animal protection around the world. The creation of responsible, alert individuals behooves any civil society, and that in turn fosters communication across borders and cultural divides. Both animal protection and conservation easily prompt international awareness, and all the more so when both are seen as possible, important, and relevant to local and international communities that together comprise the larger world. In such an environment, a healthy dialectic between local specifics and worldwide generalities can exist.

An example that illustrates such cross-fertilization comes from an area that some think "backward" in its understanding and employment of the insights that drive both the conservation and animal movements. This is the world of religion. Getting religious communities involved in each of these social

movements is important, of course, for a very high percentage of humans today (well above 80 percent around the world) self-identify as religious. So it is
instructive that a powerful "religion and ecology" movement has emerged, a
most outstanding example of which is Yale University's Forum on Religion and
Ecology (http://fore.research.yale.edu/). At the American Academy of Religion
(an association of more than 10,000 scholars and teachers), there is a robust
commitment today to making practical, sustainability-focused practices and
workshops a part of the group's annual conference. A high profile workshop
held at the 2011 annual meeting provided a revealing example of how the conservation movement and animal movement easily surmount any differences
in these social movements. Sponsored by the American Academy of Religion's
sustainability group, the workshop focused on teaching about nonhuman animals in the classroom as a win-win approach that deepens the driving insights
of each movement and thereby benefits the larger community.

These two movements share the insight that our own and our children's
character development are impacted by more than our treatment of *human*
others—this is the point of all humane education and the driving insight of
the compassion/anticruelty traditions that began originally in religious traditions but have been sustained in a wide range of secular cultures as well. It is,
in fact, our treatment of nonhuman animals, our local environment, and the
Earth as our largest community that shapes our character as individuals and
as a species. This is the practical insight driving the following passage (Waldau
2011, xv) about ways that animal protection offers benefits to humans.

> When humans experience others—again, it matters not whether these "oth-
> ers" are human or members of some other species—paradoxically this experi-
> ence of getting beyond the self allows humans to become as *fully human* as we
> can be, that is, human in the context of a biologically rich world full of other
> interesting living beings. As Viktor Frankl (1982, 115) wrote in his influential
> *Man's Search for Meaning*, "self-actualization is possible only as a side-effect of
> self-transcendence." This is true not only for human individuals but also for the
> human species as a whole. This has in fact been the message of many religions,
> many ethical systems, and various wisdom traditions anchored in small-scale
> societies.

There are, then, many reasons to emphasize the particular value of animal
protection and conservation working together to notice and take seriously,
not ignore, many facets of the more-than-human world that is our larger community. The following section invites recognition of how both conservation
and the animal movement foreground an invitation to understand the crucial
problem of violence in our shared world.

Seeing Violence Better

Because our human species has long had a deep concern over how often we war on each other, our talk of violence often centers on nation-against-nation wars. There is of course good reason to be concerned. The historians Will and Ariel Durant (1968) observed that over the last 3,421 years, only 268 had been without a war. Few believe today that human behavior in recent decades gives any hope that this pattern will change.

We all know that in addition to war there are also many other human-on-human oppressions that need our fullest attention. So even though it is often assumed that nothing could be worse than war, in 2009 Daniel Jonah Goldhagen published *Worse Than War: Genocide, Eliminationism, and the Ongoing Assault on Humanity*. The author explains his title in this way (xi):

> Our time, dating from the beginning of the twentieth century, has been afflicted by one mass murder after another, so frequently and, in aggregate, of such massive destructiveness, that the problem of genocidal killing is worse than war.

Indeed, for Goldhagen, "mass murder and eliminations are among our age's defining characteristics." One particularly tragic set of problems has been the harm visited upon the indigenous, small-scale societies still in existence. Although some people try to soften this tragedy by directly or indirectly claiming the problem occurred in *past* centuries, Goldhagen reminds us, "More groups of indigenous peoples have likely been destroyed during our age than in any other comparable time period."

If one is tempted to amend the "nothing worse than war" adage to "nothing worse than war or genocide," there remains that astounding fact that war-based and genocide-based accounts of mass killing are rivaled by stories of yet another, even larger form of human-on-human oppression. On August 23, 2009, the *New York Times Magazine* published a cover story announcing, "women's rights are the cause of our time." The article included excerpts from a book (Kristoff and WuDunn 2009, xvii) describing extraordinary harm done to women around the world.

> The global statistics on the abuse of girls are numbing. It appears that more girls have been killed in the last fifty years, precisely because they were girls, than men were killed in all the wars of the twentieth century. More girls are killed in this routine "gendercide" in any one decade than people were slaughtered in all the genocides of the twentieth century.

Beyond our human-on-human violence, of course, there is much other violence. This is important because, as suggested above, the persistence of any one form of violence strengthens our ability to commit further violence, just as curtailing any form of violence develops perspectives and character that can help us see other forms of violence better.

Advocates of conservation know of the problem of violence beyond the species line in a number of ways, such as the accelerated rate of species extinction caused by human acts, or the relentless mass destruction of wildlife and its habitat. Enter again the importance of protecting other-than-human animals—we can solve some forms of human-on-human violence better if we also recognize that there is an astonishing amount of violence that we do beyond the species line. Conservation and animal protection share this concern. But while conservation-minded citizens naturally are concerned with violence to wildlife and its habitat, the animal movement has a particular concern for identifying the full range of violence done intentionally to nonhuman animals.

One important feature of the animal movement, then, is that its advocates are willing to talk about forms of violence against sentient living beings that are ignored even as they are in our midst—what we do in factory farms, in some zoos, and research laboratories. Although as suggested above, some conservationists today still wish to ignore these practices, disingenousness with regard to any form of violence strengthens our ability to rationalize our right to violence. And once one is used to violence, what does it matter that we harm yet more habitat, push yet another species into extinction, erase yet another human community that must yield to the human juggernaut?

Conclusion—The Future Will Be Chosen

In order to invite the reader to eliminate any risk that minor differences might continue to tyrannize the cooperation between conservation activists and the advocates of the animal movement, I end here with an insight that could equally be described as conservationist wisdom or animal protectionist ethics—"Every choice we make can be a celebration of the world we want" (Frances Moore Lappe). This insight was the basis for my concluding sentence in a book-length discussion (Waldau 2011, xv) of the moral, legal, and environmental implications of animal protection.

> Through writing this book I came to understand that animal rights, as most people described it to me, is about connecting to the meaning of life.

Conservation insights are also driven by our great need to connect to the meaning of life. The two movements discussed in this chapter can, when working together, offer a very special hope—namely, that we *now* live in a time in which we can name war, genocide, habitat destruction, countless unnecessary murders of living beings, and global climate change as our heritage and predicament, but also as *problems* we can choose to face squarely. In a world in which our political systems have slipped into an appalling lack of civility, and religious traditions struggle to gain the spiritual character to promote peace rather than division, the prospect of conservation's protective, constructive, healing insights being linked to the animal movement's power to instill and nurture individuals' ethical character is a soothing one. This can happen if the active citizens in each of these movements will choose to work together. This is one choice *we* can make as a way to celebrate the world we want to live in and leave for our children.

References

Abram, David. 1996. *The Spell of the Sensuous*. New York: Pantheon.

Armstrong, Karen. 2006. *The Great Transformation: The Beginning of Our Religious Traditions*. New York: Knopf.

Beers, Diane. 2006. *For the Prevention of Cruelty: The History and Legacy of Animal Rights Activism in the United States*. Athens: Swallow Press/Ohio University Press.

Durant, Will, and Ariel Durant. 1968. *The Lessons of History*. New York: Simon and Schuster.

FAO/Food and Agriculture Organization of the United Nations. 2006. "Livestock's Long Shadow: Environmental Issues and Options." www.fao.org/docrep/010/a0701e/a0701e00.HTM.

Frankl, Viktor E. 1975. *The Unconscious God: Psychotherapy and Theology*. New York: Simon and Schuster.

Goldhagen, Daniel. 2009. *Worse Than War: Genocide, Eliminationism, and the Ongoing Assault on Humanity*. New York: PublicAffairs.

Gore, Albert. 2006. *An Inconvenient Truth: The Planetary Emergency of Global Warming and What We Can Do about It*. New York: Rodale Press.

Kristof, Nicholas D., and Sheryl WuDunn. 2009. *Half the Sky: Turning Oppression into Opportunity for Women Worldwide*. New York: Alfred A. Knopf.

Label Networks 2006. "Humanitarian Youth Culture Study." Retrieved August 7, 2006, from http://69.93.14.237/. The abstract is available at http://www.humanespot.org/content/humanitarian-youth-culture-study-2006.

Lappe, Frances Moore, and Anna Lappe. "The Online Home of Frances Moore Lappe and Anna Lappe," http://www.smallplanet.org.

Orr, David. 1994. *Earth in Mind: On Education, Environment, and the Human Prospect*. Washington, DC: Island Press.

Pew Commission on Industrial Farm Animal Production. 2008. "Putting Meat on the Table: Industrial Farm Animal Production in America." A Project of The Pew Charitable Trusts and Johns Hopkins Bloomberg School of Public Health.

Pontecorvo, Gillo. 1966. Director, *The Battle of Algiers*. Original title: *La battaglia di Algeri*. Rialto Pictures.

Reiss, Diana. 2011. *The Dolphin in the Mirror: Exploring Dolphin Minds and Saving Dolphin Lives*. Boston: Houghton Mifflin Harcourt.

Scully, M. 2002. *Dominion: The Power of Man, the Suffering of Animals, and the Call to Mercy*. New York: St. Martin's Press.

Trivers, Robert. 2011. *The Folly of Fools: The Logic of Deceit and Self-Deception in Human Life*. New York: Basic Books.

Waldau, Paul. 2011. *Animal Rights*. New York: Oxford University Press.

3

Ecocide and the Extinction of Animal Minds

Eileen Crist

IN THE LAST FEW DECADES two momentous realizations have presented themselves to humanity. One, we are in the midst of an anthropogenic crisis of life—an extinction spasm and ecological unraveling that is heading the biosphere into an impoverished biogeological era. And two, in the course of history, especially the history of domination-driven Western culture,[1] humanity has tended to deny or underestimate the mental life of animals. Besides the coincidence of their timing, the coming into knowledge of biodiversity's collapse and of the hitherto-unrecognized richness of animal minds appear entirely unrelated events. Yet there is an urgent connection between the contraction of life's diversity and the dawning appreciation of animal minds: just as we are beginning to recognize that we share the Earth with beings of extraordinary physical *and* mental complexity, we are losing that shared world.

In this chapter I explore conceptual and historical links between the unraveling of life and the denigration of animal minds—links that have foreshadowed the present historical moment of grave loss. The exploitation of the biosphere and the deprecation of animals stem from the same source: the separatist regime humanity has created in which we have entitled ourselves to unlimited access to the planet on the (tacit or declared) grounds of self-ascribed superiority over other species in general, and animals in particular. But the connection between the destruction of biological wealth and the belittlement of animals goes deeper than the obvious resonance of colonizing the natural world while denigrating its nonhuman indigenes. I argue that the long-standing denial or disparagement of animal minds is *causally* implicated in the devastation of the biosphere. Through the portrayal of animals as inferior beings, and eventually even as mechanical entities, the objectification of the natural world and its transformation into a domain of resources was vastly facilitated. As animals became successfully represented in dominant discourses as devoid of agency and experiential perspective—thereby becoming construable as means for human ends—a fortiori the (apparently) nonsentient domains of forests, rivers, meadows, oceans, deserts, and mountains (in fact, of any landscape or seascape) were made accessible to the human race without accountability or restriction.

In our time, the interface between ecocide and animal minds is tragic and ironic. Just as humanity is beginning to acknowledge and document a largely unknown world—the inner world of animals—that very world, in its diversity of forms of awareness, is coming undone. Even as human beings are becoming more receptive to the viewpoint of human-animal evolutionary continuity (not just of physical plan but of mental structure as well), we are collapsing the biosphere whose plenum of beings we might appreciate and experience through this newfound understanding. We are in danger of physically constructing a world that is as indigent in minds as René Descartes conceptually constructed it to be. But the reality and significance of this event—of losing not only physical manifestations of diversity but diverse manifestations of mind as well—eludes most of humanity. We live in a time tipping us into a planetary physical-cum-consciousness monoculture, yet most people continue to ignore Nature, failing to recognize this event horizon and remaining endlessly distracted by the noise of personal, cultural, economic, and political dramas.

I

In the late 1970s life scientists embarked on the systematic investigation of the effects of our colonization of the biosphere (understatedly known as "our ecological footprint"). An odd forty years later, the amassed knowledge is considerable and the picture it presents is grim. Extinction of species is occurring at a rate that exceeds background (nonanthropogenic) extinction by orders of magnitude: in the absence of the human impact, a handful of species would be vanishing each year; by contrast to this rate, thousands of species are driven to extinction yearly. The biosphere, moreover, is not only losing kinds of life (species and subspecies), it is also losing its abundance of wild places and creatures. The once enormous flocks, schools, and herds of animals are vanishing, and so are their migrations and the dynamic communities their migrations supported. Populations of top predators—tigers, lions, jaguars, wolves, bears, sharks, and others—are a fraction of what they were even a century or a few decades ago. Species and populations of fish (freshwater and ocean) are plummeting. On land, the once-abundant boreal, temperate, and tropical forests of the Holocene are vanishing or being displaced by biologically diminished second (third, fourth, etc.) growth or by tree plantations (which are not forests). Half of the world's biologically fabulous wetlands were lost in the twentieth century alone. Both on land and sea these losses signify the unraveling of complexity—of the interactions and cycles that connect life forms in intricate relationships. Ecological impoverishment is intensified by alien (mostly generalist) species invasions, occurring via the globalized trade and travel of a growing, consumer-oriented population that is biohomogenizing the Earth. Agricultural, chemical, and emission pollutants are exacerbating the crisis of life immensely—through the direct killing of chemical runoffs, the (in)direct killing of climate change, and the not-yet-fully-understood perils of ocean acidification. The unique qualities of Earth's places are disappearing. The diverse tree of life is being turned into a stump.

The collapse of biological wealth is escalating because human population increase, economic growth, consumption patterns, and reckless technologies are impacting a finite planet that cannot resist the scale of this onslaught. The underlying driver of these trends is the human supremacist mindset that has enshrined a no-limitation way of life—including no limitation on reproduction, no limitation on consumption and economic growth, and no limitation on the kind of technologies unleashed in the world. The sheer cumulative weight of a no-limitation civilization is ruining whatever it touches. Wherever we turn we find the living world coming undone. No sooner do we process news of the amphibian crisis, but we learn that turtles, lizards, butterflies, and birds are also in jeopardy. As we struggle to come to grips with honeybee collapse, we hear that bats are dying off. Bushmeat subsistence and trade are obliterating the animals in Africa and elsewhere. Our assault on the oceans—with the depredations of industrial fishing, fish factory farming, slaughter of marine mammals, dead zones, trash, and adverse consequences of atmospheric and climatic change—is beyond the ken of conscience. Forests are being chewed up to be replaced by tree plantations, agro-industrial monocultures (of soybeans, palm oil, and so on), and ranching and pasture ventures. The most fecund places everywhere have long come under the plow.

As tragic as each Earth news item is, the devastation lies in the whole picture. We are dismantling the very qualities that constitute the biosphere: diversity of life forms, complexity of life's interrelationships, and abundance of indigenous beings and unique places on Earth. These interconnected attributes form the cauldron of Earth's life-generating creativity: they are the foundation of life's evolutionary power, fecundity, and resilience. As a unity, I have referred to the qualities of diversity, complexity, and abundance as *the flame of life* (Crist 2004). In the wake of the havoc humanity has unleashed, life will almost certainly persist, but the flame of life is being extinguished.

Biological collapse is the historical reckoning of human *exemptionalism* and its refusal to even countenance the idea of a limited habitat niche for humanity within the biosphere. On the other hand, the denial of the richness of animals' lives goes hand in hand with human *exceptionalism*—the elevation of our species' consciousness as a unique and superior kind. Exemptionalism and exceptionalism are two sides of one coin: humanity has exempted itself from many of the biophysical limits that regulate and check animals, on the conviction that our ingenuity and technological savvy can (and should) successfully negotiate virtually any challenge. And over the course of history, colonizing the Earth and disparaging animal life have reinforced each other: the belief in human supremacy has legitimated the exercise of power over, and limitless access to, the natural world (including animals), while the seemingly triumphant domination of Nature has cast an aura of truth over the belief of human supremacy.

The colonization of the biosphere and the legislation of human-animal apartheid are thus historically and culturally intertwined. The hierarchical divide between humans and animals has been erected, in large part, through the erasure of animal minds. By the erasure of animal minds (or animal subjectivity), I refer to the historically dominant, discursively elaborated (see below), attenuated regard of animals as merely existing and reacting, as opposed to acting meaningfully and experiencing being-in-the-world. With concepts like "mind," "awareness," "subjectivity," "numinous," and "inner life," in this paper I am invoking a phenomenological understanding of mind associated with lived qualities of agency, experience, and meaning (see Abram 1996; Crist 1999). In philosophy this conception of mind is called "phenomenal consciousness" (see Hurley and Nudds 2006; Allen 2010). Novelist John (J. M.) Coetzee's regard of animals as "filled with being" and animal rights philosopher Tom Regan's plea for the animal as "subject-of-a-life" allude to phenomenal consciousness in more evocative and down-to-earth ways (Coetzee 1999; Regan 2001). Western canonical representations have suppressed or vitiated these rich dimensions of animals' lives: this is the target of my critique.

While it is dawning on more and more people that we are irrevocably losing biological diversity, complexity, and abundance of free beings and native nature, we still fail to recognize that these losses implicate the contraction of "species of mind" (see Allen and Bekoff 1997). Along with the obliteration of biological wealth in the sense of loss of species, subspecies, populations, ecosystems, and gene pools, the destruction of the diversity of animal minds (of modalities of aware perceiving, being, and experiencing) is impoverishing the

Earth (and ourselves) in ways we do not even begin to comprehend and can barely imagine. Such a comprehension is blocked by a history of disparaging animals that continues to cast its long shadow on the human mind.

Western intellectual traditions have shored up the belief in an essential divide between humans and animals, lending it philosophical, theological, scriptural, and (natural- and social-) scientific elaborations (Rodman 1980; Steiner 2005; Cavalieri 2006). Indeed, the dominant view of Western thought, from antiquity to the present, has been that human cognitive, character, and emotional virtues distinguish us from animals and that animals are *lacking*. Animals have been deemed as lacking (depending on historical and intellectual context) in the possession of reason, language, free will, religion, culture, morality, history, perfectibility, technology, complex emotions, self-consciousness, being, metacognition, and/or personhood. Representations of human nature, and of human-animal comparisons, have almost invariably been framed in terms of "difference." From the classical era forward, Western thought has repeatedly rehearsed the refrain of human superiority by consolidating and elaborating discourses of difference. Environmental thinker John Rodman put it memorably: there has been an "almost universal tendency," in the moral and political philosophical canon, "to assume the Differential Imperative as self-evident" (1980, 54).

Why have discourses of human-animal difference ruled Western traditions? What way of life has "the Differential Imperative" served? One obvious response to these questions is that positioning animals beneath the human realm produced a conceptual scheme that legitimated their exploitation. As Stuart Elden writes, establishing "a distinction from animals becomes a way of ordering, regulating, controlling, and exploiting them" (2006, 284). Such reasoning is cogent: by representing animals as inferior beings—as "subhuman," to cite a weird concept—they could be used, without negative sanction, for food, work, clothing, sport, amusement, experiment, and so forth, and they could be killed or displaced, without reservation, if they happened to foil human interests or be in the way. And yet when contemplated in the context of the broader relationship between human beings and the natural world, the denigration of animals has even more profound import than buttressing the use and killing of animals alone: corralling animals into a (separate-from-the-human) lower realm has arguably been *critical* in enabling an instrumental relation with the natural world in toto—a relationship that eventually degenerated into Nature's full-blown constitution, since the dawn of the industrial era, as a domain of "resources." Establishing a distinction from animals became a key way to stabilize the instrumentalization of the entire natural world. For if animal life had been placed on an existential par with human life—or animals

been recognized as subjects of their lives—then their ruthless exploitation and
that of their homelands would have been rendered morally unfeasible. Conversely, the representational subordination of animals vastly facilitated the total
liquidation of Nature.

In other words, while it seems unambiguous that the disparagement of animals enabled their subsequent exploitation, the historical connection may well
have been, more fundamentally, the other way around: the relentless drive to
dominate the natural world (including animals)—a drive reaching back into
the earliest histories of empire—may have *required* that animals be categori-
cally (in both senses of the word) demoted. But how, given that animals bear
the signs of owning agency, experience, individuality, and meaningful perspective? How could they be (completely or substantially) shorn of such qualities
and attributes, when the human pretheoretical understanding of animals often
concedes (and at times has honored) their mindful lives?

The categorical demotion of animals was a tricky feat: it has been accomplished by the historical creation and solidification of discourses of human-
animal difference, and especially through the refutation or downgrading of
animals' mental lives—a central mission of such discourses. The discursive
constitution of animals that excluded them from the realm of subjectivity had
to trump a nontheorized, open-ended, and relationship-bound understanding of animal being. The frameworks that were elaborated served to overlay,
blindside, intellectually best, and/or deride the natural human ability to discern
and to intuit the inner life of animals. In the next section I discuss a seminal
discourse of difference—the one that inducted, and has profoundly shaped,
the modern view of animal being: the Cartesian version of the Differential Imperative. Descartes succeeded in sharply demarcating the domains of human
and animal, in rendering implausible the existence of animal minds, in exult-
ing the human as the sole terrestrial incorporator of (the rational) soul, and
in promoting these ideas as authoritative representations of the natural order.

II

In the cosmology of the great chain of being, which dominated Western thought
for nearly two millennia, Creation was mapped as a grand and scaled existential order manifesting God's power, goodness, intelligence, and love. Plenitude
and continuity constituted its essential ingredients: the diversity of living and
nonliving entities were portrayed as separated but also linked within an in-
finitesimal gradation. This plenitude and continuum—brimming with variety
and abundance of living beings—was a hierarchical model, originating with
God and his retinue of angelic beings, descending through humans, animals,
plants, and minerals, and sinking all the way down to "the last dregs of things"

(quoted in Lovejoy 1976, 63). A unified model of the Creation, it was spun out of the fundamental polarity of spirit and matter: the entire cosmos stretched between these poles. The human realm was conceived as liminal: straddling spirit and matter, composed of and torn between antithetical yet also somehow fusible qualities. As Arthur Lovejoy put it in his classic work, in the great chain of being Man was "the horizon and boundary line of things corporeal and incorporeal" (1976, 79).

Within this cosmos all things putatively higher were valorized over all things putatively lower. Man flanked the realm of spirit but was also partly trapped in matter, and animals were wedged between the human and plant realms. Every level in the scale was authorized to use the entities below it: animals could use plants, plants could use minerals, and humans could use everything. Animals were especially singled out as "intended [by the divine providence] for man's use according to the order of nature" (quoted in Steiner 2005, 130). Human beings were distinguished from animals in possessing reason and language, which animals ostensibly lacked; animals were distinguished from plants in having sensory abilities and experiences. Perceptual consciousness thus separated animals from life forms below them, while they themselves were placed beneath humans in lacking reason and its affiliated states of consciousness such as self-determination and free will (Lovejoy 1976; Steiner 2005).

Descartes inherited this scheme, including its hierarchical inflexion, dualism of spirit versus matter, and predilection for the (human-animal) Differential Imperative. But Descartes renovated the metaphysics of the great chain of being in a way that further depressed the inferior lot of animals it had propounded. The great chain displayed the qualities of plenitude, continuity, and hierarchy, but Descartes chose to magnify the hierarchical aspect of Creation that served to glorify the human. The polarity between spirit and matter (across which the plenum of Creation yawned in continuous gradation) was transmuted by Descartes into the dualistic *saltus* of soul (or mind) and body—the eternal versus the perishing. This dualism was mapped *specifically* onto the human and animal worlds, and it diverged significantly from the preceding spirit-matter polarity in offering a discontinuous worldview in lieu of a graded scheme. Descartes's dualism exalted man for owning a "rational soul," which while well integrated into the human perishable body was also separate from it and immortal. Animals, on the other side of the divide, possessed a "corporeal soul," which was not a soul in any theological sense but a dimension of physiology that animated movement and organic function. Animals, on Descartes's innovations, became qualitatively distinct from humans, in being, like plants and other lower organisms, merely transient mortal entities. On the other hand, the possession of an everlasting soul situated the human in the ontological company of angels.

In a nutshell, Descartes rehashed but significantly reified the view that "man is godlike, animals thinglike" (Coetzee 1999, 23).

In allegiance with a legacy reaching back to classical and Neoplatonic philosophy, as well as Judeo-Christian doctrine, Descartes distinguished man by his possession of "reason." Reason—an attribute of soul (or mind) and affiliated with language in Descartes's thought—could not be mapped on a continuum as it might (in principle) be within a graded-chain cosmology or later from a Darwinian perspective. Being the manifest aspect of the rational soul, reason was either possessed or absent, and thought-articulating language was the criterion of its presence. Given their ostensible lack of language, animals, according to Descartes, do not act "through understanding but only from the disposition of their organs," and thus "beasts do not have less reason than men, but no reason at all" (Descartes 1989a, 140).

Having been dispossessed of even a rudimentary rational soul, Descartes exiled animals from the realm of spirit. Their perceptual organs, which previously served to flank them beneath humans on the great chain continuum, were downgraded by Descartes as mere corporeal analogues to human perceptual organs, but implying no similarity between animals and humans in experience or awareness. Having "no reason at all," animals could not cognitively process or interpret their perceptions and sensations, and were therefore construed as *not conscious* of what reaches them through their senses. Being devoid of thought, animals were deemed unable to experience their perceptions and sensations *as such*: their percepts were not meaningful and their various sensations did not coalesce as experiences—for example, *as* pain, hunger, pleasure, and so forth. "Pain," Descartes professed, "exists only in the understanding." There are, he added, "external movements which accompany this feeling in us; in animals it is these movements alone which occur, and not the pain in the strict sense" (Descartes 1991, 148).

Descartes delivered an unprecedented portrayal of animal existence as a sleepwalking modality. Once this picture was fabricated, similarities between human and animal behaviors—such as writhing in pain or fleeing at the sight of a predator—could be revamped through a Cartesian lens as only *apparent* similarities. Whereas human behavior was informed by and expressive of the reasoning capacity of the soul, the behavior of animals was orchestrated through corporeal mechanisms without the mediation of understanding or the experience of meaning. "The light reflected from the body of a wolf onto the eyes of a sheep moves[s] the minute fibers of the optic nerves," Descartes ingeniously conjectured. "On reaching the brain," he continued, this motion spreads "the animal spirits throughout the nerves in the manner necessary to precipitate the sheep's flight" (Descartes 1989b, 144). As this passage and others in his oeuvre reveal, Descartes's representation of animals can be understood as a gestalt—it conveyed a theory-laden image of the animal as a wound-up living automaton: a memorable image that could lodge itself like a splinter into the human mind.

Descartes elaborated the theory of mind as an interior, invisible, and unavailable domain. To be consistent with his own reasoning, therefore, Descartes could not actually say *for sure* that animals have no inner life—he could only wager as much. "Though I regard it as established that we cannot prove there is any thought in animals, I do not think it can be proved that there is none," he vacillated, "since the human mind does not reach into their hearts" (1991, 365). This passage has been cited as indicating that Descartes did not definitively believe that animals do not think (see Steiner 2005, chapter 6, on revisionist readings of Descartes's views of animals). But such an interpretation fails to notice that Descartes's equivocation ("the human mind does not reach into their hearts") was far more insidious in its consequences, than a self-assured avowal of animals' automaton-nature. Indeed, Descartes's most damaging legacy about animals is not that he turned them into organic machines—which few thinkers subsequently wholly conceded and which a commonplace understanding of animals mostly declines—but that he insinuated *doubt* about the existence of their minds and, of course, about the richness of their minds. In the wake of Descartes's full-blown privatization of the mind, one could perennially doubt one's judgment about the minds of others and most especially of animals. In addition to implanting skepticism, that could always sabotage one's lucid witnessing of animal subjectivity, the Cartesian gestalt of the animal-automaton might be superimposed on animal being to give rise to the uncanny sense that things in the world are, perhaps, radically different than they appear. In other words, the perennial demurral of the skeptical mindset, in conjunction with the fabricated image of the sleepwalking, merely corporeal animal, insinuated the possibility that the representation of the animal-as-automaton *might be true*. Insinuations go a longer way than brazen assertions, as indeed did Descartes's suggestion of the possible reality of the animal-machine, which, to this day, enjoys credibility.

Descartes's portrayal of animal being is not simply of historical interest. His thought inaugurated modern reductionist traditions regarding animal being, and his legacy of skepticism continues to hold sway. Nor is focus on Descartes intended to scapegoat a thinker who happened to hold an extremely disparaging view of animals. In fact, Descartes was not exceptional: he built his ideas on a long ancestry of narrativizing human superiority, expounding human-animal difference, and underlining animal lack. But Descartes's singular importance concerning the modern view of animals—and therefore concerning how we treat them and their homes—lies in the fact that he was the main conduit of a metaphysical and religious Zeitgeist of human supremacy into its modern materialist guise. By retaining the concept of "soul," Descartes remained congruous with Neoplatonic and Judeo-Christian traditions, but by producing a systematic slippage between the concepts of "soul" and "mind," and between the ideas of "rational soul" and "reason," he channeled (an even more virulent version of) the Differential Imperative into modern thought—including natural and social scientific conceptions.

Once the conceptual-pictorial meme of the animal automaton was instilled into the thought collective—the image of mechanical being lacking a conscious inner mediator and overseer—then epistemological possibilities of explaining behavior without reference to mental states became possible. Behavior could be conceptualized as either induced to occur from without or programmed to arise from within. (Or it could be the deterministic output of outer-stimulus meeting inner-program). Descartes's innovations thus prepared the conceptual territory for the frameworks of behaviorist "stimulus-response," classical ethological "fixed action patterns," behavioral-ecological "genetic programs," and mechanistically conceptualized "cognitive programs" and "neural nets." All have been aspects of explanatory theories for behavior, which, however different or even conflicting, *possess in common* the excision of consciousness from animal action-in-the-world.

By developing the idea that animal behavior can be understood strictly in terms of "corporeal processes," Descartes created a discursive placeholder for the development of concepts and theories of animal behavior that could actively eschew or tactfully avoid mind, while appearing to account for the production of behavior without residue. Behaviorism, which emerged over 250 years after Descartes's death (and which is still an influential school of thought), is comprehensively Cartesian—an operationalization of Descartes's premises and ideas and a disciplinary purveyor (openly or implicitly) of the animal-automaton image. Donald Griffin (2001) aptly described the majority of animal behavior scientists of the twentieth century as "inclusive behaviorists," because even those not working under its auspices acquiesced to the strictures of behaviorism in avoiding reference to mental attributes and conceding the view that mind is interior and invisible. Tellingly, the ways in which allusion to animal minds was frowned upon in the past century's behavioral sciences echoed Descartes's superciliousness: it was deemed an immature, sentimental, or merely folk inclination to see mind in animals—but serious, educated grown-ups should know better and cultivate a healthy dose of skepticism. ("There is not a preconceived opinion to which we are all more accustomed from our earliest years than the belief that dumb animals think," Descartes opined [1991, 365].)

One cannot overstate the damage that Descartes's gestalt wrought, not only for animals who were rendered usable and killable with impunity, but for the integrity of our understanding of animal life and for a dignified human presence on Earth. Through his ideas our kin came under perennial suspicion of being vacant—devoid of agency, experiential perspective, and meaningful lives; animals became entities without intrinsically valuable existence and with scarcely a claim for consideration of their lives and homes. The modern era in which the killing and exploitation of all animals has reached staggering proportions, and the imminent collapse of biological diversity has become reality, carried forward the Cartesian legacy of refuting and doubting animal minds. The concurrence was hardly a coincidence.

III

Ever since the invention of empire, human beings have appropriated the natural world into the world of, and for, people. Over the course of centuries, and accelerating in the last 250 years, humans have continually seized the natural world for amalgamation into our ostensibly separate, all-important realm. The path of the Earth's humanization has been forged via the conceptual, enacted, and technologically mediated transfiguration of the natural world into "resources"—an industrial-age concept that, nonetheless, had been prefigured in the classical and Judeo-Christian worldviews that everything on Earth has been sanctioned or created for human use. The totalitarian scope of planetary takeover called for (and was strengthened by) the discursive dismantling of nonhuman Nature's being-for-itself in order to be turned into being-for-people. Put differently, the ontological self-integrity of the nonhuman domain had to be silenced, so that its ways-to-be-used and people's rights-to-access could stand unrivaled by immanent existential validity claims. Discursive erasure has gone hand in hand with physical destruction.

The domains of subject and object have been archetypically tied with humans and nonliving things respectively. Animals initially appeared to fall within a gray zone: neither subjects (since that would put them on a par with exulted humanity), nor nonsentient life forms or inanimate objects. But over the historical course of the escalating transformation of animals and their places into resources, in the post-Cartesian era the gray zone was abandoned in favor of a sharper demarcation between subject (human) and object (nonhuman nature). To convert, for example, fish into fisheries, forests and trees into timber, animals into livestock, wildlife into game, mountains into coal, seashores into beachfronts, rivers into hydroelectric factories, and so forth, it has been helpful, probably necessary, to represent (and subsequently experience) living beings and their homes as object-like. Transmuting the living world into a domain of resources supervenes after pushing nonhumans and their dwellings into the on-

tological space of "nothing but mere objectivity," as Max Horkheimer and Theodor Adorno described Western culture's schematization of Nature (1972, 9).

Among all Nature's so-called resources, the most versatile and invaluable
have been animals. If the resources of air, water, and soil appear to trump animals in importance, it is only because animals require them. In the words of
The Animal Studies Group: "Almost all areas of human life are at some point
or other involved in or directly dependent on killing animals" (2006, 3; see
also Emel and Wolch 1998; Bekoff 2010.) But if schematizing something as a
resource is achieved through its objectification, then animals present a thorny
problem: because of all living beings, second only to other humans, animals
most strongly resist our objectifying maneuvers. Animals are very difficult to
force into the object box. "Holding their gaze," as Barry Lopez has written, we
"sense the intensity and clarity associated with the presence of a soul" (2002,
297). The numinous life of animals has always been available to nonrapacious
peoples and to receptive human beings—and always been a discomfiting suspicion of the rest. Animals do not simply grace landscapes and seascapes with
their stunning beauty: they electrify the world with—and, indeed, display the
world as—species of mind. Nor do animals simply inhabit our houses, fields,
and farms as animated bodies; they charge human dwellings with their forms
of awareness. To turn animals into resources has required the kind of discursive work that suppresses their subjectivity or succeeds in casting it into doubt.
Such work has been furnished in spades by Western discourses. It achieved its
apogee in the Cartesian gestalt.

On this line of reasoning, then, it is far from incidental that "the question
of animal mind" has been a marginalized idea, proscribed area of inquiry, and
belittled plane of experience. The suppression of animal mind has been the
sine qua non of objectifying animals and paving the modern highway to their
resourcification. At pains of cognitive dissonance, the experience of animals
as numinous cannot coexist with their callous slaughter and exploitation.
Animals' mode of existence thus has been, and had to be, represented in such
a way that its rich dimensions could be blocked from perception. To make the
point with a vivid example, most human beings cannot vivisect an animal and
simultaneously allow themselves to feel into the animal's response; something
has to give. But if an animal can be seen as subhuman or suspected of being,
maybe, sort-of-an-automaton, then its response to being cut open can be not-
seen or shrugged off. The same applies to factory farming, industrial fishing,
deforestation—to the myriad ways of killing animals and destroying their
habitats. Such things can be done to animals after their subjectivity has been
dismantled and their being has been reconstituted, cognitively and perceptu-
ally, as "mere objectivity."

The recognition of animal mind subverts the project of amalgamating the natural world by introducing an insoluble quandary into the world's resourcification. Thus the suppression of animal mind has been a key ingredient in the benighted mission of dominant civilization to colonize and tame the biosphere. If animal minds had *not* been erased, but animals were deferred to as beings with experiential perspectives, then turning their homes into resources—the Earth's rivers, wetlands, soils, rocks, forests, prairies, meadows, mountains, seas, and so on—would raise insuperable dilemmas. If animals had been esteemed as tribes, then the places they live would be regarded as integral to them: for, to invoke a similar example, how can one claim to respect a people, and at the same time burn down their village? Conversely, if animals *could* be successfully objectified (the path taken), then objectifying the rest of the living world would be a cakewalk (as it has been). The objectification of animals has been primarily accomplished through the erasure of their mental lives. And the erasure of animal minds has produced an auspicious climate, and reinforced a deluded humanistic worldview, for the indiscriminant use of everything on Earth as a means to human ends.

What is incredible is how little we actually know about the lives of animals, yet how much people have presumed, over the course of centuries, to know about their limited or nonexistent mental abilities. The discursive frameworks that have disparaged animals, and the Differential Imperative that has informed these frameworks and infected common sense, have kept us from such knowledge. Human arrogance toward animals—gilded in cosmological, philosophical, religious, and scientific systems—has fueled human ignorance about them.

IV

Even as life scientists in the 1970s initiated the methodical documentation of the collapse of biodiversity, a shift was instigated in the study of animal behavior that would eventually pry open scientific inquiry into animal minds. In 1976, Griffin, who had already distinguished himself in the field of animal behavior, published a short book titled *The Question of Animal Awareness: Evolutionary Continuity of Mental Experience*. Up until that time—a mere thirty-five years ago—the topic of animal awareness was all but unmentionable in science, and students of animal behavior in both laboratory and field were pointedly discouraged from raising questions about mental experiences. Despite Charles Darwin's nineteenth-century pioneering work in the field (see Darwin 1871, 1872), during most of the twentieth century the study of animal mind was effectively banned: regarded as an *alleged* phenomenon, and variously characterized as epistemologically unavailable, empirically nonexistent, scientifically naive, folk psychology, or wishful thinking. Given the proscription on the study of animal mind, Griffin's broaching of the topic was courageous. He devoted himself to the field of cognitive ethology that he inaugurated—writing, lecturing, and mentoring many scientists for the next thirty years of his life (Crist 2008). Griffin's landmark book *Animal Minds* (2001 [first published in 1992]) is the culmination of his life's written work.

While the scientific community was markedly ambivalent about Griffin's ideas, and has only slowly extended a welcome to the study of animal minds, Griffin's 1976 book (*Question of Animal Awareness*) encouraged an overdue confrontation with the Cartesian legacy: that only *Homo sapiens* is consciously aware, while all other animals probably lead a sleepwalking existence, or are at best dimly conscious. Work in the growing field of cognitive ethology, alongside (and in alliance with) equally significant developments in environmental ethics and animal rights, are challenging the received view of an unbridgeable divide separating human and animal mental experience.[2] Through their cumulative pull these endeavors may eventually have the power to lift the iron curtain on the inner lives of animals, opening space for the cultivation of a new understanding—and perhaps for a radical transformation of our way of life. In the meantime, while the historical proclivity for underestimating or denying animal minds is finally being punctured, we are also finding ourselves near planet-locked in a humanized world, which looms as the ultimate materialization of the same human arrogance that has long disparaged the numinous side of animals.

Conventional worries about the present ecological predicament deplore the loss of nature's services, the depletion of natural resources, or the forfeiting of ecotourist revenues in the wake of human-caused environmental catastrophes. These concerns, however, merely echo the same human-centered mindset that has driven the destruction of the biosphere in the first place. In fact, the turning of Nature into a human asset-domain *defines* the core catastrophe: by allowing ourselves to be seduced by the instrumental framing of the world as a resource domain, we have sponsored the demolition of animals and their homes (see Foreman 2011). We are not in danger of losing, to cite a platitude, "natural capital" for present and future generations, but on the contrary, having conceptually and physically constituted the world as natural capital, we have nearly lost a living, numinous world.

What is trampled out of existence by a no-limitation civilization is a world of diverse subjectivities—a world that manifests as, and is partly created through, innumerable aware collective and individual sentient actions. In its place, we are erecting a world carved strictly by the brutal knife of human arrogance. By assuming the world is lacking in diverse forms of consciousness, human supremacy is making it so. But the animal songs, calls, trails, burrows, dens, nests, haunts, engineering, landmarks, peregrinations, and migrations that once filled the Earth were shimmering mindscapes, not organism-molded matter. And so, alongside destroying biological kinds, natural habitats, and populations of animals, we are deleting the Earth's noumenal dimensions, elaborated through emotion, intention, understanding, perception, experience—in other words, through varieties of aware beings shaping and adorning the world-as-home. Animals are not "world-poor,"[3] but both the world and our own being are rendered poor without them.

What in social theory is called "the disenchantment of the world" describes a real event well underway: it is the sociopsychological reckoning of the consciousness monoculture that humanity is unwittingly and dimwittedly birthing. The destruction of diverse animal minds de-animates the world, making it spiritually empty—just as the forests, prairies, steppes, savannahs, freshwaters, and seas are rendered literally empty of animals. The humanized world emerging is not only depleted of ecosystems, species and subspecies, nonhuman populations, genetic diversity, and wilderness, it is the Cartesian imaginary realized. The disenchantment of the world, which Max Weber diagnosed as "the fate of our times" (1946), signifies the human soul's experience of the world made soul poor.

Our separatist regime has long been catapulting the biosphere toward the present predicament. Humanity's Earth-colonizing venture, however, was neither a necessary nor inexorable consequence of "who we are": it has been a sociocultural and historical outcome—albeit an outcome that humanity became increasingly unable to escape as people got locked into revamping the world in a way that *appeared* to reflect and reward human supremacy. Human beings treated the world as though the regime of human-animal apartheid were based on ontologically sound principles, and ended up, over time, creating a human-made ontology that appears to display just how "different" and "separate" we are.

But humanity might have chosen a different way of being: to embody the flowering of human consciousness as arising from and residing within an entangled bank of diverse life forms, and splendors of awareness, unique in the cosmos.

Acknowledgments

I would like to extend my appreciation to Phil Cafaro, David Kidner, Rob Patzig, Despina Crist, and Robert Crist for their thoughtful comments and suggestions on an earlier draft of this essay.

Notes

1. I here echo a common critical perspective that Mary Midgley, for example, articulates as follows: "Western culture differs from most others in the breadth of destructive license which it

allows itself, and, since the seventeenth century, that license has been greatly extended" (1995, 95). I would qualify this critical focus, however, in two ways. Firstly, divisions between Western and non-Western cultures hold little if any meaning within today's globalizing world, and with respect to the West-East divide there has been substantial (material and ideational) cultural exchange between them for millennia. Secondly, my critical focus is not on Western culture *as a whole*, but rather on Western culture as a culture of *empire*, which has manifested (and can potentially manifest) among other peoples. I am indebted to Jack Forbes's insightful analysis of the rapaciousness of empire as rooted in a form of mental disease (which he calls cannibalism or the "Wétiko syndrome") that is highly contagious. In concordance with his landmark analysis, I regard the erasure and destruction of animal minds (the subject matter of this chapter) as the work of the toxic and highly contagious meme of human supremacy.

2. The relevant literature is enormous. For a small sample, in addition to literature already cited, see Singer (1985); Mitchell, Thompson, and Miles (1997); Clark (1997); McKay (1999); Smuts (2001); Kidner (2001); Bekoff, Allen, and Burghardt (2002); de Waal and Tyack (2003); Jamieson (2003); Torres (2007); Bekoff (2007); Haraway (2008); Pedersen (2010); and Balcombe (2010).

3. Martin Heidegger characterized animal being with this term. It was one of Heidegger's great blunders to have thus added his "two-cents" to the canon's human-animal Differential Imperative. I call it a blunder because his essays "The Question concerning Technology" and "The Age of the World Picture" are withering and seminal critiques of the human (especially modern) subjugation of Nature and (relatedly) of Cartesian humanism. See Elden (2006); Heidegger (1977a, 1977b).

References

Abram, David. 1996. *The Spell of the Sensuous: Perception and Language in a More-Than-Human World.* New York: Pantheon Books.
Allen, Colin. 2010. "Animal Consciousness." *Stanford Encyclopedia of Philosophy,* http://plato.stanford.edu/archives/win2011/entries/consciousness-animal.
Allen, Colin, and M. Bekoff. 1997. *Species of Mind: The Philosophy and Biology of Cognitive Ethology.* Cambridge, MA: MIT Press.
Balcombe, Jonathan. 2010. *Second Nature: The Inner Lives of Animals.* New York: Palgrave Macmillan.
Bekoff, Marc. 2007. *The Emotional Lives of Animals.* Novato, CA: New World Library.
———. 2010. *The Animal Manifesto: Six Reasons for Expanding our Compassion Footprint.* Novato, CA: New World Library.
Bekoff, Marc, C. Allen, and G. M. Burghardt. 2002. *The Cognitive Animal: Empirical and Theoretical Perspectives on Animal Cognition.* Cambridge, MA: MIT Press.
Cavalieri, Paola. 2006. "The Animal Debate: A Reexamination." In *In Defense of Animals: The Second Wave,* edited by Peter Singer, 54–68. Oxford: Blackwell.
Clark, Stephen R. L. 1997. *Animals and Their Moral Standing.* London: Routledge.
Coetzee, J. M. 1999. *The Lives of Animals.* Princeton, NJ: Princeton University Press.
Crist, Eileen. 1999. *Images of Animals: Anthropomorphism and Animal Mind.* Philadelphia: Temple University Press.
———. 2004. "Against the Social Construction of Nature and Wilderness." *Environmental Ethics* 26 (1): 5–24.
———. 2008. "Donald Redfield Griffin." In *New Dictionary of Scientific Biography,* Noretta Koertge, editor-in-chief, vol. 3.
Darwin, Charles. [1871] 1981. *The Descent of Man, and Selection in Relation to Sex.* Princeton, NJ: Princeton University Press.
———. [1872] 1998. *The Expression of Emotion in Man and Animals.* Oxford: Oxford University Press.
Descartes, René. 1989a. "Discourse on the Method." In *The Philosophical Writings of Descartes,* vol. 1, translated by J. Cottingham et al. Cambridge: Cambridge University Press. First published in 1637.
———. 1989b. "Objections and Replies." In *The Philosophical Writings of Descartes,* vol. 2, translated by J. Cottingham et al. Cambridge: Cambridge University Press.
———. 1991. *The Philosophical Writings of Descartes,* vol. 3, *The Correspondence,* translated by J. Cottingham et al. Cambridge: Cambridge University Press.

Elden, Stuart. 2006. "Heidegger's Animals." *Continental Philosophy Review* 39 (3): 273–91.

Emel, Jody, and J. Wolch. 1998. "Witnessing the Animal Moment." In *Animal Geographies: Place, Politics, and Identity in the Nature-Culture Borderlands*, edited by J. Wolch and J. Emel, 1–24. London: Verso.

Forbes, Jack. 2008. *Columbus and Other Cannibals: The Wétiko Disease of Exploitation, Imperialism, and Terrorism*. New York: Seven Stories Press. First published in 1978.

Foreman, Dave. 2011. "The Arrogance of Resourcism." In *Conservation vs. Conservation*. Durango, CO: The Rewilding Institute and Raven's Eye Press.

Griffin, Donald. 2001. *Animal Minds*. Chicago: University of Chicago Press. First published in 1992.

Haraway, Donna. 2008. *When Species Meet*. Minneapolis: University of Minnesota Press.

Heidegger, Martin. 1977a. "The Age of the World Picture." In *The Question Concerning Technology and Other Essays*, 115–54. New York: Harper Torchbooks. First published in 1938.

———. 1977b. "The Question concerning Technology." In *The Question concerning Technology and Other Essays*, 3–35. New York: Harper Torchbooks. First published in 1955.

Horkheimer, Max, and T. Adorno. 1972. "The Concept of Enlightenment." In *Dialectic of Enlightenment*, 3–42. New York: Continuum. First published in 1944.

Hurley, Susan, and M. Nudds. 2006. "The Questions of Animal Rationality: Theory and Evidence." In *Rational Animals?*, edited by S. Hurley and M. Nudds, 1–83. Oxford: Oxford University Press.

Jamieson, Dale. 2003. *Morality's Progress: Essays on Humans, Other Animals, and the Rest of Nature*. Oxford: Oxford University Press.

Kidner, David. 2001. *Nature and Psyche: Radical Environmentalism and the Politics of Subjectivity*. Albany: State University of New York Press.

Lovejoy, Arthur. 1976. *The Great Chain of Being: A Study of the History of an Idea*. Cambridge, MA: Harvard University Press. First published in 1936.

Lopez, Barry. 2002. "The Language of Animals." In *Wild Earth: Wild Ideas for a World out of Balance*, edited by T. Butler, 296–305. Minneapolis: Milkweed Editions. First published in 1998.

McKay, Robert. 1999. "Veg(etari)animal." Unpublished conference paper. Animals in History and Culture, Bath Spa University, UK, July.

Midgley, Mary. 1995. "Duties concerning Islands." In *Environmental Ethics*, edited by R. Elliot, 89–103. Oxford: Oxford University Press.

Mitchell, Robert, N. Thompson, and L. Miles. 1997. *Anthropomorphism, Anecdotes, and Animals*. Albany: State University of New York Press.

Pedersen, Helena. 2010. *Animals in Schools: Processes and Strategies in Human-Animal Education*. West Lafayette, IN: Purdue University Press.

Regan, Tom. 2001. *Defending Animal Rights*. Urbana: University of Illinois Press.

Rodman, John. 1980. "Paradigm Change in Political Science: An Ecological Perspective." *American Behavioral Scientist* 24 (1): 49–78.

Sandler, Ronald, and P. Cafaro. 2005. *Environmental Virtue Ethics*. Lanham, MD: Rowman and Littlefield.

Singer, Peter, ed. 1985. *In Defense of Animals*. New York: Perennial Library.

Smuts, Barbara. 2001. "Encounters with Animal Minds." *Journal of Consciousness Studies* 8 (5–7): 293–309.

Steiner, Gary. 2005. *Anthropocentrism and Its Discontents: The Moral Status of Animals in the History of Western Philosophy*. Pittsburgh: University of Pittsburgh Press.

The Animal Studies Group. 2006. *Killing Animals*. Urbana: University of Illinois Press.

Torres, Bob. 2007. *Making a Killing: The Political Economy of Animal Rights*. Oakland: AK Press.

de Waal, Frans, and P. Tyack. 2003. *Animal Social Complexity: Intelligence, Culture, and Individualized Societies*. Cambridge, MA: Harvard University Press.

Weber, Max. 1946. "Science as a Vocation." In *From Max Weber: Essays in Sociology*, edited by H. H. Gerth and C. W. Mills, 129–56. New York: Oxford University Press.

4

Talking about Bushmeat

Dale Peterson

IN CENTRAL AFRICA, I visited several of the markets and talked to some of the hunters, traders, cooks, and consumers of wild animal meat—*bushmeat*, as it's called. I was traveling with a photographer friend, Karl Ammann, and on the day I'm remembering now, we were in Libreville, Gabon, accompanied by a French veterinarian named David Edderai. David was a volunteer with Vétérinaires sans Frontières, working on an experimental project to see if any wild species could be bred in captivity and sold commercially.

At noon we sampled the fare at an ordinary bushmeat restaurant. And after a bit of blue duiker, bush-tailed porcupine, and cane rat tempered with mashed greens and boiled manioc and washed down with Régab beer, we rode in David's truck over to the Mont Bouet market, a warren of passageways in loops and dead ends with crowds of buyers and sellers, gawkers and hawkers. We wandered through pockets of smells and past a procession of spices, peppers, eggplants, tomatoes, onions. We passed a hanging garden of garments and proceeded through regions of cloth, shoes, eyeglasses, a cornered tailor and his sewing machine, a dozen haberdasheries, and an exotic netherworld of women's underwear. We turned a corner, passed a bent old woman selling flip-flops, squeezed down an alleyway, were pushed farther along until suddenly we emerged into a dark barn full of chickens in stacked cages. We came to an open room and large butchery: domestic meat chopped into bright red chunks and slabs. And after more twists and turns, we pressed into the bushmeat section, a cavernous area of dead animals displayed on wooden tables arranged end to end to form corridors. Some of the meat was chopped and anonymous. Much was not anonymous.

I saw a python transformed into wheels. I saw bushpig parts, dismembered duikers and bush-tailed porcupines, whole cane rats and giant rats. I noted half a sitatunga with the hair still on. I observed three baboons with the hair singed off, exposing a cream-colored skin and puffy faces. David spoke to one woman, chatting about prices as she chopped off the heads of duikers with a cleaver. And in the area of monkey parts, three young women, laughing and clearly delighted at the comic spectacle the three of us offered, thrust monkey heads at our faces and cried out: "Eat this! It's good! It's good!" But their laughter expressed a gentle teasing, so it seemed, and we laughed back—or tried to.

We found an elephant trunk sliced into wheels and were told that each wheel cost the equivalent of a few dollars per kilo. Karl negotiated the price down to less than that for a two-kilo wheel. It was about ten inches in diameter, with a gristly inner core surrounded by a radiating circle of red meat and then an outer rind of thick skin. The vendor wrapped it up in a sheet of paper and handed it to Karl, who said he intended to photograph it.

At that point we noticed an interesting piece farther down one aisle: a big hairy leg. We moved in to examine it. Not an antelope leg. Far too broad and muscular. Not a monkey or baboon leg. Too big. And the fingered foot at the end was a dead giveaway. It was an entire chimpanzee leg, severed at the hip and stretched out across the table, looking huge and muscular, like a gorilla arm, and the fingered foot was curled into a half-clenched fist.

But David, meanwhile, had become engaged in another spectacle elsewhere. The spectacle was mostly him, as he squatted inside a wire-mesh pen where a male duiker the size of a small deer lay on his side, breathing fast and bound with some twine wrapped tightly around his four legs. Best way to keep meat fresh: still alive. He was a bay duiker, David said, adding that he needed male bay duikers for the breeding project.

He huddled over the animal, felt the legs, one by one. Slowly, he unwrapped the legs from the binding—the duiker was quiet, looking up at him with large dark eyes. David again ran his hands carefully over each leg, in turn. "It's not good. This leg is cold," he said. The duiker may have been bound for a couple of days, David thought. Too long. But he massaged those legs slowly and gently until the creature emitted a long, rasping groan. Then another. And another. David picked him up, carefully tucked his legs under while holding him upright, and the duiker began shifting his muscles and testing his weight. Soon he was hobbling and then, obviously fearful, had stumbled over to the side of the pen where he began bashing his head against the wire mesh.

By now a couple dozen people were standing around, engaged as an audience. One man asked if we intended to kill the animal. Others suggested we should. Still, David quietly negotiated with the owner, bought the duiker, and then, having gently wrapped him in a bit of cloth, located a *carteur*—freelance wheelbarrow worker—and gingerly placed the wrapped animal into the wheelbarrow. The three of us followed our carteur back out of the market, Karl carrying his newspaper-wrapped elephant trunk slice under one arm.

We look like fools, I thought. As we followed our carteur up this corridor and down that alleyway, around one corner and through another passage, as we sometimes stumblingly negotiated the streaming crowds of shoppers and sellers, occasionally I would hear the voice of someone trying out his or her high school English: "Hello! How are you? Where are you from? What are you doing?" There was commentary, gesturing, laughter, studied indifference—but little antagonism, I thought, mostly amusement, tolerance, curiosity. Possibly mild contempt.

Why Should We Talk about Bushmeat?

Most Americans and Europeans like their meat disguised. They want it packaged and refrigerated so that it looks and smells like soap. They prefer to remain ignorant about where the meat comes from, and they will spend their lives never thinking about the transition from feed to fed to food. They're disgusted by the sight of a dead animal, and they turn up their noses at the prospect of good organ meat. They insist on taking their meat from a tiny spectrum of boring species—cows, domestic pigs, and the like—and they would never dream of experimenting with different cuts or new species.

The big city markets in Central and West Africa have plenty of the boring domestic meats, if you happen to prefer that. But if you want meat that will recall when your family lived out in the village, meat with real flavor from an interesting wild animal, meat taken by skilled hunters from the forest and often cured by a flavorful process of smoking, meat with a real story behind it—*bushmeat*, in other words—you will have to pay a premium, as you would for any other luxury item.

Bushmeat is a common source of protein throughout West and Central Africa. Think *seafood*, a widely loved form of food consumed around the world, and an important traditional source of protein for people living near the sea. Seafood is just one finny form of bushmeat: consisting of wild critters taken in astonishing variety and in forms that might seem odd to the outsider but are normal and often marvelously delicious to the initiated. And seafood is, like all other bushmeat, too often harvested unsustainably.

The idea of *sustainability* evokes an image of wise farming. The wise farmer harvests each year a crop that is replaced a year later by an equally good crop. Nature herself smilingly provides this free bounty: one that exploits the free services of sun and rain, the steady maintenance of fertile soils through natural or intelligently promoted fertilization, and the continuous retention of seed crop. This reassuring image leads theorists to speak of hunting and killing wild animals as "harvesting," and it brings them to imagine that a forest can provide the wise hunter and his family a steady bounty of animal protein.

Some scientists estimate that African tropical forests can sustainably produce as much as 200 kilograms of wild animal biomass per square kilometer per year. A more likely figure, according to ecologists John Robinson and Elizabeth Bennett, would be around 150 kilograms of animal biomass per square kilometer per year. Considering that around 65 percent of an animal's weight becomes edible meat, this production should provide about 100 kilograms of edible meat per square kilometer per year. People living in Central Africa's Congo Basin actually consume an average of 47 kilograms per person per year (while people in the industrial nations consume around 30 kilograms per person per year), which suggests that these forests might support, sustainably, around two persons for every square kilometer. That may have been the level of human presence in Central Africa a century ago. But people in the Congo Basin today are living at densities of around five to twenty persons per square kilometer, with their numbers doubling every generation. So we can see that the forests of the Congo Basin cannot possibly keep up with the present demand, while projecting into the future suggests a time of devastating food shortages (Robinson and Bennett 2000a; see also Barnes and Lahm 1997).

An approaching food shortage is the first reason to talk about bushmeat. A second is the promise of biodiversity loss—extinction—that threatens some species more immediately than others. Species with higher reproduction levels will ordinarily tolerate hunting better than those with lower levels. In other words, hunters who are wiping out the animals of a forest opportunistically will eliminate species in a progression, starting with the slower reproducers and finishing with the faster. Apes are among the slowest reproducing mammals of all (Marshall, Jones, and Wrangham 2000), so the African apes—bonobos, chimpanzees, and gorillas—although they constitute only around 1 percent of the total bushmeat take, are also among the most endangered by the recent explosion in bushmeat commerce. Somewhere between 5,000 and 50,000 bonobos are still alive in the African forests. Perhaps 115,000 gorillas of all types remain, and roughly around twice that number of chimpanzees are left (Butynski 2001). These are small numbers, especially as we compare them to those of *Homo sapiens*: with a current population of well over six billion and increasing by about eighty million per year.

The third reason to talk about bushmeat concerns public health. Eating any meat risks disease, but domestic meat can be monitored much more thoroughly than wild animal meats. In addition, the various disease-causing organisms known to inhabit the bodies and blood of primates (including many species of monkeys and the great apes), having already adapted to live in an old primate host, a stable reservoir, are preadapted to find a welcoming home in a new primate host, such as the human one. What does this preadapted relationship mean more specifically? Virologists recognize the West African monkey called the sooty mangabey as the source of HIV-2 in humans, while chimpanzees are now known to be the source of HIV-1. HIV-2 accounts for 1 percent of all AIDS cases, past and present; HIV-1 accounts for the remaining 99 percent. We might regard the zoonotic transformations of SIV-1 and SIV-2 into HIV-1 and HIV-2 as a historical event: past and better forgotten. However, a number of monkey species living in Africa are reservoirs for several Simian Immunodeficiency Viruses (SIVs) closely related to the immediate ancestors of HIV-1 and HIV-2 (Hahn et al. 2002; Gao et al. 1999; also, Weiss and Wrangham 1999). A 2002 survey in Central Africa tested the blood of 788 monkeys sold as meat or kept as pets and found more than one-fifth of them positive for the presence of several strains of SIV, including five previously unknown lines (Peeters et al. 2002). These and other dangerous pathogens in the blood of many different primate species (Winter 2004) amount to a hidden brew that could one day produce some devastating new pandemic among humans, perhaps caused by a recombinant virus that becomes HIV-3. That's one reason why I would never eat a monkey or an ape.

The other reason why I would never eat a primate is the fourth reason we should be talking about bushmeat. Monkeys are our close relatives, while apes are our closest relatives. Apes are intelligent enough to make and use tools, to learn sign language, to anticipate the future and show creativity in problem solving, to recognize themselves in the mirror, and so on. These are signs of high intelligence, and behind that intelligence is a set of emotions and some degree of awareness or cognition—a *psychological presence*, if you will—that for me places the apes outside the category of food. Simply put, I won't eat an animal that is significantly aware or has a clear psychological presence. To some degree the same idea might be argued for a large number of animals, but surely we can imagine it to be true for the larger-brained and more highly encephalized mammals, including the cetaceans, proboscideans, and primates.

Why Aren't We Talking about Bushmeat?

Experts have quietly commented on the consumption of bushmeat and its apparent unsustainability for some time. As early as 1974, Emmanuel Asibey, chief wildlife officer of Ghana, described the catholicity of taste for meat in his country, noting that the markets offered ants and anteaters, birds of most kinds, fruit bats, giant snails, hares, maggots, monitor lizards, rodents (including mice, cane rats, house rats, giant rats, porcupines, squirrels of all kinds), snakes (Gaboon vipers, pythons, puff adders), and tortoises and turtles of every species. Among the primates, all native species—from bush babies to monkeys to chimpanzees—were also presented as food. Asibey estimated then that in Accra, Ghana's capital, one market had sold 155 tons of bushmeat over seventeen months between 1968 and 1970 (Asibey 1974).

Writing in *Biological Conservation*, Asibey expressed the concern that such levels of consumption would produce shortages and ultimately ecological collapse. But in spite of Asibey's prescient early warnings, the comments of a few contemporaries (such as Sabater Pi and Groves 1972), and the studies of a number of biologists during the 1980s and early 1990s (see Robinson and Bennett 2000b), it was not until 1997 that the first North American popular magazine focusing on natural history or conservation published an article addressing the subject (McRae 1997). By then the trade in bushmeat had gone from a generally rural subsistence activity to a full-fledged commerce spread broadly across the middle of Africa and concentrated in the big cities. The commerce has become an important part of national economies in West and Central Africa, and it has taken on international dimensions, with major markets emerging in Paris, Brussels, London, Montreal, Toronto, New York, Chicago, and Los Angeles (Brown 2006; Cheng and Okello 2010). The industrial exploitation of bushmeat now removes as much as five million metric tons of wild animal biomass each year out of the Congo Basin forests (Pearce 2002).

The bushmeat industry, an indigenous one that profits and serves Africans, has become possible in its present form since the arrival of modern logging in the Congo Basin. Modern logging is largely an exogenous industry, profiting Europeans and Asians as well as, to a lesser degree, an elite class of urban Africans. Starting in the second half of the twentieth century and expanding rapidly in the 1980s and 1990s, foreign logging companies brought in some enormously powerful new technologies for cutting and pushing and picking, and within a few years—an instant on anyone's biological timescale—they had opened up enormous portions of the world's second largest rain forest. This event has created an endless flow of wood out of the African interior and across the seas to consumers in Europe and Asia. My estimate for the size of this industry is ten million cubic meters of wood per year, with an average exchange value of around $100 per cubic meter. I'll let economists quibble over the details of those figures, which obviously will vary from year to year, but they are fair estimates, and they have the additional virtue of making our arithmetic simple: Logging in Central Africa amounts to a $1 billion per year business (Forests Monitor 2001; Global Forest Watch 2000; also "Brazzaville Signs" 2000).

Some large conservation organizations, such as the World Wildlife Fund, instead of resisting the destruction of some of our Earth's final stands of virgin rain forest, have supported the loggers in their public relations by spending millions of dollars supposedly helping them become "greener" and closer to "sustainable" (Carroll 2001; Apele 2002). I consider tropical forest logging to be Exhibit A in demonstrating how debased the word *sustainable* has become. Simply put, tropical forest logging, which concentrates on a small number of commercially valuable species and operates in a return cycle of a few decades, breaks open ecosystems that are tens of millions of years old, and it routinely cuts trees that are hundreds of years old. A random sample of Sapele boles logged in northern Congo were radiocarbon-dated at four to nine hundred years (Assessment 1996, 36–39). Yes, trees sometimes grow back—but ecosystems do not. And by creating vast networks of roads and trails, logging has for the first time in history opened these forests to professional hunters and the army of bushmeat traders who follow.

I've stood at the edge of Gabon and watched the ceaseless flow of wood—on trucks, trains, barges—as it reaches the Atlantic, and I am inclined to think of the industry that produces this flow as akin to mountaintop-removal mining or, maybe, something more like "rape." But the World Bank and other donor NGOs call it "development," and they have assisted it with mega-million-dollar loans and deals. Whatever we call it, we can all agree that a lot of money is involved—a billion dollars yearly for the logging industry, hundreds of millions for the bushmeat business—and money is the first reason why people find it hard to talk about bushmeat.

Let me express this idea with an accusation posed as a question that I sometimes hear: *How dare you, a rich westerner, talk of limiting development, which represents the rightful economic advance of impoverished peoples in the Third World?* Responding to that accusation posed as a question completely would require a book, rather than the few paragraphs I have, but let me start by agreeing that it implies, fairly, that westerners who want to influence conservation elsewhere in the world should be prepared to pay.

The accusation posed as a question also implies, unfairly, that the value of "development" in the dollars-and-cents form of timber income and bushmeat commerce actually improves the lives of most ordinary Africans. It does not. True, logging provides temporary jobs for some people who are mostly outsiders brought into the forest. The same can be said for bushmeat hunting and the trade associated with it. It's also true that logging brings to national governments money in the form of fees, taxes, and bribes, while the bushmeat industry engages comparable distributions. But most of the big money from logging concentrates in the hands of a comparatively few individuals from the urban centers, while the vast majority of poor people in rural areas, where logging removes trees and bushmeat hunting removes animals, are witnessing the disintegration of their traditional subsistence wealth. The grand Sapele trees of northern Congo, to consider one of many possible examples, have been hard hit as a favored timber species among loggers. Since Sapele is remarkable for its unusual combination of lightness, hardness, strength, and water resistance, the local people depended upon it for making dugout canoes and roof beams. Local people also relied on Sapele bark for analgesic and anti-inflammatory medicines, and the mature trees were the only source of *Imbrasia oyemensis*, a tree-living caterpillar relied upon as a fallback food when hunting and fishing were unproductive (Lewis 2001). I speak of these uses of Sapele in the past tense, because loggers have significantly moved the trees into the past tense (SGS Internal 2000).

This reverse Robin Hood scheme—stealing subsistence wealth from the rural poor to give monetary wealth to the urban elite—is nowhere more obvious than among the most dispossessed of all, the forest specialists sometimes known as Pygmies. In Cameroon, I visited a Baka Pygmy encampment consisting of a campfire and some beehive-shaped huts woven out of leaves and branches. Dressed in rags, these people looked as if they were suffering from malnutrition, and they discussed their illnesses and asked for soap. Occasionally, I would look up to watch a giant Mercedes-Benz truck with thousand-dollar boles of timber chained to the carriage rattle down the road, raising a fine film of dust. Meanwhile, less glamorous trucks move down that same dusty road and carry out the animals these people rely on as food (Peterson 2003, 196, 197).

A second reason people find it hard to talk about bushmeat is related to the first, but it focuses on food. Here's the idea expressed by the kind of accusation posed as a question I'm most familiar with: *How dare you, a well-fed westerner, criticize the eating habits of hungry Africans?* It's a hard question to answer mainly because the people who pose it tend to be convinced of their own moral rectitude. But it misses the point.

The bushmeat I would talk about does not involve the meat you find inside the cooking pot in a Pygmy village or in a village of non-Pygmy people living semisubsistence lives in the countryside. Rather, the bushmeat I would talk about has passed through a chain of commerce that professionally removes it from the forest and industriously moves it into the big cities, where it sells at a premium. What sort of premium? I asked an African, skilled at bargaining and familiar with the markets in Yaoundé, Cameroon, to bargain for different cuts of meat. This informal yet well-documented experiment yielded the following results, in Central African francs (CFA):

Chimpanzee hand: 5,000 CFA.

Chimpanzee head: 12,000 CFA.

Smoked chimpanzee arm: 2,500 CFA.

Smoked chimpanzee hand: 2,500 CFA.

Big piece (at least one kilo) of boneless beef: 1,500 CFA.

Big piece of beef with bone: 1,300 CFA.

Big piece of boneless pork: 1,300 CFA.

Other more methodical surveys demonstrate a similar pattern, which can be summarized simply: In the big cities of Central Africa, choice, not hunger, sustains the bushmeat industry.

Yet that choice underlies so many people's reluctance to talk about bushmeat. Choice means preference. In talking about bushmeat in Central Africa, we are discussing a commerce with deep cultural roots, and talking critically about bushmeat amounts to criticizing a culture. It's like criticizing whaling in Norway, or challenging Apartheid in the South Africa of a few years ago. We must be willing to enter such conversations. Cultural criticism is no more and no less than the dynamic exchange of ideas among people, which is always a useful thing.

How Can We Talk about Bushmeat?

We can talk about bushmeat in two ways: by invoking human self-interest and by invoking human other-interest.

Self-interest is the easy one. Human self-interest is easy because the emotions it engages transcend trends and traditions. Whether we live in a forest in Africa or a suburb in North America, the first three problems presented by the Central African bushmeat industry—future food shortages, a coming wave of extinctions, some serious threats to public health—can be described in terms of a human self-interest that should resonate everywhere.

We might think of self-interest as a kind of universal language in the way that money is, and, indeed, the two are often indistinguishable. So, for example, we can pull down the accountant's visor and speculate on the enormous costs likely to be associated with a new viral pandemic emerging from the rain forests of Central Africa and passing zoonotically into the human species. We can estimate the costs of future food shortages and the expense of dealing with the consequence. And likewise with extinctions. You might imagine that extinctions are beyond pricing, but since people regard nature as valuable for self-interested reasons—for aesthetic pleasure or emotional satisfaction—we can place a value estimate on individual species' continued existence. Economists John Loomis and Douglas White have concluded, based on their own extensive surveys, that the average American would pay from $6.25 to $7.63 per year to rescue Atlantic salmon from oblivion, from $24.00 to $35.96 per year to save grizzly bears, from $22.07 to $33.07 for whooping cranes, and so on (Whipple 1996). This kind of economic thinking is useful, I believe, in large part because anyone anywhere can appreciate it.

Other-interest is the hard one. Yes, of course, we all feel a certain other-interest—when the other is another person. But what about another species? Ordinarily, when we begin talking about the interests of other species, we risk ridicule. We might be dismissed as *animal rightists* or *animal welfare fanatics*. Other-interest is a moral stance, and when we take such a stance regarding animals or plants or ecosystems, we are in danger of being dismissed and disrespected or even of becoming irrelevant, since this kind of moral stance is not universal or transcultural.

The universality of a human concern for other humans is easy enough to demonstrate. You can go anywhere in the world and discover routine examples of human-to-human concern. Yet during those same travels you will simultaneously discover an utter chaos of human responses to nonhumans. Dogs are loved in one place and eaten in another. Cats are revered here and feared there. Elephants are wonderful or terrible, depending on where or who you are. In places like Central Africa, moreover, where cultures can be very well defined and highly localized, people's attitudes about animals may change in the brief time it takes to cross a river. Raft across the Luo River in north central Democratic Republic of Congo and you can pass from the land of the Mongo people to the land of the Mongandu. You will simultaneously move from a land where bonobos are nearly extinct, because they are hunted, to a land where bonobos are plentiful because they are protected by tale and taboo (Kano 1992).

The moral concern that some people, including myself, say they feel for nonhuman animals appears to be neither universal nor innate. It is not so much a product of nature, I believe, as of nurture acting upon nature. We *learn* to care for other creatures, although we do so through extending our innate capacity to care for other people.

Some people like to speak of such learning to care for animals as *breaking the species barrier*, and they think of that barrier as a negative bias called *speciesism*. That recently coined term evokes an analogy with racism and sexism, and it promotes the idea that good people might overcome this *ism* as readily as good people have moved to overcome the other *isms*. But the analogy fails to recognize the depth of our own species narcissism. We can overcome racism and sexism because we already possess all the moral and emotional equipment to treat other members of our own species with greater fairness and kindness or, as we sometimes say, greater humanity. Can we overcome this speciesism with the same level of coordination or commitment? I think not. *Breaking the species barrier* is a convenient but simplistic concept unlikely to be replicated in real life, while *speciesism* is a misleading term that fails to recognize the complexity of our attitudes toward other humans and other animals.

Nor will we learn a greater moral concern for our fellow creatures by somehow "reconnecting" with nature or following the noble example of indigenous peoples. Most Africans living in the Congo Basin are closer to nature than the average urban and suburban North American. Yet most Central African urbanites and virtually all of the Central African hunters I have met show little compassion for animals. Animals are a commodity that can be exchanged for other, more appealing commodities. This utilitarian attitude is not merely limited to hardened professional hunters; it seems to be a more general cultural perspective that is reflected in languages across the continent. The Lingala language of Central Africa uses *eyama* to mean both *animal* and *meat*, while the Swahili of East Africa uses *nyama* similarly. Krio, the lingua franca of Sierra Leone, has *bif* signify both *animal* and *meat*. Nigerian Hausa employs *nama* for the same double signification, while francophone Africa applies *viande* similarly (Peterson 2003, 63, 64; Matthiessen 1991, 36, 198). To break the legs of a quivering, liquid-eyed forest antelope might seem cruel to you and me, but to a hunter in Central Africa, it's a reasonable way to keep the animal in place and the meat fresh simultaneously (Hennessey 1995).

Expressing an other-interest in nature, taking a moral stance about animals (or plants or ecosystems), will not resonate with everyone, and it is not a magical solution to anything. But I feel it's right, and I believe that it's pragmatically important for conservation. Taking a moral stance about nature means that we understand human need is not the sole criterion for value. What is morally valued cannot be monetarily valued: not commodified or bought and sold. In taking a moral stance, then, we remove certain items from the marketplace. We might permanently protect virgin stands of tropical forest from logging or forever strip away the price tag from certain meats, such as chimpanzee and gorilla meat.

Yes, taking a moral stance about animals (or plants or ecosystems) is inconveniently complicated. How does one describe or conceptualize it? We can foolishly insist that all animals have an equal moral value, in which case we find ourselves unable to move our feet for fear of crushing an ant. We might organize our moral concern based on an animal's degree of evolutionary closeness to humans. Or, as a third alternative, we might base our concern on the idea of a species' *psychological presence*: a certain level of intelligence or cognition predicted by a certain brain size or complexity or an encephalization ratio, or some other assessment.

Moral concern based on evolutionary closeness to humans versus one based on psychological presence: What's the difference? One difference is that a morality based on psychological presence would protect those very large-brained and highly intelligent animals who happen not to be close evolutionary relatives of *Homo sapiens*, such as elephants and whales. Another difference is that evolutionary closeness to humans is easy to measure, while psychological presence is a discouragingly nebulous concept—hard to measure.

Yet another difference between the two approaches may be the most important, as we return to the subject of bushmeat. Many African traditions already accept the idea that certain animals are protected by cultural taboo from being hunted, killed, or eaten because they are imbued with a special closeness to people. The Mongandu people forbid the eating of bonobos because, so tradition says, when these animals believe no one is watching, they walk upright on two legs in the style of people (Kano 1992). The Oroko of Cameroon say that humans can sometimes become apes, and thus for a hunter to find an ape is propitious, while killing one brings bad luck. For the Kouyou of Congo both gorillas and chimps are taboo because they are too close to human (Peterson 1993, 74, 75). African traditions protecting apes are in the minority (and for some traditions, unhappily, the very human resemblance of apes adds to their symbolic value as fetish or food [Peterson 1993, 76–78; see also Cormier 2002]). Nevertheless, Central Africans in general are not unfamiliar with the argument that apes are special and deserve special protection of a moral sort because they are close to human.

Talking about bushmeat should begin by focusing on particular kinds of bushmeat, especially the primates—and most especially the apes. Apes amount to only around 1 percent of the average hunter's take, which means that removing apes from the menu should challenge the self-interest of comparatively few people. Removing apes from the menu, moreover, might also be presented as a matter of other-interest, a moral issue, which could engage people's moral emotions in resonance with existing traditions.

By "talking about bushmeat," of course, I mean having a cross-cultural conversation about it. But let me be clear what kind of cross-cultural conversation I envision. An American psychologist spent time trying to "convert" (his word) a Cameroonian gorilla hunter. As part of that brave effort, the psychologist gave the gorilla hunter a photographically illustrated children's book called *Koko's Kitten*, which is the true story of a sign-language-using captive gorilla who had and loved her own pet kitten. The story is a moving one, and it nicely illustrates the intelligence and emotional depth of a gorilla. For someone who has only seen gorillas fleeing ignominiously from the far end of a shotgun, it must have been enlightening. But in the end, I believe, the gorilla hunter was effectively converted from his former occupation by a steady job that didn't involve mud or blood. Meanwhile, one day I went out to find the hunter's old camp, and I found in its place a new camp of the pair who had replaced him. In other words, speaking airily in terms of other-interest is not enough, while self-interest remains a compelling motivation in a place where people are economically desperate.

More to the point, perhaps, when I think of having a cross-cultural conversation about bushmeat, I imagine something more encompassing than a mere conversation between two individuals, with one trying to convert the other. I imagine a larger and more multifarious conversation involving mass audiences and mass media—soap opera on the radio or television comes to mind—or a conversation where the main speakers are internationally admired. Soccer or film or music stars, for example. Such a conversation, then, would amount to a deliberate attempt to move cultural attitudes in a certain direction by respecting the past, honoring the present, and looking to a future that people are already, in some ways, prepared for.

References

Apele, Sarah. 2002. "Dangerous Bedfellows." Privately circulated report.

Asibey, Emmanuel A. O. 1974. "Wildlife as a Source of Protein in Africa South of the Sahara." *Biological Conservation* 6 (January): 32–39.

Assessment of the CIB Forest Concession in Northern Congo. 1996. Gland, Switzerland: IUCN.

Barnes, R. F., and S. A. Lahm. 1997. "An Ecological Perspective on Human Densities in the Central African Forests." *Journal of Applied Ecology* 32: 245–60.

"Brazzaville Signs Logging Contract with Foreign Firms." 2000. Panafrican News Agency, February 28.

Brown, Susan. 2006. "West Develops a Taste for Primates." *New Scientist* (July 8): 8.

Butynski, Tom. 2001. "Africa's Great Apes." In *Great Apes and Humans: The Ethics of Co-Existence*, edited by B. A. Beck et al., 3–56. Washington, DC: Smithsonian Institution Press.

Carroll, Richard. 2001. "Summary of Organization's Interest and Involvement in the Bushmeat Issue." In *BCTF Collaborative Action Planning Meeting Discussion Papers*, 81–88. Silver Spring, MD: BCTF.

Cheng, Maria, and Christina Okello. 2010. "Tons of Bushmeat Smuggled into Paris." Associated Press, June 18.

Cormier, Loretta Ann. 2002. "Monkey as Food, Monkey as Child: Guajá Symbolic Cannibalism." In *Primates Face to Face: The Conservation Implications of Human-Nonhuman Primate Interconnections*, edited by Agustín Fuentes and Linda D. Wolfe, 63–84. Cambridge: Cambridge University Press.

Forests Monitor. 2001. *Sold Down the River*. Cambridge: Forests Monitor.

Gao, Feng, et al. 1999. "Origin of HIV-1 in the Chimpanzee *Pan troglodytes troglodytes*." *Nature* 397 (February): 436–41.

Global Forest Watch. 2001. *An Overview of Logging in Cameroon*. Washington, DC: World Resources Institute.

Hahn, Beatrice, et al. 2000. "AIDS as Zoonosis: Scientific and Public Health Implications." *Science* 287 (January 28): 607–14.

Hennessey, A. Bennett. 1995. *A Study of the Meat Trade in Ouesso, Republic of the Congo*. Wildlife Conservation Society and GTZ.

Kano, Takayoshi. 1992. *The Last Ape: Pygmy Chimpanzee Behavior and Ecology*. Stanford, CA: Stanford University Press.

Lewis, Jerome. 2001. "Utilization of Forest Resources and Local Variation of Wildlife Populations in Northeastern Gabon." In *Sold Down the River*. Cambridge: Forests Monitor.

Marshall, Andrew J., James Holland Jones, and Richard W. Wrangham. 2000. *The Plight of the Apes*. Cambridge, MA: Briefing for U.S. Representatives Miller and Saxton.

Matthiessen, Peter. 1991. *African Silences*. New York: Random House.

McRae, Michael. 1997. "Road Kill in Cameroon." *Natural History* (February): 36ff.

Pearce, Fred. 2002. "Death in the Jungle." *New Scientist* (March 9): 14.

Peeters, Martine, et al. 2002. "Risk to Human Health from a Plethora of Simian Immunodeficiency Viruses in Primate Bushmeat." *Emerging Infectious Diseases* 8 (May): 451–58.

Peterson, Dale, 2003. *Eating Apes*. Berkeley: University of California Press.

Robinson, John G., and Elizabeth L. Bennett. 2000a. "Carrying Capacity Limits to Sustainable Hunting in Tropical Forests." In *Hunting for Sustainability in Tropical Forests*, edited by J. G. Robinson and E. L. Bennett, 13–30. New York: Columbia University Press.

———, eds. 2000b. *Hunting for Sustainability in Tropical Forests*. New York: Columbia University Press.

Sabater Pi, Jorge, and Colin Groves. 1972. "The Importance of Higher Primates in the Diet of the Fang of Rio Muni." *Man* 7 (June): 239–43.

SGS Internal Audit Programme: Associated Documents. 2000. Unpublished report.

Weiss, Robin A., and Richard Wrangham. 1999. "From *Pan* to Pandemic." *Nature* 397 (February): 385, 386.

Whipple, Dan. 1996. "Congress May Get Bill to Save Rare Species." *Insight on the News*, July 1.

Winter, Stuart. 2004. "Health Workers Fear 'Foaming Monkey Virus' Could Trigger New AIDS Epidemic." *Sunday Express*, March 21.

5

Conservation, Animal Rights, and Human Welfare

A Pragmatic View of the "Bushmeat Crisis"

Ben A. Minteer

Ethical Fault Lines in Conservation?

IN POPULAR DISCOURSE there is often little distinction made between conservation and animal rights/welfare; both are seen as expressing an ethical concern for nature, whether this attitude is directed specifically at whales or wetlands, rhinos or rain forests. While there are probably many explanations for this generalized lumping of ethical regard, the undifferentiated view of animal and conservation ethics is doubtless supported by the galvanizing force of particular cases of wildlife abuse and destruction. For example, conservationists and animal rights/welfare proponents of every stripe would presumably condemn the brutal slaughter of dolphins in Taiji, Japan, depicted in the gripping 2009 documentary, *The Cove*; the wanton 2007 killing of ten mountain gorillas in war-torn Virunga National Park (Democratic Republic of Congo [DRC]); the ongoing destruction of tigers for the illegal wildlife trade; and so forth. While different underlying reasons may be offered for such judgments (reflecting varying ethical orientations to individual animals, or marine mammals, or primates, or perhaps endangered species, etc.), these deeper philosophical differences clearly do not always preclude conservationists and animal rights proponents from holding the same normative view that the killing of the dolphins, gorillas, and tigers is morally wrong.

This "compatibilist" understanding of environmental and animal ethics, moreover, finds additional reinforcement in the language of those wildlife conservation organizations that combine animal welfare and conservationist messages. The Sea Shepherd Conservation Society and Defenders of Wildlife, for example, often make impassioned appeals to stop various acts of cruelty to wild animals alongside more traditional conservation arguments geared toward the protection of biological populations, species, and habitats. Similarly, many animal welfare groups, such as the Humane Society of the United States, have developed programs that overlap with key aspects of the conservation agenda, such as antipoaching campaigns and efforts to slow (or halt) wildlife habitat loss. In the academic community, recent high-profile events such as the "Compassionate Conservation" conference held in the United Kingdom in 2010—sponsored by the Oxford-based wildlife conservation research unit WildCRU and Born Free (a hybrid animal welfare/conservation organization)—is further evidence of this accommodating view toward the well-being of individual animals and the conservation of threatened species and ecosystems.

Yet despite this there has long been a deep division between those who identify primarily with the ethical convictions and policy goals of conservation and those who adopt the ethical view and agenda of animal rights/welfare. A good illustration of this rift may be seen in a recent interchange over the animal welfare–conservation relationship in the journal *Conservation Biology*, the flagship journal for conservation science and management in the United States. There, Perry and Perry (2008) argued for greater cooperation among animal rights supporters and wildlife conservationists, pointing out that both groups are committed to promoting animal well-being, even if they emphasize different understandings of this good (i.e., individuals versus populations/species). They suggested, furthermore, that both groups have important policy goals in common, including preventing the introduction of invasive species (thus avoiding negative ecological impacts and the need for lethal control), and regulating more stringently the international exotic pet trade, which is widely seen as a major conservation problem and a welfare issue (Perry and Perry 2008, 32). Yet in his reply to their paper, Michael Hutchins (of the Wildlife Society) knocked away the olive branch:

> It would be wonderful if we could all get along, but it is time to recognize that some ideas are superior to others because they clearly result in the "greatest good." As a conservationist, I reject animal rights philosophy. . . . It is time to face up to the fact that animal rights and conservation are inherently incompatible and that one cannot be an animal rights proponent and a conservationist simultaneously. To suggest otherwise only feeds into the growing public confusion over animal rights, welfare, and conservation and their vastly different implications for wildlife management and conservation policy. (Hutchins 2008, 816)

Hutchins correctly notes that there is a significant distinction to be made between the relatively more moderate claims of "animal welfare" advocates, who typically seek to reduce animal suffering in domestic and wild contexts (often looking to balance overall harms and benefits rather than to allow individual interests to "trump" the good of the many), and "animal rights" proponents, who can take a far more abolitionist line on the fair treatment of individual animals in a manner analogous to certain core ethical notions of human personhood (see, e.g., Regan 2004). Yet the larger message is nonetheless clear—namely, that right-thinking conservationists should reject or at least significantly de-privilege the zoocentric (i.e., individual animal-centered) claims coming from many defenders of the interests and dignity of individual animals, especially in cases where these positions run counter to the traditional population-centered, species-centered, and ecosystem-centered goals of wildlife managers and con-servation scientists. This position is not that surprising given that the dominant scientific, ethical, and policy orientation in biological conservation typically focuses on promoting the viability of populations and species, or the health, integrity, and/or resilience of ecosystems.

The philosophical differences between conservation or environmental ethics and animal rights/welfare were laid bare in the early 1980s when environmen-tal ethicists like J. Baird Callicott (1980) argued that "true" environmentalists could not be animal liberation supporters. Callicott's case against animal ethics was joined by subsequent work by environmental philosophers, including Mark Sagoff (1984) and Eric Katz (1991), who also underscored the policy divergence of animal-centered approaches and more ecologically oriented environmental policies. At the root of the philosophical dispute is the distinction between in-dividualism and holism in both ethical accounting and managerial concern: the population-system-process orientation of conservation policy and environmen-tal management appears to run in the opposite direction of the individualistic orientation of zoocentric ethics and activism, which again we may organize into two primary strains: (a) animal welfare/liberation; and (b) animal rights.

The animal welfare position is primarily concerned with the human inflic-tion of suffering on individual animals able to experience states of pleasure and pain, or those creatures that are sentient. According to the leading philosopher of the movement, Peter Singer, sentient beings have *interests*; these interests must be taken into account when we make decisions affecting their well-being (to not do so would be to demonstrate a morally arbitrary preference for hu-man interests over the equivalent interests of nonhumans). Singer has argued that a pervasive "speciesism" grips modern society, a discriminatory attitude parallel to racism or sexism that underlies the ethically indefensible neglect of animals' interests simply because they are the interests of animals rather than humans (Singer 2002).

A more stringent animal-centered or zoocentric view may be found in the "animal rights" approach of Tom Regan, whose position shares Singer's ethical individualism but finds its philosophical grounding in the rights tradition rather than utilitarianism (the moral framework that informs Singer's animal welfare views). For Regan, those animals that are self-conscious and are able to form beliefs and desires deserve direct moral consideration (i.e., they are not "mere means" as Kant would have put it). According to Regan, such individuals are "experiencing subjects of a life" (a class including humans but also all "mentally normal mammals of a year or older"), and are thus "ends-in-themselves" that deserve respect; they are not to be valued as sources of human satisfaction or amusement (Regan 2004). Regan's position is potentially more demanding than Singer's given that the rights-based approach resists the method of utility maximization, which could allow a position such as Singer's to support harming animals if such actions would be expected to promote, on balance, greater benefit than harm, all things—or rather, all interests—considered.

While it is clear that strong philosophical and managerial distinction between conservation and animal rights/welfare remains compelling for some, as mentioned above it has never accurately captured the reality of many organizations' and individuals' ethical and policy commitments to both the reduction of animal suffering and the conservation of species and landscapes. Perhaps because of this, subsequent philosophical work following the early "line in the sand" drawn by environmental philosophers has softened the conflict to some degree, and we have seen the emergence of conciliatory projects that have tried to build bridges between the views and goals of animal advocates and environmental holists (e.g., Varner 1998; Jamieson 1998). Even Callicott, for example, would eventually retract his more aggressively antizoocentrist arguments in an effort to embrace an accounting of animal interests under his own multitiered model of environmental holism (Callicott 1998).

Converging Values in the Bushmeat Crisis

The intersection of animal rights/welfare and conservation ethics is particularly intriguing in the case of what has become known as the "bushmeat crisis," a subject of increasing concern in both the nature conservation and development communities (Bennett et al. 2007; see also Peterson, this volume). The bushmeat problem raises an intricate complex of ecological, economic, cultural, and, most fundamentally, ethical challenges regarding the survival of species and the welfare of animals, as well as the health and livelihood of some of the poorest and most vulnerable peoples on the planet. "Bushmeat" is an African word for the meat of terrestrial wild animals harvested for household consumption and/or for commercial sale in local and regional markets. It is a nonspecific term that covers a wide array of hunted species, including duikers (forest antelopes), cane rats, wild pigs, monkeys, chimpanzees, and gorillas. The level of off-take of wild animals for meat varies by ecological type, nation, and continent, but the highest levels of harvest take place in the tropical forests of West-Central Africa, with significant exploitation of forest-dwelling species for wild meat consumption also occurring in many parts of Asia and South America (Brown and Davies 2007, 1).

Comparatively inexpensive and plentiful, bushmeat is a primary source of animal protein in Central Africa; in the half-dozen countries that comprise the Congo Basin, for example, approximately 80 percent of animal protein is derived from wildlife and as much as one million metric tons of bushmeat is consumed each year, the equivalent of almost four million cattle (http://www .bushmeat.org). Demand for bushmeat in Central and West Africa, moreover, is growing, especially among urban populations—a trend resulting from the confluence of the lack of alternative protein sources and cultural tastes for wild meat, among other drivers. This has stimulated the emergence of a large and often lucrative trade in wild meat in Africa, South America, and Asia, with estimates of the economic value of bushmeat across countries in West and Central Africa ranging from US$ 42–205 million (Davies 2002) and more than US$ 175 million per year in the Amazon Basin (Rao and McGowan 2002). In many areas, hunting for trade in urban markets is an important source of household income, especially in the absence of alternative livelihood opportunities (Kümpel et al. 2010).

Although people have been hunting and consuming bushmeat for thousands of years, in recent times these practices have put greater pressure on tropical wildlife as a result of rapid (human) population growth in Africa and Southeast Asia, as well as the loss of habitat via agricultural conversion and settlement, rampant road building that hastens access to forest interiors, and a host of advances in hunting technology (e.g., shotguns, outboard motors, etc.) (Bennett 2006). Indeed, the encroachment of traditional industrial forces in tropical forests such as logging and oil exploration has been linked to the surge in the commercial bushmeat trade as these activities not only create access to remote forest wildlife via road construction but also facilitate transportation of bushmeat to urban markets—and stimulate local sales at or near the harvest point as hunters sell bushmeat directly to resource industry staff in concessions (Thibault and Blaney 2003; Peterson 2003; Poulsen et al. 2009). These developments, and the rise of a commercial market in bushmeat that has dramatically magnified harvest rates beyond those that would characterize subsistence off-take, have been an increasing source of concern over the past two decades among conservationists who have come to see hunting as a major, if not primary, threat to wildlife populations across the tropics (Bennett et al. 2002).

Precise measures of the sustainability of wildlife harvests in bushmeat regions are difficult due to both the bioecological complexity surrounding the calculation of sustainable exploitation rates and the lack of adequate data for many wildlife populations of concern. Nevertheless, conservation scientists and advocates generally agree that hunting rates exceed sustainable levels across many parts of the tropics, especially in African rain forests, with dozens of populations already declining or in danger of going locally extinct (e.g., Robinson and Bennett 2002; Fa, Peres, and Meeuwig 2002; Nasi et al. 2008). In the Congo Basin, an estimated 60 percent of mammalian species are hunted unsustainably (Chivian and Bernstein 2008, 43). A recent study conducted by TRAFFIC, the wildlife trade monitoring network, suggests that the nation of Cameroon appears to be exceeding an estimated sustainable off-take of wildlife by more than 100 percent, with Gabon and the Republic of the Congo also approaching this level (Ziegler 2010).

Furthermore, the relatively low level of animal production of tropical forest ecosystems magnifies the impact of hunting in these areas. Intact tropical rain forests, for example, produce only one-sixth of the mammal biomass of tropical grasslands; most of that production is in primates (Bennett 2006). Primates tend to be scarce and breed slowly, making them particularly vulnerable to bushmeat hunting and at great risk of local extinction (Bennett et al. 2007; Nasi et al. 2008). Although up 80 percent of the population of certain small, fast-breeding animals, such as elephant shrews and agoutis, can be harvested sustainably every year, less than 4 percent of primates can be taken without a significant risk to population viability (Bennett 2006, 108). The recent emergence of a "luxury bushmeat" market in lemur meat, taken from Madagascar's iconic primate, has only added to conservationists' concern about the dramatic biological and ecological costs of the trade (Barrett and Ratsimbazafy 2009).

In addition to the decline of particular populations and species targeted for hunting and market, the overharvest of forest wildlife can also produce wider ecological effects that are of growing concern to conservationists. The bushmeat trade's impact can be observed well beyond targeted wildlife. Intensive harvesting of bushmeat in tropical forests, for example, produces what some biologists have referred to as the "empty forest syndrome" (e.g., Redford 1992), in which largely intact tree cover masks significant biodiversity loss following the reduction of large animals for the wild meat trade—as well as the secondary loss of predators that prey upon them (Bushmeat Crisis Task Force, n.d.). There is also the concern that overhunting of forest wildlife can ultimately affect forest vegetation by reducing seed dispersal and tree growth (Stoner et al. 2007; Brodie et al. 2009). This may in turn have implications for the provision of ecological services, as some ecologists have suggested overhunting may be indirectly reducing an important global sink in degrading the carbon sequestration potential of tropical forests (Brodie and Gibbs 2009; though see also Jansen, Muller-Landau, Wright 2010).

The human impact of the bushmeat trade, too, has become a source of increasing anxiety within the development, public health, and conservation communities. Consumption of wild meat by humans increases the risk of zoonotic transmission of disease; as Peterson (this volume) notes, virologists have linked HIV-1 and HIV-2 to wild African primates, and there is fear that related Simian Immunodeficiency Viruses have found a reservoir in African monkey populations. The 2003 SARS epidemic is thought to have originated in the human exposure to an infected wild Himalayan palm civet (*Paguma larvata*) in a live-animal meat market in China (Chivian and Bernstein 2008, 43–44). A 2005 survey of bushmeat hunters in Cameroon found two viruses (HTLV-3 and HTLV-4) that researchers believe came from primates — and that have been linked to neurological disease (Marris 2005).

Adding to such anthropocentric concerns is the economic issue of declining livelihoods and the economic costs of dwindling populations of harvested game, a trend that threatens the food security of poor, landless peoples in tropical regions (Rao and McGowan 2002, 580). Fa, Currie, and Meeuwig (2003) predict that if current extraction levels in the Congo Basin continue, "there will be a significant decline in available wild protein by 2050, and there will be insufficient non-bushmeat protein produced to replace the amounts supplied by wild meats" (Fa, Currie, and Meeuwig 2003, 75). What is more, the distributional pattern of this decline promises to hit the most vulnerable communities the hardest. As Bennett and colleagues (2007, 885) write, the livelihood impact and loss of wild meat as the available protein source declines in these regions will be especially problematic for marginalized groups and indigenous peoples who lack alternative sources of income and opportunities to enter the labor market.

Principle and Pragmatism in Bushmeat Policy

The environmental, animal, and human impact of the commercial trade in bushmeat would seem to point toward a common policy response, at least in principle. Whether the concern is primarily the survival of threatened wildlife populations (especially primates), the reduction of seed dispersers and prey species for large carnivores, or the increased risk of exposure to zoonotic diseases, clearly there are compelling reasons to attempt to develop effective bushmeat policies and practices that will address the broad range of conservation and development values (Bennett et al. 2007). Furthermore, the bushmeat problem appears to be a case in which both animal rights/welfare and a strong nature-centered ethic of conservation would be supported by a strict ban on bushmeat harvest and trade, the establishment of more tightly managed (for biodiversity preservation) protected areas in bushmeat regions, and increased enforcement and interdiction efforts.

The ban-and-enforcement approach would certainly comport well with the canonical work in environmental ethics, which has largely been defined by the search for a nature-centered or nonanthropocentric ethical system that will motivate unswerving preservationist plans and policies (e.g., Taylor 1986; Rolston 1988, 1994; Westra 1994; Katz 1997). A key part of this project has been to engender a profound change in worldview—and a radical ethical transformation—within individuals and societies that many environmental ethicists believe is essential to protecting wildlife and plant species, preventing the degradation and destruction of ecological systems, and generally "doing right" by nature. A global ban on bushmeat harvest and trade, if effectively enforced, would thus seem to be justified by appeal to a strong nonanthropocentric principle requiring the preservation of species and natural systems for their own sake and safeguarding these species and systems against all forms of human encroachment and degradation.

Such a policy of prohibition would also be strongly supported by the animal ethics perspective given that the goal would be to protect targeted and nontargeted animals (considered individually) from being harmed and/or killed for a growing market in bushmeat. This approach would presumably satisfy both animal welfare "sentientists" concerned with minimizing the pain and suffering of conscious beings and animal rights proponents who would apply a stronger "no use" standard that respects the dignity of "ends-in-themselves." Finally, by reducing contact with and consumption of wild meat (especially primates) via enforcement of a hunting and trade ban, health risks would be lowered, thus addressing a significant anthropocentric concern about the bushmeat enterprise.

Yet as mentioned earlier, this broad convergence of values and interests in addressing the bushmeat problem—including the ethical and programmatic concerns of animal rights/welfare and species/ecosystem conservation—will in practice be more complicated to achieve and maintain, especially as we begin to consider specific strategies and tactics to manage the bushmeat trade and the implications of a prohibition policy on other critical interests, especially human livelihoods. Indeed, the overlapping interests and ethical arguments pointing toward curbing the unsustainable harvest of and trade in wild meat have the potential to pull in different directions at the planning and policy level. This is especially true to the degree that these ethical principles and agendas are articulated in absolutist and ideological ways that preclude efforts to achieve pragmatic compromises and workable solutions that attempt to engage the full range of values at stake in the bushmeat dilemma.

One potential area of tension in formulating bushmeat policy, for example, exists between strict nature preservationist views and more sustainable development and human livelihood goals. Although human health benefits from a prohibitive policy response would likely ensue with respect to reduced risk of contracting zoonotic diseases, other anthropocentric interests, such as (short-term) protein availability, the income-generating potential of the trade in wild meat for poor, marginal human communities, and the cultural value of bushmeat hunting and consumption, may be in direct conflict with the strong preservationist response. Over the last decade, conservationists and development professionals have engaged in a series of debates on the broader question of whether poverty alleviation or biodiversity protection should dominate international conservation efforts, a conversation that has at times divided more nonanthropocentric "nature protectionists" advocating a strong protected areas and biodiversity-centered strategy for conservation from more (broadly) anthropocentric "social conservationists" who view biodiversity (and nature generally) as vehicles for sustainable development and the improvement of human welfare over the long run (see, e.g., Adams et al. 2004; Sanderson and Redford 2003; Roe and Elliot 2004; Roe 2008; Miller, Minteer, and Malan 2011). Proposals to address the bushmeat problem that focus narrowly on the biodiversity conservation goal—when articulated at the expense of human livelihood and welfare interests—thus run the risk of clashing with the development agendas of social conservationists, pitting wildlife and ecological protection against the interests of rural communities who may have few alternative protein sources available and a limited range of economic opportunities.

This conflict becomes even more significant given the real-world conditions that will constrain any practicable bushmeat policy. Despite its attractiveness to doctrinaire conservationists (and animal rights proponents), most conservation and development planners do not see a global ban on bushmeat as the most realistic or effective approach to addressing the problem. Indeed, frank assessments of the viable policy options for addressing the bushmeat problem seem to concur that a strict prohibition-and-enforcement-based policy will be unsuccessful, especially if these pressing human welfare needs are not addressed (Bennett et al. 2007; Nasi et al. 2008). These judgments take a pragmatic view of the possibilities for biodiversity conservation within the particular governance, economic, and cultural contexts of bushmeat regions, which are often beset by inadequate administrative resources as well as weak formal traditions of natural resource monitoring and management, and limited overall institutional capacity. They also account for the undeniably powerful incentive structures propelling bushmeat harvest and trade, including the lack of available protein substitutes, the low barriers to entry in the enterprise (compared to the capital requirements of agriculture), and varying cultural preferences for wild meat over that from domestic animals (Brown 2007; Kümpel et al. 2007; Hurst 2007). And they recognize the imperative to balance competing values and interests in policy response to the bushmeat problem, including ethical regard for human welfare as well as for wildlife conservation and ecological health.

What emerges from these more pragmatic analyses is the need to develop nuanced, adaptive, and context-specific bushmeat policies balancing the sustainable use of wildlife in bushmeat regions with the legal protection of listed (threatened) species, rather than advocating a universal and undifferentiated preservationist policy that prohibits wildlife exploitation altogether. Along these lines, some of the more promising efforts to reconcile biodiversity protection with sustainable use and development goals in the bushmeat case focus on the need to integrate conservation and development interests via coordinated spatial planning. Bennett and colleagues (2007, 886), for example, propose an integrated, landscape-level approach that employs a range of land-use strategies, including the designation of protected areas (managed for biodiversity conservation), production forests for resource production (including hunting of nonthreatened species), and "farm bush" areas devoted to sustaining local livelihoods—and which could provide both bushmeat and agricultural products. The development of alternative sources of protein, such as intensively bred cane rats, cattle, and/or farmed fish, could certainly play a role in reducing the need for wild meat in rural communities and relieving hunting pressure on overexploited populations, though the success of these alternatives will depend upon a range of biological, economic, and political factors (e.g., Bowen-Jones and Pendry 1999; East et al. 2005; Wilkie and Godoy 2001; Wilkie et al. 2005). These efforts, moreover, will likely be most effective when joined by a host of additional measures, including the use of economic incentives and sanctions such as increased taxation, the imposition of fines on bushmeat traders engaging in unsustainable practices, and increased monitoring and enforcement efforts (Wilkie and Carpenter 1999).

Among other challenges, these sorts of pragmatic attempts to reduce unsustainable bushmeat harvest and trade—while also increasing protection for threatened species—require building effective alliances among the various stakeholders in bushmeat regions, including rural communities, conservation and development specialists, local and national officials, and the extractive industrial sector (i.e., logging, oil drilling, and mining). The emerging consensus appears to be that these alliances will require a great deal of flexibility and ethical accommodation, especially among more nonanthropocentrically minded conservationists who rightly view extractive enterprises as posing some of the greatest threats to wildlife health and ecological integrity. But there is also reason to believe that these partnerships have the potential to productively reshape hunting and market practices, at least at the local scale.

For example, as mentioned above, logging companies are a significant factor in the commercial bushmeat trade (just as they are in ecological destruction); road building for timber harvest fragments and degrades wildlife habitat and also stimulates hunting by providing a demand for wild meat at logging concessions and a key transportation route linking remote forest interiors to urban bushmeat markets (Nasi et al. 2008, 29–30; Poulsen et al. 2009). For that very same reason, some conservation organizations have established partnerships with logging companies and government agencies in bushmeat regions to develop collaborative wildlife management systems that encourage sustainable hunting, protect wildlife populations and minimize habitat destruction, and promote rural livelihoods. The Project for the Management of Ecosystems in the Periphery of the Nouabale-Ndoki National Park (DRC) is perhaps one of the better-known cases of this sort of multisector, collaborative approach. A joint effort of the Wildlife Conservation Society, a Congolese logging company, and the national government, the project established wildlife use zones to control access and increase protection of bushmeat species. It also developed a program of conservation education for company managers and local residents, enhanced wildlife regulations in company policies, and pursued the development of affordable protein alternatives (i.e., bush farms and importation of beef), among other practices (Poulsen, Clark, and Mavah 2007; Nasi et al. 2008). Although the program has been criticized for not holding the timber company to a high enough standard with respect to wildlife conservation and forest sustainability, supporters view it as an important effort in shaping subsequent national-level policy requiring all logging concessions in Northern Congo to pay for wildlife protection and practice wildlife management as part of their operations (Aviram, Bass, and Parker 2003, 11).

The upshot is that a feasible, effective, and ethically inclusive policy response to the bushmeat dilemma will require balancing a complex of values and interests as well as accommodating diverse stakeholders in workable, multilevel partnerships that can reduce human impact on wildlife species and tropical forest systems while improving the food security and livelihood prospects of poor rural people. It will also require that specific policy agendas and management regimes demonstrate great context sensitivity given the cultural and institutional variability across bushmeat areas, the differing degrees of biological vulnerability of wildlife populations, and varying levels of productivity and options for achieving sustainable harvest rates within particular ecosystems (Robinson and Bennett 2004). Indeed, as Nasi and colleagues (2008, 40) conclude, the bushmeat problem is simply not amenable to universal solutions, but rather must be "nation, site, and context-specific, be based on a detailed knowledge of hunting patterns and the ecology of the hunted species and be tailored to local cultural, socio-economic, and political conditions" (see also Secretariat of the Convention on Biological Diversity 2011).

Yet once more the incorporation of sustainable use principles within an integrative bushmeat policy program will not be agreeable to more strict nature preservationists, both in environmental ethics (mentioned above) and in biodiversity conservation, who worry that these strategies will open the door to further exploitation and destruction of populations and ecosystems (e.g., Terborgh 1999). This approach will also not appease most animal rights proponents: a pragmatic bushmeat policy combining sustainable use with protection of threatened populations will obviously still harm and kill individual animals. But special protections for threatened and/or vulnerable species such as primates (see, e.g., Ape Alliance 2006) and increased monitoring and enforcement efforts—to the degree they are effective—should reduce harm and death within these populations. Furthermore, efforts to curtail the unsustainable commercial trade in bushmeat for urban consumers would also be expected to lead to a decrease in the overall number of animals harmed and killed beyond subsistence consumption. Both animal rights/welfare and conservation proponents, moreover, could support greater restrictions on unnecessarily harmful and indiscriminate bushmeat hunting techniques, such as the use of snares that often results in prolonged animal suffering of target and nontarget animals (Bowen-Jones and Pendry 1999).

All the same, it is true that a realistic and balanced policy response to the bushmeat problem will require a significant number of concessions and compromises from the more zealous advocates of animal protection, just as it will from the more preservationist oriented conservation scientists, ethicists, and advocates. The moderate elements of the animal welfare community, which seek to reduce animal suffering balanced with other interests and concerns, may be far more likely to support a sustainable harvest-protection bushmeat policy than the more ideological factions of the animal rights community, which take a harder line on human exploitation of animals—regardless of whether such exploitation is justified on conservation grounds. And again, the more pragmatic wing within the nature conservation camp will presumably be more open to the mixed, managed use-protection model than the preservationist wing, which typically prefers a "fortress conservation" approach focused on expanding protected areas and limiting human access to biological resources.

But banning hunting and consumption of bushmeat, despite the appeal such a policy would hold for strong nature protectionists and animal rights advocates, is simply not realistic in light of the economic, cultural, and political factors described above. Even more seriously, it is a policy that would run afoul of a host of powerful ethical obligations to promote human welfare in impoverished regions and would violate core principles of procedural and social justice in conservation decision making and resource allocation. This is an especially important point given that conservationists have at times treated these commitments cavalierly (see, e.g., Brockington 2002; Dowie 2009). Top-down, heavy-handed attempts to impose a wild meat ban would likely have disastrous short-term consequences for human well-being in bushmeat regions—not to mention potentially undercut conservation efforts by disenfranchising local people and putting even greater pressure on threatened populations and protected areas (Brown 2007).

Conclusion: Conservation without Ideology

As biodiversity scientist John Robinson (2011) writes, ideological stances in conservation—which may be found among those championing species protection and human livelihoods, as well as animal rights—frequently create intractable and polarizing ethical dilemmas in practice. The solution, he argues, is to relax adherence to absolutist principles and convictions and pay more attention to the critical role of sociocultural and ecological contexts in conservation planning and policy, including efforts to enhance project sustainability and effectiveness over the long run. Among other things, Robinson's more pragmatic outlook requires an embrace of ethical and strategic pluralism in conservation projects, as different conservation contexts will necessitate different approaches and ethical justifications. These ethical convictions and strategies will be continuously revised and clarified over time as the conservation community learns from its failures and successes and adjusts its priorities in light of new information and changing social, cultural, and ecological conditions (Norton 2005).

The attempt to recognize and understand this value pluralism, including the search for points of policy convergence and common ground when possible, is a key feature of what some are calling "ecological ethics," a pragmatic model of ethics for ecologists and biodiversity scientists that tries to accommodate the widest possible range of conservation and human values in management decisions and policy making (see, e.g., Minteer and Collins 2005a, 2005b, 2008). Problems like the bushmeat crisis demonstrate the need to see the connections and synergies among these various realms of value—and adopt creative and effective methods of coalition building and problem solving among stakeholders—rather than coming to a philosophical agreement on the final and universal goal for conservation (see Vucetich and Nelson, this volume, for a defense of the latter approach). To be "pragmatic" in conservation planning and policy making is to acknowledge that there are many potentially valid ends to be pursued within the conservation agenda, including species protection, ecological resilience, and human and animal well-being. The challenge is thus to determine which combinations of values and goals are possible and desirable within particular conservation plans and projects.

As the bushmeat case illustrates, however, hard choices will still have to be made — often to the dissatisfaction of more doctrinaire voices in conservation and animal advocacy. But there remains considerable value in pursuing more inclusive and integrative strategies in conservation ethics and policy and not presuming that different underlying philosophical perspectives will always result in conflicting policy preferences and agendas, even if trade-offs will ultimately need to be addressed at the project level (Leader-Williams, Adams, and Smith 2010). By not defining certain stakeholders or positions in advance as lacking a "true" conservation ethic, or as acting on improper or misguided motives, the pragmatist approach reflects a deeper faith in the possibility of building diverse coalitions around specific conservation and development problems. And it compels us to engage demanding conservation challenges — such as the bushmeat dilemma — in a cooperative, experimental, and nonideological fashion, whether we ultimately care more about animals, ecosystems, or human livelihoods.

References

Adams, W. M., R. Aveling, D. Brockington, B. Dickson, J. Elliott, J. Hutton, D. Roe, B. Vira, and W. Wolmer. 2004. "Biodiversity Conservation and the Eradication of Poverty." *Science* 306: 1146–48.

Ape Alliance. 2006. *Recipes for Survival: Controlling the Bushmeat Trade*. London.

Aviram, R., M. Bass, and K. Parker. 2003. *Extracting Hope for Bushmeat: Case Studies of Oil, Gas, Mining, and Logging Industry Efforts for Improved Wildlife Management*. Report available at http://www.bushmeat.org/docs.html.

Barrett, Meredith A., and Jonah Ratsimbazafy. 2009. "Luxury Bushmeat Trade Threatens Lemur Conservation." *Nature* 461: 470.

Bennett, Elizabeth. 2006. "Consuming Wildlife in the Tropics." In *State of the Wild 2006: A Global Portrait of Wildlife, Wildlands, and Oceans*, edited by S. Guynup, 106–13. Washington, DC: Island Press.

Bennett, E. L., E. Blencowe, K. Brandon, D. Brown, R. W. Burn, G. Cowlishaw, G. Davies, H. Dublin, J. E. Fa, E. J. Milner-Gulland, J. G. Robinson, J. M. Rowcliffe, F. M. Underwood, and D. S. Wilkie. 2007. "Hunting for Consensus: Reconciling Bushmeat Harvest, Conservation, and Development Policy in West and Central Africa. *Conservation Biology* 21: 884–87.

Bennett, E. L., E. J. Milner-Gulland, M. Bakarr, H. E. Eves, J. G. Robinson, and D. S. Wilkie. 2002. "Hunting the World's Wildlife to Extinction." *Oryx* 36: 328–29.

Bowen-Jones, Evan, and Stephanie Pendry. 1999. "The Threat to Primates and Other Mammals from the Bushmeat Trade in Africa, and How This Threat Could Be Diminished." *Oryx* 33: 233–46.

Brockington, Dan. 2002. *Fortress Conservation: The Preservation of the Mkomazi Game Reserve*. Bloomington: Indiana University Press.

Brodie, J. F., O. E. Helmy, W. Y. Brockelman, and J. L. Maron. 2009. "Bushmeat Poaching Reduces the Seed Dispersal and Population Growth Rate of a Mammal-Dispersed Tree." *Ecological Applications* 19: 854–63.

Brodie, Jebediah F., and Holly K. Gibbs. 2009. "Bushmeat Hunting as Climate Threat." *Science* 326: 364–65.

Brown, David. 2007. "Is the Best the Enemy of the Good? Institutional and Livelihoods Perspectives on Bushmeat Harvesting and Trade—Some Issues and Challenges." In *Bushmeat and Livelihoods: Wildlife Management and Poverty Reduction*, edited by G. Davies and D. Brown, 111–24. Malden, MA: Blackwell.

Brown, David, and Glyn Davies. 2007. Introduction. In *Bushmeat and Livelihoods: Wildlife Management and Poverty Reduction*, edited by G. Davies and D. Brown, 1–10. Malden, MA: Blackwell.

Bushmeat Crisis Task Force (BCTF). n.d. *Bushmeat: A Wildlife Crisis in West and Central Africa and around the World.* Document accessed online at http://www.bushmeat.org/bushmeat_and_wildlife_trade/what_is_the_bushmeat_crisis.

Callicott, J. Baird. 1980. "Animal Liberation: A Triangular Affair." *Environmental Ethics* 2: 311–38.

———. 1998. "Back Together Again." *Environmental Values* 7: 461–75.

Chivian, Eric, and Aaron Bernstein. 2008. "How Is Biodiversity Threatened by Human Activity?" In *Sustaining Life: How Human Health Depends on Biodiversity*, edited by E. Chivian and A. Bernstein, 29–74. New York: Oxford University Press.

Davies, Glyn. 2002. "Bushmeat and International Development." *Conservation Biology* 16: 587–89.

Dowie, Mark. 2009. *Conservation Refugees: The Hundred-Year Conflict between Global Conservation and Native Peoples.* Cambridge, MA: MIT Press.

East, T., N. F. Kümpel, E. J. Milner-Gulland, and J. M. Rowcliffe. 2005. "Determinants of Urban Bushmeat Consumption in Río Muni, Equatorial Guinea." *Biological Conservation* 126: 206–15.

Fa, J. E., D. Currie, and J. Meeuwig. 2003. "Bushmeat and Food Security in the Congo Basin: Linkages between Wildlife and People's Future." *Environmental Conservation* 30: 71–78.

Fa, J. E., C. A. Peres, and J. Meeuwig. 2002. "Bushmeat Exploitation in Tropical Forests: An Intercontinental Comparison." *Conservation Biology* 16: 232–37.

Hurst, Andrew. 2007. "Institutional Challenges to Sustainable Bushmeat Management in Central Africa." In *Bushmeat and Livelihoods: Wildlife Management and Poverty Reduction*, edited by G. Davies and D. Brown, 158–71. Malden, MA: Blackwell.

Hutchins, Michael. 2008. "Animal Rights and Conservation." *Conservation Biology* 22: 815–16.

Jamieson, Dale. 1998. "Animal Liberation Is an Environmental Ethic." *Environmental Values* 7: 41–57.

Jansen, P. A., H. G. Muller-Landau, and S. J. Wright. 2010. "Bushmeat Hunting and Climate: An Indirect Link." *Science* 327: 30.

Katz, Eric. 1991. "Defending the Use of Animals by Business: Animal Liberation and Environmental Ethics." In *Business, Ethics and the Environment: The Public Policy Debate*, edited by W. M. Hoffman, R. Frederick, and E. S. Petry Jr., 223–32. New York: Quorum Books.

———. 1997. *Nature as Subject: Human Obligation and Natural Community.* Lanham, MD: Rowman and Littlefield.

Kümpel, N. F., T. East, N. Keylock, J. M. Rowcliffe, G. Cowlishaw, and E. J. Milner-Gulland. 2007. "Determinants of Bushmeat Consumption and Trade in Continental Equatorial Guinea: An Urban-Rural Comparison." In *Bushmeat and Livelihoods: Wildlife Management and Poverty Reduction*, edited by G. Davies and D. Brown, 73–91. Malden, MA: Blackwell.

Kümpel, N. F., E. J. Milner-Gulland, G. Cowlishaw and J. M. Rowcliffe. 2010. "Incentives for Hunting: The Role of Bushmeat in the Household Economy in Rural Equatorial Guinea." *Human Ecology* 38: 251–64.

Leader-Williams, N., W. A. Adams, and R. J. Smith, eds. 2010. *Trade-Offs in Conservation: Deciding What to Save.* Oxford: Wiley-Blackwell.

Marris, Emma. 2005. "Monkeys Infect Bushmeat Hunters." *Nature News* (online), http://www.nature.com/news/2005/050516/full/news050516-2.html.

Miller, T. R., B. A. Minteer, and L-C. Malan. 2011. "The New Conservation Debate: The View from Practical Ethics." *Biological Conservation* 144: 948–57.

Minteer, Ben A., and James P. Collins. 2005a. "Ecological Ethics: Building a New Tool Lit for Ecologists and Biodiversity Managers." *Conservation Biology* 19: 1803–12.

————. 2005b. "Why We Need an 'Ecological Ethics.'" *Frontiers in Ecology and Environment* 3: 332–37.

————. 2008. "From Environmental to Ecological Ethics: Toward a Practical Ethics for Ecologists and Conservationists." *Science and Engineering Ethics* 14: 483–501.

Nasi, R., D. Brown, D. Wilkie, E. Bennett, C. Tutin, G. van Tol, and T. Christophersen. 2008. *Conservation and Use of Wildlife-Based Resources: The Bushmeat Crisis.* Secretariat of the Convention on Biological Diversity, Montreal, and Center for International Forestry Research (CIFOR), Bogor. Technical Series no.33.

Norton, Bryan G. 2005. *Sustainability: A Philosophy of Adaptive Ecosystem Management.* Chicago: University of Chicago Press.

Perry, Dan, and Gad Perry. 2008. "Improving Interactions between Animal Rights Groups and Conservation Biologists." *Conservation Biology* 22: 27–35.

Peterson, Dale. 2003. *Eating Apes.* Berkeley: University of California Press.

————. This volume. "Talking about Bushmeat."

Poulsen, J. R., C. J. Clark, and G. A. Mavah. 2007. "Wildlife Management in a Logging Concession in Northern Congo: Can Livelihoods Be Maintained through Sustainable Hunting?" In *Bushmeat and Livelihoods: Wildlife Management and Poverty Reduction,* edited by G. Davies and D. Brown, 140–57. Malden, MA: Blackwell.

Poulsen, J. R., C. J. Clark, G. Mavah, and P. W. Elkan. 2009. "Bushmeat Supply and Consumption in a Tropical Logging Concession in Northern Congo." *Conservation Biology* 23: 1597–608.

Rao, Madhu, and Philip J. K. McGowan. 2002. "Wild-Meat Use, Food Security, Livelihoods, and Conservation." *Conservation Biology* 16: 580–83.

Redford, Kent H. 1992. "The Empty Forest." *BioScience* 42: 412–22.

Regan, Tom. 2004. *The Case for Animal Rights.* Updated ed. Berkeley: University of California Press.

Robinson, John G. 2011. "Ethical Pluralism, Pragmatism, and Sustainability in Conservation Practice." *Biological Conservation* 144: 958–65.

Robinson, John G., and Elizabeth L. Bennett. 2002. "Will Alleviating Poverty Solve the Bushmeat Crisis?" *Oryx* 36: 332.

————. 2004. "Having Your Wildlife and Eating It Too: An Analysis of Hunting Sustainability across Tropical Ecosystems." *Animal Conservation* 7: 397–408.

Roe, Dilys. 2008. "The Origins and Evolution of the Conservation-Poverty Debate: A Review of Key Literature, Events, and Policy Processes." *Oryx* 42: 491–503.

Roe, Dilys, and Joanna Elliot. 2004. "Poverty Reduction and Biodiversity Conservation: Rebuilding the Bridges." *Oryx* 38: 137–39.

Rolston, Holmes, III. 1988. *Environmental Ethics: Duties to and Values in the Natural World.* Philadelphia, PA: Temple University Press.

————. 1994. *Conserving Natural Value.* New York: Columbia University Press.

Sagoff, Mark. 1984. "Animal Liberation and Environmental Ethics: Bad Marriage, Quick Divorce." *Osgoode Hall Law Journal* 22: 297–307.

Sanderson, Steven E., and Kent H. Redford. 2003. "Contested Relationships between Biodiversity Conservation and Poverty Alleviation." *Oryx* 37: 389–90.

Secretariat of the Convention on Biological Diversity. 2011. Livelihood Alternatives for the Unsustainable Use of Bushmeat. Report prepared for the CBD Bushmeat Liaison Group. Technical Services No. 60, Montreal, SCBD, 46 pages.

Singer, Peter. 2002. *Animal Liberation.* Rev. ed. New York: HarperCollins.

Stoner, K. E., K. Vulinec, S. J. Wright, and C. A. Peres. 2007. "Hunting and Plant Community Dynamics in Tropical Forests: A Synthesis and Future Directions." *Biotropica* 39: 385–92.

Taylor, Paul W. 1986. *Respect for Nature: A Theory of Environmental Ethics.* Princeton, NJ: Princeton University Press.

Terborgh, John. 1999. *Requiem for Nature.* Washington, DC: Island Press.

Thibault, Marc, and Sonia Blaney. 2003. "The Oil Industry as an Underlying Factor in the Bushmeat Crisis in Central Africa." *Conservation Biology* 17: 1807–13.

Varner, Gary E. 1998. *In Nature's Interests? Interests, Animal Rights, and Environmental Ethics.* Oxford: Oxford University Press.

Vucetich, John A., and Michael P. Nelson. This volume. "The Infirm Ethical Foundations of Conservation."

Westra, Laura. 1994. *An Environmental Proposal for Ethics: The Principle of Integrity.* Lanham, MD: Rowman and Littlefield.

Wilkie, David S., and Julia F. Carpenter. 1999. "Bushmeat Hunting in the Congo Basin: An Assessment of Impacts and Options for Mitigation." *Biodiversity and Conservation* 8: 927–55.

Wilkie, David S., and Ricardo A. Godoy. 2001. "Income and Price Elasticities of Bushmeat Demand in Lowland Amerindian Societies." *Conservation Biology* 15: 761–69.

Wilkie, D. S., M. Starkey, K. Abernethy, E. Nstame Effa, P. Telfer, and R. Godoy. 2005. "Role of Prices and Wealth in Consumer Demand for Bushmeat in Gabon, Africa." *Conservation Biology* 19: 268–74.

Ziegler, Stefan. 2010. "Application of Food Balance Sheets to Assess the Scale of the Bushmeat Trade in Central Africa." *Traffic Bulletin* 22: 105–16.

PART TWO

CONSERVATION BEHAVIOR AND "ENLIGHTENED MANAGEMENT"

GUIDELINES FOR RESTORING, RECREATING, AND REDECORATING NATURE

MANY OF THE PROBLEMS we face have originated and grow in severity because we ignore nature. The essays in this part raise a large number of issues that are central to this general theme, including the consequences of not paying attention to what we routinely and habitually (some might say obsessively) do. Biology, economics, and politics all enter the scene. There are many lessons from past attempts to save individuals, species, and ecosystems and their fragile, precious, and irreplaceable webs of nature, but it's not clear that we learn much from our successes and failures. Reintroduction projects have played a large role in our attempts to recreate or restore ecosystems and are discussed as examples of what can be done but how difficult it is to achieve our goals.

Daniel Blumstein's chapter tackles head on how we ignore evidence and by doing so jeopardize nature conservation. He also considers how science is done, the peer-review process, and how the practice of doing science influences the evidence we have available for making decisions. Arguing that the best conservation emerges from an explicitly evidence-based approach, Blumstein discusses several ways that evidence can be used to improve conservation outcomes. He concludes we must encourage stakeholders to use, not ignore, evidence. Echoing what others have written here and elsewhere, Blumstein stresses that scientific evidence should hold a special place in conservation and management, and decisions *must* involve all stakeholders. He also considers the sticky question of when we should move ahead even when we lack sufficient evidence that supports this decision but have a good idea that inaction could have negative consequences. Other essays in this volume also ponder how we must balance evidence and intuition when we try to deal with problems whose resolution depends on input and consensus from diverse parties, including biologists, psychologists, economists, and politicians. Cultural differences also play a role and complicate the process. Pluralism, not absolutism, has to be the strategy of choice, but once again we're faced with a question of which guiding principles can be compromised and which cannot.

Joel Berger follows up Blumstein's ideas by showing how we fail to learn from past mistakes when we ignore complex relationships among various species, in this case interactions among grazing sheep, coyotes, and rabbits. We also ignore the fact that webs of nature are dynamic, not static. Considering that rabbit drives used to kill these animals in mass because in Wyoming and Idaho they're considered predators or varmints, Berger notes that killing rabbits might actually increase coyotes' predation on sheep. There also are unintended consequences that shift ecological relationships in a given area that might negatively affect human health. Berger cautions that while we *can* learn from past mistakes we often do not. There are no substitutes for long-term research that takes into account the intricate and dynamic relationships among different species and the areas in which they live. But even when we get a more or less complete picture for one situation, the application of these findings may not apply to other circumstances. Once again things can appear hopelessly complex and insoluble, but that's the real world with which we need to deal.

Moving us into an area that is becoming more and more important—namely, how we interact with animals who enter into our homes and directly influence our lives—Camilla Fox shows that ignoring nature leads to ineffective predator management. Considering practical matters of dealing with coyotes, Fox shows that the long-term wanton and rampant killing of these incredibly adaptable mammals by the US government and others has not had much of an effect on their numbers or predation on livestock, extremely small that it is. Indeed, recent figures for nonpredator deaths of cattle and calves provided by the United States Department of Agriculture (USDA 2011) show that more than 95 percent are due to disease and other factors, not active predation. When we pay attention to the biology of coyotes and also use a community-based approach that stresses coexistence and compassion, livestock-predator conflicts can be more readily solved. Once again, give and take is the name of the game.

Marco Festa-Bianchet echoes the messages of the preceding three essays by arguing that evolutionary biology is important for conservation and the development and implementation of sustainable management. In his wide-ranging essay he argues that human activities become a selective pressure, and we should expect evolutionary changes as a consequence of those activities. We also have a sort of "Walt Disney" expectation that a "balance" (between predators and prey, herbivores and vegetation, parasites and hosts) will be the norm at any temporal and spatial scale. We expect small reserves to have a predator-prey equilibrium, when in reality equilibriums may only exist over very large spatial scales. We expect the consequences of human impacts (overharvests, habitat destruction, exotic species) to be fixed quickly by restoration projects, but return to "normal" may take generations or be impossible. Festa-Bianchet also notes that society emphasizes the "here and now" and cannot plan over the long term. He concentrates on harvesting fish where there is good evidence that unsustainable overfishing selects for reproductive strategies that are maladaptive under natural situations. To quote Festa-Bianchet, "My argument is simple: if some traits make an individual less likely to be harvested, and if harvest pressure is high, then if those traits have a genetic component they should become more common over time. Humans will then shape the evolution of harvested species, sometimes with results that may be detrimental to both the species and the harvesters." He concludes that sustainable compromises are possible as we learn more about the mating ecology of species. We need evolutionary and ecologically enlightened management, but before it is widely accepted it will have to break numerous cultural, economic, and political barriers.

Philip Seddon and Yolanda van Heezik write about what might be called "enlightened management." They focus on reintroduction projects. Seddon and van Heezik also weave in what we know about the importance of children having direct experiences with nature, the absence of which leads to a ratcheting down of expectations as people don't realize what has been lost but accept the highly modified and depauperate environment that surrounds them as normal. They argue convincingly: "We need to reset public baselines through the restoration of biodiverse natural environments in which people can reconnect with nature and renew their expectations of what is normal. We need actively to put native species back into areas from which they have been lost." Reintroduction projects can help put species back into areas where they once lived but there's been a taxonomic bias in that they tend to concentrate on charismatic iconic or poster species. While these model species are needed for garnering public support for these projects, Seddon and van Heezik stress: "There needs to be a shift in focus away from species chosen solely for their public appeal and toward restorations that include keystone species in an explicit attempt to restore ecosystem function." If we ignore the importance of less iconic species then we're ignoring their importance in ecosystem integrity. Seddon and van Heezik's essay raises many questions that need to be given serious attention in future reintroduction projects. They also stress that public support is much needed if they are to be undertaken and succeed. People get accustomed to the absence of animals in their surroundings and have concerns about economic and other impacts on their lifestyles. How to reengage the public and get people involved in biodiversity conservation is a huge task but can be made easier with community-based projects and also by getting youngsters to have direct experiences with nature. Gene Myers also writes about this in his essay concerned with children, animals, and conservation (part 4).

Following up on Seddon and van Heezik's essay and the importance of reintroduction projects for biodiversity conservation and for reengaging people with nature, Sarah King writes more specifically about reintroduction projects involving Przewalski horses and red wolves and how they and other reintroductions might be made more successful if we know about the details of the behavior of the animals under consideration. She notes that despite the many differences between Przewalski horses and the red wolf and their reintroductions, there are also many similarities: both were reintroduced from populations that had been captive for several generations owing to their extinction in the wild, and both reintroduction projects need to address problems of hybridization, inbreeding, and conflicts over land use. Thus knowledge of the animals' behavior will help managers mitigate these problems and can potentially aid persistence in the wild of current populations.

While King's conclusion may seem very obvious, detailed behavioral studies are often lacking in reintroduction efforts and conservation studies in general. She and others suggest they would be more successful if we knew more about the animals with whom we're working. These projects can be extremely time consuming and expensive and success rates are very low, but we're ignoring nature if we don't pay attention to the details.

We also ignore nature if we don't consider the behavior of the individual animals we're studying or trying to reintroduce or in whom we're otherwise interested. This is why Liv Baker's essay on the importance of understanding individual differences among animals is so important and builds nicely from King's paper. We know that age and gender differences play a large role in how individuals behave both in a group and when alone. Baker notes that behavioral ecologists are now interested in the fitness consequences of different personalities within a population. We also know, for example, that mortality, breeding, dispersal, partner preferences, and disease risk are correlated with individual differences in personality. Baker notes there also are ethical considerations when we try to conserve threatened species in translocation projects. Individuals can experience varying levels of stress from trapping injuries, capture and transport myopathy, weight loss, and disease when being moved from one place to another, and a concern for their well-being and individual differences in response to being moved must be taken into account. Many translocation projects are unsuccessful and perhaps they'd be more successful if individual differences were given more consideration. We like to say there's no place like home and if individuals are to survive when being moved around we must be sure they wind up in a place in which they feel safe, a location that's considered home.

Other researchers have also been interested in how individual differences influence behavior and survival. Consider for example the work of Samantha Bremner-Harrison and her colleagues on how individual differences in the behavior of captive-bred swift foxes influenced survival after their release (Bremner-Harrison, Prodöhl, and Elwood 2004). They assessed variation in boldness in these carnivores and discovered that those foxes who died in the six months following release were the boldest individuals. In the presence of novel stimuli in captivity these foxes left their dens more quickly, approached stimuli more closely, and were less fearful than did those foxes who survived. These individuals were less suited for release.

On a different scale and going beyond the individual differences among the animals who interest us, we also must pay attention to what we've done in the areas from which we've removed animals. For example, when wolves were moved to Yellowstone National Park to reintroduce these native carnivores, a program that has enjoyed great success, nothing was known about the changes in social dynamics in the packs from which wolves were removed or the ecological consequences of their removal. When I asked this question at a meeting, I was met with silence, not with scorn. Nobody knew. Were we "robbing Peter to pay Paul"? In our efforts to "recreate" or "restore" one ecosystem were we compromising another?

We also ignore nature when we allow animals such as wolves to be killed because they've become "pests," because they do on occasion kill livestock. If we're going to reintroduce animals such as wolves we can't ignore who they are, lest we wind up in the future trying to reintroduce them again because they've been exterminated for behaving as the predators they are. We can't continue to employ these kill, reintroduce, and kill cycles (Bekoff 2002, 2006, 2010). They're economically unsustainable and unfair to the animals and perhaps others who would benefit from a shift in our priorities.

Clearly, we need to know more about the individual animals. While funding for long-term observational and descriptive studies is lacking and ever decreasing, perhaps the field notes others have taken during their research could be useful (e.g., Canfield 2011). Over a period of twenty or more years I've gone back to and shared with others field notes from our long-term study of the social ecology of coyotes and the research we conducted on various birds. While it's difficult to generalize from one study to another, it's better than nothing. Sharing might help us along as we attempt projects that traditionally haven't been very successful because of a lack of basic information on the behavior and behavioral ecology of the species under consideration. We also need to convince funding agencies that these data are essential and that patience is needed, because these sorts of studies take time. If support isn't forthcoming we might have to admit we don't know enough either to begin or continue a project and reestablish priorities.

References

Bekoff, M. 2002. "The Importance of Ethics in Conservation Biology: Let's Be Ethicists Not Ostriches." *Endangered Species Update* 19 (2): 23–26.

———. 2006. *Animal Passions and Beastly Virtues: Reflection on Redecorating Nature.* Philadelphia: Temple University Press.

————. 2010. *The Animal Manifesto: Six Reasons for Expanding Our Compassion Footprint*. Novato, CA: New World Library.

Bremner-Harrison, Samantha, Paulo Prodöhl, and Robert W. Elwood. 2004. "Behavioural Trait Assessment as a Release Criterion: Boldness Predicts Early Death in a Reintroduction Programme of Captive-Bred Swift Fox (*Vulpes velox*)." *Animal Conservation* 7: 313–20.

Canfield, Michael R., ed. 2011. *Field Notes on Science Nature*. Cambridge, MA: Harvard University Press.

USDA. 2011. "USDA Releases 2010 Non-Predator Death Loss Report." http://beefmagazine.com/cowcalfweekly/1230-non-predator-loss/.

6

Why We *Really* Don't Care about the Evidence in Evidence-Based Decision Making in Conservation (and How to Change This)

Daniel T. Blumstein

IF WE VALUE NATURE, we should value scientific evidence to help manage and preserve it. These days, evidence-based decision making is touted as the way to improve everything from health care (Cochrane 1972), to environmental education (Keene and Blumstein 2010; Saylan and Blumstein 2011), to international developmental aid (Duflo and Kremer 2005), as well as an important way to improve conservation outcomes (Pullin and Knight 2001, 2009; Schreiber et al. 2004). In the field of conservation biology, managers talk about adaptive management, a process that ultimately uses evidence to improve management outcomes. Yet when one digs down beneath the surface, properly designed experiments that are explicitly part of adaptive management are rare, managers are reluctant to embrace the method, and many people hold that good decisions emerge from a process that has little to do with evidence. In this chapter I will explore the question of whether and when scientific evidence is important, and, when it may not be, how we can generate better conservation and management outcomes.

Why Scientific Evidence Should Be Valued

Many pundits now declare the end of science. For instance, in a *Wall Street Journal* Op-Ed, Daniel Henninger (2009) concluded that science had become postmodernist and therefore creates biased and relativistic results that should not have any special standing. From my perspective, nothing could be further from the truth.

The essence of the scientific method is to pose a testable hypothesis, collect data, and evaluate the results. Often these hypotheses are phrased in terms of a formal null hypothesis. For instance, does drug x

have no influence on blood pressure. If we find that patients taking drug x have a 50 percent reduction in their blood pressure compared to those taking controls who had no change in blood pressure, we would infer that drug x reduces blood pressure. If we don't find an effect of drug x on blood pressure we are in a bit of a quandary: the drug could work but must be given at a different dosage, it may not work, or something else might be responsible for the lack of change. We learn more by refuting our null hypotheses than we learn by not refuting them.

By asking a series of questions using this "hypothetico-deductive method" (e.g., Popper 1958) we can quickly discover what drugs may reduce blood pressure and which don't. We can discover drugs to treat cancer. We can develop stem cell therapeutic technology. If we're so inclined, we can design better weapons and renewable sources of energy. We can learn about factors that may generate or maintain biodiversity in the oceans, in the forests, and in hospital operating rooms.

Posed this way, the scientific method is a brilliant way to efficiently generate knowledge and evaluate hypotheses. Science is a way of knowing (Moore 1999). Science does not tell us what questions should be asked (Should we develop weapons? Should we develop wave generated power?), but once asked, science is a process to separate the valid from the invalid. Part of the process that allows us to trust the outcome is that scientific findings are published in peer-reviewed journals.

Peer review is a process by which results are subjected to (usually) anonymous review by skeptical experts. Many scientists are brutally competitive and will look for the slightest reason to find fault with a submitted manuscript they are reviewing. Why? Ideally because they feel that only the best results deserve publication. And when peer review works, the reviewers take their own (almost always unpaid) time and make constructive comments that serve to improve the resulting paper. Sometimes scientists unethically reject competitors' papers for no other reason than that they are competitors, but these all-too-human outbreaks are rare in a process that generally works remarkably well.

Once published, a paper is a target and its findings are subjected to continued scrutiny by scientists who seek to refute it on empirical or theoretical grounds. Why? Ideally, and frequently, because scientists seek to better explain the world around us. However, scientists are people and careers are made not only on building new ideas but also by refuting high profile ideas. Imagine how famous an evolutionary biologist would be who suddenly discovered a major fault with evolutionary theory. However, to do so, experts must evaluate the challenger's hypothesis and findings. If the peers' evaluation is positive, we've made substantial scientific progress. The fact that evolutionary theory has withstood the test of time and been constructively modified over time is a testament to its validity.

Peers are those who have sufficient background knowledge in a field to properly evaluate a paper. Peers are those who have published and developed their reputation in that field by their work and their results. Just because someone may not "believe" in evolution (not that anyone should believe in evolution because it should be a testable hypothesis), doesn't mean that they are not a valid peer. Just because someone has a PhD doesn't mean they are necessarily able to evaluate a particular finding if it's outside their field. Of course, we must rigorously guard against "group-think" and processes that completely shut out alternative opinions, but experts in a field really are those who are best qualified to evaluate a new finding before publication.

For these reasons, I believe that scientific evidence should hold a special place in conservation and management. How then is scientific evidence used to make decisions?

Adaptive Management and Decision Making

Adaptive management is a very important process by which scientific evidence is used to enhance conservation outcomes (Walters and Holling 1990). Adaptive management uses controlled experiments or uncontrolled comparisons to quantify the effects of management interventions on management outcomes (Salafsky and Margoluis 2003).

In active adaptive management, managers design formal experiments that employ a BACI—before-after, control-impact—design (Walters and Holling 1990; Underwood 1992). In a BACI design, the difference before and after a treatment is compared across a control and some impact or management intervention. Formally, this design allows managers to isolate the effect of a particular impact or management intervention because there is a control. Controls are situations in which nothing is done and thus a control makes it possible to account for annual or other factors that might simultaneously be influencing an outcome.

An example will help illustrate active adaptive management. Assume you're trying to increase the success of a captive-rearing and reintroduction program. You have identified a problem—animals are killed by predators soon after release. You hypothesize that by training them to be more aware of predators before release they will survive better upon release. You divide up your animals to be released into two groups; one group gets trained, the other gets the added handling experience without formal training. You have data from the year before where there was 0 percent survival of introduced animals. You then release these animals and compare the survival of the trained versus untrained animals. Because you had 0 survival the year before, the analysis is simply whether the trained animals survived more than the untrained ones. If so, you can conclude that training was effective. However, what if the survival of both groups increased? In this case you might conclude that there are annual effects (maybe there was no predator around that year), or that the increased handling you did for your control group enhanced survival. If the training does not specifically enhance survival, you may decide it's too costly to do. This scenario illustrates how and why it is important to have a control group; it allows managers to isolate the effect of a management intervention and see if it specifically is responsible for enhanced survival.

Selecting controls in adaptive management scenarios is essential but creates some novel issues. What, for example, is one to do if by having a control, one knows that a population may lose a substantial number of individuals from an endangered species? Indeed, many managers find using controls in these situations ethically challenging and contrary to the goals of management, which may be desperately trying to increase the abundance of a threatened or endangered species (Johnson 1999). In the above example, having a control group in which you didn't train animals might be sentencing them to death—because all prior experience pointed to 0 percent survival for untrained animals. In such circumstances managers often find it ethically and indeed practically difficult to justify having a control.

In passive adaptive management, managers compare the outcomes of uncontrolled experiments to either previous outcomes or they may employ "natural controls." Because controls are not formally designed into the comparisons, it's not possible to isolate the effect of the treatment on the outcome. Thus without a control one wouldn't know whether it was the training, or annual variation, or simply the handling that enhanced survival. Nevertheless, and in spite of these shortcomings, it is perhaps better to be making these sorts of comparisons than not making comparisons and relying simply on intuition to make management decisions. Can we do better?

Darwinian Decision Making

In some cases it is ethically or politically difficult to run proper controls. For instance, if managers know that current captive breeding practices result in 50 percent mortality, and that 50 percent mortality is unsustainable, then it is essential to increase survival. Rather than having a control (which one knows will continue to have 50 percent mortality), perhaps it is better to directly compare two (or more) alternative treatments. In the medical literature this is known as comparative effectiveness evaluation.

The shortcoming of this is that without a proper control, there may be a nagging uncertainty about whether something else changed during the experiment. However, a comparative effectiveness approach may be defensible if there are welfare costs to a business as usual control that one already knows doesn't work. In other words, if one knows a lot of animals are going to suffer or die, it may be preferable to compare two different possible solutions rather than having a control.

A comparative effectiveness study is best conducted as a BACI design, except here we are comparing the after minus before to the two alternative treatments. I have previously suggested that to come up with the best alternative treatments, experts should be consulted (Blumstein 2007). This will to some extent mimic a Darwinian process whereby a variety of alternative treatments are generated, and the best will be rapidly identified. By this means, Darwinian decision making can be an important tool in adaptive management.

Does the Best Evidence Lead to the Best Management Outcomes?

Conservation and management are political. To be convinced, simply ask why the Yellowstone wolves were considered fully recovered and thus de-listed from the Endangered Species Act when by hunting and killing wolves straying from the safety of protected areas the population would immediately decline to levels that might not be considered sustainable (Bergstrom et al. 2009). Or ask why some species are listed while others are not (Harllee et al. 2009).

Decisions made in the political sphere are not necessarily based on the best available scientific evidence. Should they be? In a compelling book, *The Paradox of Scientific Authority: The Role of Scientific Advice in Democracies*, Bijker, Bal, and Hendriks (2009) argue that scientific evidence is best evaluated by a committee, working out of the spotlight, and charged with providing the best interpretation of the scientific results as possible to political decision makers. Such high-level committees are exemplified by the US National Academy of Sciences, which creates committees tasked with providing information to Congress. Bijker, Bal, and Hendriks focused on the Health Council of the Netherlands, a committee tasked with providing the best possible scientific evaluation for politicians with respect to health and medical issues. By working out of the spotlight and behind closed doors, these committees are free to evaluate evidence with little oversight. If the committee is well chosen and diverse, then the recommendations to the policy makers should be well thought out and useful. Employing such "expert" decision makers to evaluate evidence and develop reports that enable decision makers, however, is not the normal way that decisions are made.

Often, scientific evidence is mixed with politics without going through a committee's "filter." Or the results from the scientific experts are discounted. This is because good political outcomes are often viewed as those that go through a process that involves stakeholders (Burgman 2005). Stakeholders are those that self-identify with an issue. In a representative democracy, we want to involve people who care about issues and we want their views to be understood, and, if popular, represented. Thus many management decisions involve getting stakeholders involved in a process that generates a consensus. From a managers' view, this may be the sort of outcome that is most desirable.

Deciding whether to list or de-list a species is a political decision. Deciding to kill "problem" animals is a political decision (animals aren't problems—we perceive them as problems!—Goodall and Bekoff 2002). Deciding how to allocate funds among competing conservation needs is a political decision (funds going to wolf conservation are not going to sage grouse conservation). So what then is the role of scientific evidence in decision making?

How Should Evidence Be Used in Decision Making?

I suggest that there are many management decisions that require evidence to enhance effective conservation. The Centre for Evidence-Based Conservation (http://www.cebc.bangor.ac.uk/index) was founded in 2003 and is dedicated to using systematic reviews to enhance conservation efficacy. The Centre has sponsored a variety of reviews that include a variety of topics (all reviews are posted at http://www.environmentalevidence.org/Reviews.htm): Are mammal and bird populations declining in the proximity of roads and other infrastructure? Does MHC diversity decrease viability of vertebrate populations? What are the impacts of human recreational activity on the distribution, nest occupancy, and reproductive success of breeding raptors? Are marine protected areas effective tools for sustainable fisheries management? These topics are varied and provide managers with the best-available evidence to enable thoughtful decisions, even if decisions are made in the political sphere.

While I believe that evidence should be an important part of decision making, sustainable decisions *must* involve stakeholders (Schreiber et al. 2004). That said, stakeholders *must* be charged with using the available evidence to make the best decisions. In other words, creating decision-making processes that explicitly respect the process of using data, value experimental data more than correlative data, and seek to build in data collection as part of ongoing adaptive management. Evidence, viewed this way, is an essential part of the process of making a decision. Evaluation, viewed this way, is built into both ongoing monitoring and the decision-making process.

How Should Lack of Evidence Be Handled?

In many cases lack of sufficient evidence is often used as an excuse for inaction. If the consequences of inaction are small, there may be sufficient time to collect more data. However, if the consequences of inaction are great, it is probably best to adopt the "precautionary principal" that essentially states it's better to be safe than sorry and the onus is on those who want action to demonstrate that action will not be harmful (http://www.sehn.org/wing.html). It is important to realize that many opponents to action will harp on the uncertainties involved in the decision-making process and argue about the costs of action. For instance, opponents to limiting fossil fuel use or to developing "clean" energy often point to the costs associated with changing our fuel consumption habits. In cases like this, it is only sensible to articulate the costs of inaction. If the costs of inaction are greater than the costs of action, a rational decision is to proceed cautiously. For instance, I would suggest that the ecological and environmental consequences of melting the polar ice caps and releasing methane—a potent greenhouse gas—from the thawed permafrost are extreme and probably exceed the costs to increasing conservation and developing alternative fuel sources. Regardless, data should continue to be collected and analyzed and decisions modified based on current data.

Island Fox Conservation: Two Examples of Wise Management

The island fox (*Urocyon littoralis*) is a diminutive North American canid and is endemic to Southern California's Channel Islands. Island species are especially vulnerable to stochastic events, and different islands, each with its endemic subspecies, were threatened by some different problems. Two successful, scientifically based recovery programs illustrate features that should be (and often are) modeled in other recoveries. Coonan, Schwemm, and Garcelon (2010) describe much of this.

Santa Catalina Island, the largest of the southern Channel Islands, had a bout of canine distemper that caused a dramatic decline in population size (at one point there were fewer than one hundred foxes). Scientific management that included vaccination of surviving foxes, considerable work led by stakeholders (especially the Santa Catalina Island Company), as well as captive breeding followed with reintroduction to recover the population, ultimately led to the successful recovery of this island's population. This was facilitated by having relatively few stakeholders involved (most of the island is owned and managed by the Santa Catalina Island Company), a small population of residents, the ability to control visitor behavior, and a the presence of a strong "scientific culture" for management.

Meanwhile, the Northern Channel Island populations declined precipitously because golden eagles (*Aquila chrysaetos*) self-introduced themselves to the islands. Fortuitously, the foxes on some of the islands were being studied by graduate students and monitored by government researchers, and this decline was tracked with precision.

The best available evidence suggested that an ecological phenomenon called "hyperpredation" was responsible for their decline. Bald eagles were naturally on the islands but high levels of PCBs in the fish they ate took a toll on reproduction; it was hypothesized that the vacancy left by bald eagles permitted golden eagles to self-introduce themselves. Golden eagles were primarily supported by a large feral pig population on the islands, and foxes were inadvertent victims of a growing eagle population. Foxes were brought into captivity both for their safety and to begin a captive breeding program. Managers had been working for years to remove the pigs from the islands and stepped up their efforts on this. Many eagles were live-trapped and relocated to north-central California. A captive breeding program for bald eagles (*Haliaeetus leucocephalus*) was expanded with the ultimate goal being to replace the golden eagles with bald eagles.

Eventually, in 2004, the US Fish and Wildlife Service formally listed the foxes on the Northern Channel Islands as critically endangered. This brought the US Fish and Wildlife Service into the mix of stakeholders. From a low high of several thousand easy-to-see foxes active on the islands, to a low of about seventy animals scattered across all the Northern Channel Islands, the mix of management was successful and the population grew with the removal of pigs and golden eagles. By mid-2010, there were more than 1,700 foxes populating the northern islands and the species was headed for de-listing.

Throughout, various stakeholders that included the National Park Service, the Nature Conservancy, the US military, University of California researchers and land managers, as well as zoos, and public interest groups, were actively involved in discussing and debating management options. Scientific consultants were brought in, and while the process led to no formal active adaptive management projects, scientific evidence was highly valued by all stakeholders and used throughout the process.

Other Examples of Wise and Potentially Wise Management

Nichols and Williams (2006) review the case of adaptive harvest management of mallard ducks (*Anas platyrhynchos*) in North America. Scientific-based management of duck hunting involves stakeholders, and active monitoring is an explicit part of the process. Based on annual population estimates and population trajectory, along with a survey of juvenile survival, various population models are parameterized annually and recommendations are made for harvest size. The population remains stable despite extensive hunting.

Innovative adaptive management programs abound in New Zealand. One (Armstrong, Castro, and Griffiths 2007) has focused on the hihi (*Notiomystis cincta*), a critically endangered bird that was barely surviving on a single island. Managers wanted to expand the range and incorporated a series of population models and experimental reintroductions. Regular monitoring identified factors that could be used to increase survival (experimental provision of sugar water and experimental removal of mites) and those that influenced survival but could not be controlled (the presence of a fungal spore). Ultimately, animals were successfully introduced to several islands and removed from an island with high fungal spore levels.

Management of captive giant pandas (*Ailuropoda melanoleuca*) has always involved active participation of major stakeholders and experiments conducted in captivity (e.g., Swaisgood et al. 2001), yet field research lagged behind in scientific management. Future studies are being planned in an adaptive context that involves the Chinese government working with local communities in a way to employ manipulative experiments to inform the management of wild populations (Swaisgood et al. 2011). Time will reveal the degree and role of experimental active management in the field and whether it helps inform management and results in success.

Conclusions and Recommendations

Systematic reviews, whether conducted by an individual, a research group, or a private committee, are an excellent way to provide evidence to decision makers. Decision makers must include stakeholders who have a vested interest in the outcome; sustainable solutions involve stakeholder support. Stakeholders must be charged with using evidence to make decisions. The onus is on those who oppose the evidence to build compelling arguments about why the evidence should be ignored. It should be unacceptable to not act because of insufficient evidence if the consequence of inaction is potentially great. Because evidence is often lacking in many conservation problems, it is essential to build into the decision-making process the ability to collect new data and to reevaluate decisions based on these new data. Controlled experiments should be done unless there are good reasons not to. Viewed this way, adaptive management is a process that should be embraced because it provides ongoing evaluation of conservation outcomes and is designed to improve management outcomes.

References

Armstrong, D. P., I. Castro, and R. Griffiths. 2007. "Using Adaptive Management to Determine Requirements of Re-Introduced Populations: The Case of the New Zealand Hihi." *Journal of Applied Ecology* 44: 953–962.

Bergstrom, B. J., S. Vignieri, S. R. Sheffield, W. Sechrest, and A. A. Carlson. 2009. "The Northern Rocky Mountain Gray Wolf Is Not Yet Recovered." *BioScience* 59: 991–999.

Bijker, W. E., R. Bal, and R. Hendriks. 2009. *The Paradox of Scientific Authority: The Role of Scientific Advice in Democracies.* Cambridge, MA: MIT Press.

Blumstein, D. T. 2007. "Darwinian Decision-Making: Putting the Adaptive into Adaptive Management." *Conservation Biology* 21: 552–53.

Burgman, M. 2005. *Risks and Decisions for Conservation and Environmental Management.* Cambridge: Cambridge University Press.

Cochrane, A. L. 1972. *Effectiveness and Efficiency: Random Reflections on the Health Services*. London: Nuffield Provincial Hospitals Trust.

Coonan, T. J., C. A. Schwemm, and D. K. Garcelon. 2010. *Decline and Recovery of the Island Fox: A Case Study for Population Recovery*. New York: Cambridge University Press.

Duflo, E., and M. Kremer. 2005. "Use of Randomization in the Evaluation of Development Effectiveness." In *Evaluating Development Effectiveness*, edited by G. K. Pitman, O. N. Feinstein, and G. K. Ingram, 205–32. New Brunswick, NJ: Transaction Publishers.

Goodall, J., and M. Bekoff. 2002. *The Ten Trusts: What We Must Do to Care for the Animals We Love*. New York: HarperCollins.

Harllee, B., M. Kim, and M. Nieswiadomy. 2009. "Political Influence on Historical ESA Listings by State: A Count Data Analysis." *Public Choice* 140: 21–42.

Henninger, D. 2009. "OPINION: Climategate: Science Is Dying." *Wall Street Journal*, December 9, 2009.

Johnson, B. L. 1999. "The Role of Adaptive Management as an Operational Approach for Resource Management Agencies." *Conservation Ecology* 3, no. 8. http://www.consecol.org/vol3/iss2/art8/.

Keene, M., and D. T. Blumstein. 2010. "Environmental Education: A Time of Change, a Time for Change." *Journal of Evaluation and Program Planning* 33: 201–4.

Moore, J. A. 1999. *Science as a Way of Knowing*. Cambridge, MA: Harvard University Press.

Nichols, J. D., and B. K. Williams. 2006. "Monitoring for Conservation." *Trends in Ecology and Evolution* 21: 668–73.

Popper, K. 1958. *The Logic of Scientific Discovery*. London: Hutchinson.

Pullin, A. S., and T. M. Knight. 2001. "Effectiveness in Conservation Practice: Pointers from Medicine and Public Health." *Conservation Biology* 15: 50–54.

———. 2009. "Doing More Good Than Harm: Building an Evidence-Base for Conservation and Environmental Management." *Biological Conservation* 142: 931–34.

Salafsky, N., and R. Margoluis. 2003. "What Conservation Can Learn from Other Fields about Monitoring and Evaluation." *BioScience* 53: 120–22.

Saylan, C., and D. T. Blumstein. 2011. *The Failure of Environmental Education (and How We Can Fix It)*. Berkeley: University of California Press.

Schreiber, S. G., A. R. Bearlin, S. J. Nicol, and C. R. Todd. 2004. "Adaptive Management: A Synthesis of Current Understanding and Effective Application." *Ecological Management and Restoration* 5: 177–82.

Swaisgood, R. R., F. Wei, W. J. McShea, D. E. Wildt, A. J. Kouba, and Z. Zhang. 2011. "Can Science Save the Giant Panda (*Ailuropoda melanoleuca*)?" *Integrative Zoology* 6: 290–96.

Swaisgood, R.R., A. M. White, X. Zhou, H. Zhang, G. Zhang, R. Wei, V. J. Hare, E. M. Tepper, and D. G. Lindburg. 2001. "A Quantitative Assessment of the Efficacy of an Environmental Enrichment Programme for Giant Pandas." *Animal Behaviour* 61: 447–57.

Underwood, A. J. 1992. "Beyond BACI: The Detection of Environmental Impacts on Populations in the Real, but Variable, World." *Journal of Experimental Marine Biology and Ecology* 161: 145–78.

Walters, C. J., and C. S. Holling. 1990. "Large-Scale Management Experiments and Learning by Doing." *Ecology* 71: 2060–68.

7

Cautionary Wildlife Tales

Learning to Fail or Failing to Learn?

Joel Berger

GEORGE SANTAYANA once famously said, "Those who cannot re-member the past are condemned to repeat it" (Santayana 1905). Simi-larly, when we ignore lessons from nature, we continue to suffer the fate of fools because we compromise our economies and abilities to effectively restore nature. In this short essay, I use several cases that have played out during the last one hundred years to reinforce the above point. Specifically, I describe how our reckless historic pursuit to remove hares from western USA ecosystems has had striking eco-logical consequences. While the understanding of species interactions within the context of food webs is complex and people with a love of nature and animals may be excused for failing to appreciate nuanced relationships, the time has passed not to notice what has gone wrong by failing to listen.

I was once asked by a rancher from eastern Oregon to come along on a rabbit drive. Being from Los Angles, I knew little about this prac-tice. I quickly learned. People flock together and corral the rabbits before flogging them to death. The ecological goal is to reduce rabbit densities, and in this high desert landscape agricultural values are the cultural mores. Rabbit drives are not new, having been around for more than one hundred years in "modern times" and occurring from California to New Mexico and on up into Saskatchewan.

For thousands of years before this, Native Americans used similar methods to procure rabbits and bison and pronghorn. Tibetans did similarly for their high elevation antelopes, the chiru. The difference in economic rationale between the distant past and now is virtually none—enriching one's livelihood. In the past, game, small or big, was harvested for food and clothing. Today's semirecent drives were designed for financial rewards, which involved manipulation of nature's food web for individual gains. In this case, the killing of white-tailed and black-tailed jackrabbits (which are true hares) was to reduce these "noted" pests either to minimize forage off-take to benefit livestock or to decrease herbivory on crops. Even today in both Wyoming and Idaho, jackrabbits are officially considered predators or varmints.

In his 1949 book, *A Sand County Almanac*, Aldo Leopold said, "To keep every cog and wheel is the first precaution of intelligent tinkering." The broader issue of course is what we have learned and what we fail to learn about manipulating nature, and assuredly what the knowledge we derive tells us about the land and the lives that remain upon it. Our human legacies are large, and it is we, as world arbiters, who decide what biological diversity remains for future inheritance. Desert and grassland hares offer a valuable metaphor for what we have and have not learned about interactions among species nature and its bearing on conservation.

Much scientific effort is devoted to understanding the extent to which ecosystems are regulated by top-down forces such as predation and bottom-up drivers like plant productivity. With wolves or bears removed many believe that elk or moose multiply and then overbrowse their riparian vegetation, which in turn no longer supports biologically rich communities, including migrant songbirds. With sea otters removed, urchins proliferate, leaving in their wake a decimated kelp community. The explosion of white-tailed deer has reduced acorn masts in the eastern United States and simplified ecological interactions (Estes and Terborgh 2010).

In contrast to top-down effects of carnivores, abiotic drivers such as cold, rain, and snow all shape growing seasons and plant productivity (and diversity), factors that ultimately control the abundance of life ecosystems support. Desert and Arctic environments sustain far less diversity and biomass than do temperate and tropical ones. In turn, such factors regularly dictate the abundance of top carnivores. Such interplay between top-down and bottom-up forcing depends on an area's history, disturbance regime, species composition, and nuances associated with individual species and their densities.

The key question is not what rabbit drives have to do with ecological health, though this is clearly relevant to those of us preoccupied with interfacing science with conservation. It is what our tinkering has done to the landscapes that we humans depend upon and how these changes affect processes that we, as a society, care about. In essence, setting aside a personal zest to conserve biodiversity, I adopt here a human-centric approach and ask how rabbit drives affect our human economies or societal values. Three points come to the fore.

First, on public lands in the American West, domestic sheep grazing has occurred for more than a century. It continues although the sheep industry is less robust now than in the past. Both US government and state efforts still target coyotes through predator control programs because coyotes have been and remain important predators of domestic sheep (Berger 2006). The amount of predation is inversely related to jackrabbit abundance, at least in the northern Great Basin desert; with fewer black-tailed jackrabbits, more sheep are killed (Knowlton and Stoddart 1992). While it is unclear how widespread this pattern is and despite the millions of dollars spent in eradication campaigns, these relationships have not been studied elsewhere. It is possible that with other factors equal, the killing of jackrabbits inadvertently exacerbates predation on domestic sheep because hares, as alternative, semilarge prey, are scant.

Second, an interesting dynamic links coyotes and jackrabbits with an ecological process of societal interest—long distance migration. In and around Grand Teton National Park in Wyoming, pronghorn move from the park's summering grounds to spend their winters at distant sites in the Upper Green River Basin. This is the longest terrestrial mammal migration between Canada and Tierra del Fuego, with some animals moving about 700 kilometers round-trip (Berger, Cain, and Berger 2006). Because of its length and national prominence, this migration has been afforded federal protection through approved national forest management plans (Cohn 2010).

Federal policy, however, does not necessarily assure long-term conservation of this unusual migration because of the nuanced yet complex interactions between coyotes and jackrabbits that was set in motion through human actions a century ago. We now know that where wolves have been eradicated, coyote abundance seems to increase, in part because wolves dampen coyote densities (Berger and Gese 2007). Coyotes in the Grand Teton region account for up to 80 percent of pronghorn fawn mortalities (Berger, Gese, and Berger 2008), an issue of concern if sustaining migration is a goal as adequate fawn recruitment is requisite to maintain a population. Of note is that white-tailed jackrabbits once occurred within Grand Teton but they are now considered extirpated (Berger 2008). Why jackrabbits no longer occur is unknown.

Although speculative, it is possible that—as in the above described sheep model—coyote predation on fawns intensifies because white-tailed jackrabbits are no longer available as prey and, hence, coyotes switch to neonatal ungulates instead (Berger 2008). Whether the forty-seven historic leases that enabled cattle grazing within the confines of what is now Grand Teton National Park have had a long-term impact that aggravated this predator-prey dynamic is conjectural. Only a comparative study elsewhere will facilitate knowledge about ecosystem-level effects of hare extirpation and its consequent impact on the migrations humans strive to protect.

Third, rabbit drives are likely to have had unintended consequences that shift a system's ecology. The pummeling of hares a century ago creates an interesting thought experiment with possible consequences for human health and further finances, specifically the possible spread of malaria (Livingston 2010). In California's Fresno County more than 43,000 were killed during a two-month period in 1892 (Palmer 1896). By 1915, to the south in Kern County, residents in the town of Bakersfield called for the creation of a tax-funded mosquito abatement program, suggesting mosquitoes had grown worse. While ecological relationships are rarely so simple and many confounding variables are likely involved, if—lacking abundant hares as prey—mosquitoes switched to humans, then we indirectly created a food web impact that affected human livelihoods in an unanticipated fashion.

There are obviously direct and indirect effects that stem from human actions, some of which lead to a misunderstanding about how nature functions and some of which do not. When tsetse flies are poisoned in areas of Africa with low human densities, the habitat has been subsequently rendered more habitable for people. Conscious decisions were put in place to manipulate local ecologies with a goal to increase human occupancy. People move in, and it's not unexpected that wildlife declines. These are direct effects of humans.

Indirect effects stem from a spate of interactions set in motion by human action when a third species or intermediary is involved. In such cases and especially in our past, little foresight appears to have been given to the longer-term ecological consequence of removing or altering a key species. This is understandable given our lack of knowledge of ecological dynamics. More recently, we seem to pay the price. In the case of rabbit drives, this might involve exacerbated predation on sheep. It might involve affecting a long distance migration through influences on juvenile recruitment. It might involve human health. Can we learn from the past—sure. Do we? It's clear that we continue to suffer from an unkind past and some present pursuits in the name of making a better world. Until we pay attention to, rather than ignore, lessons from nature, George Santayana may have gotten it right.

References

Berger, J. 2008. "Undetected Species Losses, Food Webs, and Ecological Baselines: A Cautionary Tale from Yellowstone." *Oryx* 42: 139–43.

Berger, J., S. L. Cain, and K. Berger. 2006. "Connecting the Dots: An Invariant Migration Corridor Links the Holocene to the Present." *Biology Letters* 2: 528–31.

Berger, K. M. 2006. "Carnivore-Livestock Conflicts: Effects of Subsidized Predator Control and Economic Correlates on the Sheep Industry." *Conservation Biology* 20: 751–61.

Berger, K. M., and E. M. Gese. 2007. "Does Interference Competition with Wolves Limit the Distribution and Abundance of Coyotes?" *Journal of Animal Ecology* 76: 1075–85.

Berger, K. M., E. Gese, and J. Berger. 2008. "Indirect Effects and Traditional Trophic Cascades: A Test Involving, Wolves, Coyotes, and Pronghorn." *Ecology* 89: 818–28.

Cohn, J. 2010. "A Narrow Path for Pronghorns." *Bioscience* 60: 428.

Estes, J. A., and J. Terborgh, eds. 2010. *Trophic Cascades: Predators, Prey, and the Changing Dynamics of Nature*. Washington, DC: Island Press.

Knowlton, F. F., and L. C. Stoddart. 1992. "Some Observations from Two Coyote-Prey Studies." In *Ecology and Management of the Eastern Coyote*, edited by A. Boer, 101–21. Fredericton, NB: Wildlife Research Unit, University of New Brunswick.

Livingston, C. 2010. "Rabbit Roundup: Kern County's Hare Raising Battle." *Bakersfield Magazine*, December 7, 22–23.

Palmer, T. S. 1896. "The Jack Rabbits of the United States." US Department of Agriculture, Bulletin 8. Washington, DC: Government Printing Office.

Santayana, G. 1905. *The Life of Reason*. New York: Scribner.

8

Coyotes, Compassionate Conservation, and Coexistence

Why Ignoring Nature Means Ineffective "Predator Management"

Camilla H. Fox

PERHAPS NO OTHER wildlife "management" program in the United States reflects such a blatant disregard for nature as the federal government's predator "control" program. This program, euphemistically called "Wildlife Services" and functioning under the United States Department of Agriculture (USDA), has been criticized as arcane, ecologically destructive, and indiscriminate by professional scientists worldwide and by the highly esteemed American Society of Mammalogists. The program is a perfect example of how ignoring nature—who the predators are, how they live, and why they are of central importance in the ecosystems in which they live—results in the inhumane slaughter of millions of animals, and that indiscriminate killing has not, does not, and cannot solve the problems at hand.

Under their predator management program, the USDA Wildlife Services killed more than five million animals in the United States in 2010—including 113,000 mammalian carnivores. Predators targeted include coyotes, wolves, bobcats, badgers, mountain lions, foxes, and bears; many nontarget animals fall victim to the leghold traps, snares, and poison baits employed. Much of this killing takes place on public lands throughout the West. State and county governments are provided incentives to contract with Wildlife Services through matching cooperative funding agreements. Taxpayers foot the bill for this carnage; in 2010, more than $126 million was spent to fund Wildlife Services.

The program functions under the antiquated federal Animal Damage Control Act (ADC Act), which was signed into law in 1931. The ADC Act authorizes the Secretary of Agriculture to "conduct campaigns for the destruction or control" of animals considered threats to agriculture or ranching operations. Never before had Congress authorized such absolute annihilation of nature—codifying the federal government's role in killing predators in the service of private economic interests while demonstrating a complete disregard for wildlife's critical functional ecological role. Under this arcane law, government agents continue to trap, snare, poison, and shoot any animal who "may" harm livestock, aquaculture, or agricultural crops. If ever an act of Congress demonstrated a total disregard for nature, it was passage of the ADC Act.

Given the green light by the act, Wildlife Services conducts a quiet, relentless war against North America's wildlife. Few Americans have heard of the Wildlife Services program. Even fewer know that their tax dollars pay federal agents to shoot wolves, coyotes, and other predators from low-flying aircraft and to set poison bait and snares to trap and kill them.

Case Study: What's Wrong with This Approach?

In 2003 Dennis Slaugh was exploring the wild lands near his home in northeastern Utah when he stopped to examine a possible surveyor's marker embedded in the ground. However, it wasn't a surveyor's marker; it was a spring-loaded M-44 "coyote getter" designed to explode in the mouth of a curious coyote that tugged on baited meat.

As Slaugh dusted off the metal knob, it exploded in his face, spraying him with sodium cyanide. Slaugh was hospitalized and forced to retire early from his job as a heavy equipment operator, but he survived. Today, Slaugh still has trouble breathing and suffers constant nausea and weakness—all symptoms of sodium cyanide poisoning. As terrible as his encounter was, he was lucky to have run into sodium cyanide rather than Compound 1080.

Compound 1080 (sodium fluoroacetate) is a highly toxic, slow-acting, odorless, colorless, tasteless poison with no antidote. It is so deadly that just one teaspoon is toxic enough to kill one hundred human adults. Used by the USDA in rubber bladders tied around the necks of sheep and goats, the "livestock protection collars" (LPCs) are meant to poison attacking predators. However, the pouches are just as easily punctured by vegetation and barbed wire, leaking Compound 1080 into the environment where grazing animals can be poisoned from eating the contaminated forage.

And what is the fate of the predators who do attack and pierce the poison-filled collars? They suffer a prolonged and painful death, sometimes running and convulsing up to ten hours before they die. Government reports indicate that fewer than 10 percent of the bodies of poisoned animals are recovered, which leaves 90 percent to enter the ecosystem as food for exploring badgers, bobcats, crows, bears, and pets. Scavenging leads to the secondary poisoning of thousands of innocent companion animals and unoffending wildlife, including threatened and endangered species, each year. Of the 113,000 native carnivores killed in 2010, close to 15,000 were killed with deadly poisons.

Coyote: The Most Persecuted Carnivore in North America

The primary target of the federal government's poison and lethal predator control program is the coyote (*Canis latrans*). The coyote is the most persecuted predator in North America today. It's estimated that on average at least half a million coyotes are killed every year in the United States alone—one per minute—by federal, state, and local governments and by private individuals. The USDA's Wildlife Services program kills more than 80,000 coyotes each year. Despite scientific evidence suggesting this approach is misguided and ultimately ineffective, the emphasis on lethal coyote control persists. Coyotes are also killed for their fur, for "sport," and in "body-count" contests where prizes are awarded for killing the most and/or largest coyotes. Most states set no limit on the number of coyotes that may be killed, nor do they regulate the killing method. They are poisoned, snared, "denned" (the practice of killing coyote pups in their dens), and hounded (where packs of hounds are set upon and often mutilate a single coyote in the wild or in an enclosure in a practice known as "penning"). If ever an animal embodied what it means to ignore nature it is the coyote—America's iconic native *Song Dog*.

In the face of persecution, coyotes have expanded their range continent-wide largely in response to human alterations of the environment and the killing of wolves—their greatest competitor. The eradication of wolves in much of the lower forty-eight states opened up new territories that coyotes quickly filled. Unlike wolves, coyotes have adapted to living in close proximity to people and now inhabit even the most densely populated metropolitan cities from Boston to San Francisco, Austin, and Seattle. Communities are often ill equipped to deal with the presence of coyotes, and conflicts arise when an uninformed populace intentionally or unintentionally feeds coyotes. Moreover, state wildlife agencies are often cash and staff strapped, so that urban wildlife issues and public outreach are not priorities. Far too often the approach to coyote conflicts—whether in agricultural areas or urban landscapes—is lethal.

Why Lethal Coyote Control Is Ineffective

Coyotes, like all carnivores, are self-regulating. They often establish home ranges and defend territories, which means they ensure against overabundance in any given area. Efforts by humans to reduce their populations have largely been unsuccessful because coyotes exhibit strong compensatory responses to lethal control. While lethal control may result in short-term reductions in the number of coyotes in a specific area, the vacuum is soon filled by coyotes migrating from surrounding areas and by increased pup survival in remaining populations. Females in exploited populations tend to have larger litters because competition for food is reduced and more unoccupied habitat is available. Lethal control also often selects for coyotes that are more successful, wary, nocturnal, and resilient—what some biologists call a "super coyote."

Marc Bekoff (2010), who studied coyotes for more than thirty-five years, stresses that "killing does not and *never* has worked. When a space opens where a coyote had lived another individual simply moves in. Usually the offending coyote is not identified. And it is ethically indefensible to wantonly go out and kill coyotes because they try to live among us, arrogant big-brained invasive mammals who have redecorated the homes of coyotes and other animals and then conveniently decided that they have become 'pests' when we don't want them around any longer. . . . Confrontations with coyotes can almost always be traced to irresponsible human actions including allowing dogs to run free off leash and feeding the coyotes, either intentionally or unintentionally. And, it's pretty easy to clean up all of these problems and coexist peacefully with coyotes."

Hence, "coyote control" is a bit of an oxymoron as coyotes have responded to our attempts to control them with sheer perseverance to survive. Love them or not, coyotes are here to stay. They are, as William Bright, in his superb collection of stories titled *A Coyote Reader* notes, "the trickster par excellence."

The Role of North America's Native *Song Dog*

Coyotes are a vital component of healthy ecosystems and play an important ecological role in keeping rodent and small mammal populations in check. Studies conducted in the fragmented habitats of coastal Southern California showed that the absence of coyotes and/or their removal allowed smaller predators such as foxes and feral cats to proliferate, leading to a sharp reduction in the number and diversity of native ground-nesting birds. Similar findings have been made in more rural areas where coyote removal negatively affects songbird and waterfowl diversity. Coyotes have also been found to help control Canada goose and white-tailed deer populations on the east coast. Hence, in areas where coyotes are the apex mammalian predator, their removal can precipitate ecological chain reactions that lead to profound degradation of the health, integrity, and diversity of the ecosystem. They are also efficient scavengers and offer many natural services we may not fully appreciate.

Although coyotes are the most maligned and misunderstood native carnivore in the United States, Native Americans once revered them. Known as the "trickster" and "creator" figure in many tribal lore stories, coyotes were highly respected for their keen intelligence, devoted parenting, and adaptable ways.

In honor of the coyote's resourcefulness, intelligence, and rightful place in the ecosystem, the Navajo called the species "God's Dog." Coyotes have much to offer us, not only by keeping ecosystems healthy but also by providing inspiring examples of ingenuity and adaptability in an ever-changing world.

An Alternative Approach:
Compassionate Conservation and Coexistence

Public controversy over the USDA's predator control program and testing of the deadly poison Compound 1080 on coyotes in the Northern California county of Marin led to a decision by public officials to take an alternative approach to livestock-predator conflicts. In 2000 the Marin County Board of Supervisors voted to cease contracting with the federal agency and to instead adopt an alternative community-based program known as the Marin County Strategic Plan for Protection of Livestock and Wildlife. The program provides cost-share funds to assist ranchers with implementation of nonlethal predator deterrent methods including livestock guard dogs and llamas, improved fencing, and night corrals.

According to Marin County Agricultural Commissioner Stacy Carlsen, who oversees implementation of the nonlethal cost-share program, "Over six years . . . losses have fallen to 2.2 percent—and the program costs over $10,000 a year less than the old one," as reported in *Bay Nature* magazine. "For the first couple of years we couldn't tell if the [loss] reductions were a trend or a blip. Now, we can say there's a pattern. . . . In a few years we'll be a model without anyone questioning our success" (quoted in Agocs 2007). This innovative model sets a precedent for meeting community needs and values where both agriculture and protection of wildlife are deemed important.

A New Paradigm: Paying Attention to Nature

Greater understanding of the ecological importance of native carnivores and increasing public opposition to lethal "control" have led to growing demand for humane and ecologically sound conservation practices. Despite shifting public attitudes and values, however, traditional predator and wildlife management techniques persist, leading to increasing tension between conservationists and management institutions. This tension is reflected in increased litigation, legislation, and public ballot initiatives.

I founded Project Coyote in 2008 to foster a new approach to the way coyotes and other predators are viewed and "managed" in the United States. Project Coyote provides a voice for native carnivores in wildlife management policy and practice, fostering compassionate conservation and coexistence. Project Coyote also aims to support research into humane, practical, and ecologically based approaches to coyote-human conflicts and to promote these alternatives. We need a new paradigm in the way we coexist with native carnivores and other wildlife, one that recognizes their important ecological role and their intrinsic worth as beings who share finite space and time on this planet. Ignoring nature never has and never will work as we strive for coexistence with other beings.

For more information, visit http://www.ProjectCoyote.org.

References

Agocs, Clifford. 2007. "Making Peace with Coyote." *Bay Nature* (January 1). http://baynature.org/articles/jan-mar-2007/making-peace-with-coyote/?searchterm=carlsen.

Bekoff, Marc. 2010. "Coyotes: Fascinating Animals Who Should Be Appreciated, Not Killed." *Psychology Today* (May 12). http://www.psychologytoday.com/blog/animal-emotions/201005/coyotes-fascinating-animals-who-should-be-appreciated-not-killed.

9

Why Evolutionary Biology Is
Important for Conservation

Toward Evolutionarily
Sustainable Harvest Management

Marco Festa-Bianchet

HUMANS HARVEST many wild animal and plant species for consumption, to obtain material goods, or for cultural, religious, and recreational reasons. Here I will address how the harvest of wild vertebrates by people can be a powerful selective force, shaping the evolution of harvested species. My argument is simple: if some traits make an individual less likely to be harvested, and if harvest pressure is high, then if those traits have a genetic component they should become more common over time. Humans will then shape the evolution of harvested species, sometimes with results that may be detrimental to both the species and the harvesters.

Although overharvest continues to be a problem for many species, especially those whose products have a commercial value, modern wildlife and fisheries management has made considerable progress toward developing models of sustainable harvest. In most cases we have the ecological knowledge to establish sustainable quotas. Those quotas are generally derived from demographic models that predict how many individuals of different sex-age classes can be harvested from a population without causing a decline. Information on age- and sex-specific survival and reproductive rates is necessary to develop these models. We now know that much higher yields can be obtained if harvest is concentrated on sex-age classes that make a low contribution to population recruitment, typically young individuals and males. Some sport fishing regulations now prohibit the harvest of large fish, in recognition of their high reproductive value. Much progress has also been made in our understanding of the relationships between wild vertebrates and various factors that may affect their population growth, such as habitat quality, weather, predation, parasites, and disease. Those can all be taken into account in harvesting programs, so that, for example, a population in habitat with high productivity and predictable weather can sustain a higher harvest rate than one in a poor habitat with unpredictable weather, and a population without natural predators can be harvested at a higher rate than one in a pristine habitat with abundant predators.

Research in wildlife and fisheries management has mostly addressed questions in population demography and habitat characteristics. Given a habitat of a certain productivity, predation rate, and parasite prevalence, a given demographic structure and some measure of variability in these characteristics over time, we can compute a "safe" level of harvest. We now often have the knowledge required for ecologically enlightened consumptive management. Whether or not we apply that knowledge is not my topic here. That knowledge has accumulated over many decades of research and continues to advance. Recently, however, researchers have begun to consider that selective harvests favoring a certain kind of individual (young or old, prereproductive or mature, large or small, fast- or slow-growing, bold or shy) may have an evolutionary impact on the harvested population (Allendorf et al. 2008). That is an important concern because evolutionary change may eventually affect the ability of a population to sustain harvests or even to persist.

Research on the potential evolutionary impact of human harvests has focused on two main types of artificial selective pressures. First, harvests reduce the average life expectancy in comparison to unexploited populations. That may create an advantage for individuals that reproduce early, often, and invest a large proportion of available resources into reproduction, even at the cost of reduced survival (Allendorf and Hard 2009). Second, selective harvests may disadvantage individuals with a particular morphological or behavioral characteristic (Coltman 2008).

There is now abundant evidence that both commercial and sports fisheries can affect life-history evolution of exploited fish (Sharpe and Hendry 2009). Many populations of commercially harvested fish are severely depleted, and concerns have been raised about their persistence (Hutchings and Festa-Bianchet 2009). The high mortality caused by fishing can favor fish that reproduce earlier than in unexploited populations. The reproductive strategy of many long-lived fish favors a delay in age of maturity, mostly because larger females are more fecund, and early breeding attempts reduce growth. If size is also correlated with survival, early reproduction may also increase natural mortality. Therefore, under natural selection a fish belonging to a long-lived iteroparous species would have higher fitness if it delayed reproduction until it reached a certain size. Because fish have undetermined growth, in general the older they are the larger and the more fecund they become. Fishing, however, increases the risk of mortality. A delay in first reproduction, advantageous under natural selection, may lead to zero fitness if there is a high chance of being harvested. Under those circumstances, it may be better to breed at a small size and young age, even if that carried a higher risk of mortality (because of depleted energy reserves, or greater predation risk) or lower future reproductive potential (because of stunted growth as a result of precocious reproduction). These costs are unlikely to be an important selective force under heavy fishing pressure because most fish will be killed before they can experience the negative fitness consequences of early reproduction. For the severely depleted stock of northern cod (*Gadus morhua*) along the Atlantic coast of Canada, it has been suggested that this "maladaptive" reproductive strategy imposed by artificial selection through decades of overfishing may now impede population recovery despite lowered fishing pressure (Swain, Sinclair, and Hanson 2007). Theory and experimental work both suggest that recovery from human-induced "adaptations" may take much longer than the time required for those artificial adaptations to evolve. That is because natural selective pressures are typically weaker than the artificial selective pressures caused by high levels of harvests (Allendorf and Hard 2009; Walsh et al. 2006). Artificial evolution does not happen in isolation, and little is known of what genetically correlated characters may be affected by selection for early reproduction, or how this change in life-history strategy may affect ecological relationships with, for example, potential prey and predators.

For terrestrial mammals, evidence that human harvest has affected life-history evolution remains sparse. Although theoretical approaches suggest that high hunting mortality should select for earlier primiparity, records from harvested red deer (*Cervus elaphus*) in Norway do not suggest any trend in age of primiparity (Mysterud, Yoccoz, and Langvatn 2009). In Norway, however, a large proportion of sport-hunted red deer are calves, so that it is unclear whether the added mortality on breeding-age females (which would select for earlier maturity) may be counterbalanced by high harvest mortality of pre-reproductive deer (which would select for a more conservative reproductive strategy). Our ability to predict the possible evolutionary impact of sport harvests of ungulates is limited by the lack of precise data on sex- and age-specific mortality of harvested population. We often have data on what sex-age classes are harvested but rarely on the availability of each sex-age class and therefore on the relative harvest mortality.

For animals that have been harvested over millennia, such as many species of large herbivores, human harvest may have shaped the evolution of reproductive strategies over much of their evolutionary history. In that regard it is useful to compare mortality patterns caused by large predators and by humans.

Natural predation often affects sex-age classes that are most vulnerable and that also contribute relatively little to population growth, including very young and very old individuals. For example, wolves (*Canis lupus*) mostly kill young of the year or senescent elk in Yellowstone (Wright et al. 2006) and moose (*Alces alces*) in Isle Royale (Peterson 1999), but sport hunters mostly take prime-aged female elk from the same Yellowstone population. Similar numeric impacts could have very different effects on population dynamics: suppose hunters and wolves each remove one hundred female elk, but while hunters avoid shooting calves and take females based on their availability (mostly those aged two to ten years), wolves kill almost exclusively calves and senescent females. The impacts on population growth are thus very different: because young elk have low survival, and might wait two years before they start reproducing, whereas senescent elk are near the end of their lifespan, the elk that wolves remove would have contributed much less to population recruitment than those removed by the hunters. This consequence can be calculated based on knowledge of age-specific survival and reproductive rates. What is much more difficult to quantify is the possible effect of hunting on life-history evolution. We can suppose that under a "normal" ecological situation, a female elk that survived to her first reproduction could expect to have five to ten breeding opportunities before reaching senescence, when she would again be vulnerable to wolf predation as well as to intrinsic causes of mortality (Gaillard et al. 2000). Under those conditions her best reproductive strategy would involve substantial restraint within any one reproductive episode, to avoid compromising her residual reproductive potential (Martin and Festa-Bianchet 2010). If, however, sport-hunting mortality meant that she could only reproduce one to three times before being shot, then there should be a strong pressure on her to make sure that those few reproductive opportunities propagated her genes in the population. She may breed as soon as possible and devote a much greater proportion of her available resources to maternal care, possibly at the risk of lowering her own survival chances. Increased mortality of prime-aged adults should favor a riskier reproductive strategy. Harvest mainly directed at young of the year could have the ecological consequence of allowing the take of a higher number of animals while diminishing its impact on population growth (Milner et al. 2006), and the evolutionary consequence of selecting for a more conservative reproductive strategy by increasing juvenile mortality over adult mortality (Carlson et al. 2007).

A similar reasoning could apply to males, although male reproductive costs typically involve mating effort rather than parental care. In many hunted populations of cervids (Mysterud, Solberg, and Yoccoz 2005), few males survive beyond the age of three or four years, because of the heavy harvest pressure from sport hunting. With such a short life expectancy, a strategy of allocating resources to body growth rather than investing all available energy in attempting to mate would not be favored. Males lucky enough to survive to a rutting season should do all they can to obtain some matings. Heavy male-biased harvests would also remove most of their competitors, further selecting against a strategy based mostly on achieving large size to win combats with other males. The overwhelming selective pressure will be to avoid getting shot.

I have presented some speculative arguments. So far, the only strong evidence of alterations in reproductive strategy apparently caused by artificial selective pressure comes from fish. Few studies, however, have looked for evidence of artificial selection in other species. Comparative analyses of heavily hunted and protected populations may offer some interesting insights, but one practical problem is that the predicted ecological consequences of simply reducing density through hunting are similar to the predicted evolutionary consequences: higher reproductive success, earlier age at primiparity, and greater reproductive effort (Allendorf and Hard 2009; Festa-Bianchet 2003).

One possible source of artificial selection is trophy hunting, where the desirability of an animal as a target depends on the size of secondary sexual ornaments such as horns, antlers, or tusks. These ornaments are typically used as weapons in male-male combat, and for some species there is evidence that their size affects reproductive success. For example, presenescent male red deer with large antlers and mature bighorn rams with large horns obtain high mating success (Coltman et al. 2002; Kruuk et al. 2002). In both of these species, male-male combat involves clashing or pushing with the horns or antlers. Weapon size, in addition to body size and physical strength, is likely a strong factor in determining male dominance status. Dominant males control access to estrous females and have higher fitness than subordinates. In bighorn sheep, a detailed study of the survival and reproductive success of rams with known horn size and pedigree showed that thirty years of unrestricted selective hunting favored small-horned rams. Rams had phenotypically and genetically smaller horns than their ancestors in the same population a few decades earlier.

Bighorn sheep biology and hunting regulations conspired to ensure artificial selection in this isolated population. Any Alberta resident can buy a "trophy" sheep tag, and harvest is only limited by the definition of "legal" ram, based on a minimum horn curl. Rams with large horns can be taken, but small rams are protected. About 40 percent of "legal" rams were harvested each year, removing many competitors for "illegal" rams before the rut. In bighorn sheep, large horns increase a male's mating success only if he survives to about six to seven years of age and can reach a body mass that will help him assert his dominance over other rams. A male with rapidly growing horns (that may have become highly successful had he survived to six to seven years of age) was at risk of being harvested at four years. Combining hunting and natural mortality, that ram only had about a 10 percent chance of being alive to rut as a seven-year-old, compared to a 70 percent chance for a ram only exposed to natural mortality. Rams can be "legal" as four-year-olds or die at twelve years with "illegal" horns, and a "legal" five-year-old will typically be subordinate to an "illegal" nine-year-old who may weigh 20 percent more. Since that study, evidence suggesting artificial selection for small-horned rams has been found in bighorn sheep in British Columbia and in European mouflon (*Ovis aries*) in France.

Artificial selection through harvests may lead to maladaptation by countering natural selection, but it can also have two other undesirable consequences. First, it may decrease genetic variability, as reported for trophy-hunted populations of red deer in Europe, because phenotypes that are seen as undesirable for trophy production may be associated with rare alleles that could easily disappear from the population. On the other hand, in some polygynous systems, removal of large dominant males may increase the genetic variability of subsequent cohort by reducing the level to which a few males may monopolize matings. Although maintaining genetic variability is often a goal of conservation programs, we know very little about the long-term consequences of redistributing reproductive success among subordinate males. Dominant males may carry genes that confer fitness benefits to their offspring. For example, ability to survive to a mature age and phenotypic traits associated with winning male-male contests may partly be genetically determined. Females may increase fitness through selection of dominant mates. Selective hunting may also affect the evolution of harvested populations because of genetic correlations. For example, in bighorn sheep several fitness-related traits are genetically correlated with horn size in rams. Thirty years of artificial selection based on horn size in the Ram Mountain population led to a decrease in body mass that was partly determined by a positive genetic correlation between mass and horn size. Rams with genes for large horns also tended to have genes that allowed them to reach large body mass. Trophy hunting targeting horn size led to a correlated decrease in body mass. The daughters of large-horned rams tend to be larger and to wean larger lambs, which in turn have higher survival than small lambs. Under these circumstances, by removing the best breeding males, trophy hunting could diminish the genetic quality of a population.

Having established that harvest may have undesirable selective effects on wildlife, one can then ask the question of how those effects can be eliminated or minimized. Sport hunting has a rich cultural tradition and is part of several conservation programs in many parts of the world. Trophy hunting in particular can provide substantial revenues that can be used for conservation (Whitman et al. 2004) and is the only form of economic activity that can protect some habitats and ecosystems as an alternative to more destructive human uses. Ecotourism is typically limited to areas that are relatively safe and easy to reach and where people have a good opportunity to see the animals they seek. Sport hunters, and trophy hunters in particular, will instead go to very remote locations and hunt animals that are difficult to see, in addition to being willing to pay large sums of money to pursue their activities. Although many current trophy-hunting programs provide no tangible benefits to conservation, properly managed ones could be instrumental in the conservation of many species and their habitats. For example, in central Asian mountains, large mammals and their habitats are threatened by poaching for meat and luxury products and by competition and disease transmission from increasing densities of domestic livestock. Ecologically and evolutionary sustainable trophy hunting could contribute substantially to the conservation of biodiversity in these regions, because hunters are willing to pay tens of thousands of dollars to harvest a single trophy animal (Harris and Pletscher 2002). In many parts of Europe and North America, well-regulated sport-hunting programs, including varying levels of trophy hunting, have contributed to conservation of ecosystems where numbers of both ungulates and their predators have increased in recent decades. Although many areas of controversy divide hunters and conservationists, properly managed sport hunting provides a desirable alternative to many forms of industrial development, urbanization, or agricultural activities that threaten habitats.

It is therefore important to consider under what circumstances sport hunting, and trophy hunting in particular, may be a threat to biodiversity. Wildlife managers and ecologists have identified ways to ensure ecologically sustainable harvests, and there is no reason why we could not develop ways to ensure evolutionarily sustainable harvests.

First, we need to ask if artificial selection is necessarily a bad thing? Some changes in phenotype or behavior driven by human activities may not have particularly detrimental effects. If moose in Europe have less palmate antlers than in North America because centuries of selective hunting favored males with deerlike antlers, or if brown bears and wolves in Europe are "better" at living with humans than those in North America, that is not necessarily something that we should worry about. In fact, research into why European wolves can survive in areas with much greater human activity than in North America, or why brown bear aggressiveness toward humans varies across populations, may at some point reveal that these differences in behavior partly stem from centuries of artificial selection and may have contributed to the persistence of some populations. That research could provide valuable insight in the management of large carnivores in North America.

If trophy hunting simply selected for males with smaller horns or antlers, that would not be something that we need to fret about, in comparison to poaching, habitat destruction, or introduced disease. Unfortunately, however, artificial directional selection against fitness-related traits can have unwanted correlative effects on other traits (Coltman et al. 2005). Therefore, our first concern should be whether or not the trait under artificial selection has important fitness consequences. Recent research in different species of ungulates will illustrate this problem. A trophy's value mostly depends on its size: hunters seek males with large horns or antlers. In bighorn sheep, rams with large horns have high reproductive success, if they survive to seven to eight years of age, and large horns are genetically correlated with fitness-related traits. Therefore, strong selection against large-horned rams caused by unrestricted trophy hunting has possible negative consequences for genetic diversity and population dynamics. Based on similarity in behavioral ecology—namely, apparent natural selection for large horns; no compensatory horn growth (males with rapid early horn growth tend to have greater horn growth also later in life); combats partly decided by horn/antler size and body mass; data suggesting substantial monopolization of matings by a few mature and dominant males—I would expect similar results in all other wild sheep (*Ovis*), wild goats (*Capra*), and cervids of the genera *Cervus*, *Dama*, and *Rangifer* (Coltman et al. 2002; Kruuk et al. 2002; McElligott and Hayden 2000; Willisch and Neuhaus 2009). In chamois (*Rupicapra*), and mountain goats (*Oreamnos*), on the other hand, there is little evidence that horn size affects male mating success (Mainguy et al. 2009; Rughetti and Festa-Bianchet 2010). In these species, male-male combat apparently depends mostly on body size and possibly strength and agility: horns are used not to wrestle or clash with the opponent, but to injure him. Male chamois and mountain goats need horns to fight, but having longer horns than their opponent does not increase their fighting ability. In these species, males with rapid early horn growth reduce growth later in life, so that early horn growth is a poor predictor of asymptotic horn size. Body mass also varies more than horn size, again suggesting that being large is more important than having large horns. Trophy hunting of these species may have little impact on their evolution because the trait targeted by hunters (horn size) apparently plays a limited role in male reproductive success (Rughetti and Festa-Bianchet 2010). I would expect similar results in other ungulate species where fighting among males does not involve direct clashes with horns or antlers, such as *Capreolus* and some territorial antelopes (Lundrigan 1996). Because we know very little about the determinants of male reproductive success in most ungulates, however, it is difficult to determine in which species selective hunting may result in a strong evolutionary response.

The intensity of artificial selection will affect its evolutionary consequences, especially when it acts opposite to natural selection. Predicting the outcome of selection in the wild is notoriously difficult because of the large number of selective pressures acting contemporaneously, changes in the strength of selection from year to year, interactions with environmental variables, the effects of migration and mutation, and possible genetic correlations among traits subject to different selective pressures (Postma, Visser, and van Noordwijk 2007). Genetic correlations that differ in strength according to sex and age add to the list of variables that may modulate artificial selective pressure (Poissant, Wilson, and Coltman 2010). Clearly, however, selection of overwhelming strength is likely to have major effects. Very high harvest rates, such as those for overexploited fish, some ungulates, and carnivores (Fryxell et al. 1999; Toïgo et al. 2008), likely mean that avoiding human harvest, or adopting a reproductive strategy adapted to extremely high adult mortality, becomes the main factor in determining fitness and therefore driving evolution. Unfortunately, we rarely have reliable data on harvest rates, particularly in cases where harvest tends to be concentrated on specific sex-age classes. In the Ram Mountain study, as typical for sheep hunting in North America, less than 2 percent of the population was harvested each year. For rams that fit the legal definition of a harvestable ram, however, harvest rate was about 40 percent. Where mountain sheep are hunted using a minimum-curl definition of "legal" ram and an unlimited number of permits, it is likely that harvest rates of 30 to 50 percent of "legal" rams are common, and possibly even higher in areas with good access by motorized vehicles. Apparently sustainable trapping-induced mortality of nearly 40 percent has been reported for marten (*Martes americana*) (Fryxell et al. 1999). In sport-hunted populations of ungulates where less than 5 percent of males survive beyond three to four years of age (Mysterud, Solberg, and Yoccoz 2005; Toïgo et al. 2008), avoiding harvest may well be the single greatest determinant of male reproductive success. We should expect evolution of behavioral and morphological traits that are associated with surviving the hunting season.

Migration from unharvested populations would considerably dampen the effects of artificial selection (Tenhumberg et al. 2004). Migration of individuals that escaped artificial selection could be particularly important for trophy hunting, because males in many species may roam over several female groups looking for mating opportunities (Hogg 2000). Because the hunting season is typically before the rut, males moving out of protected areas would have a high mating success in hunted areas where many of their competitors will have been shot. Migration from protected areas may dampen the effects of artificial selection, but it may also reduce the effective population size within the protected areas themselves, as gene flow will be mostly unidirectional. Nevertheless, a network of unharvested areas and policies aimed at protecting migration corridors should reduce the intensity of artificial selection. Marine protected or no-fishing areas are often advocated as a means to increase fish populations, especially for species where most recruitment originates away from exploited areas (Allendorf and Hard 2009). They may also have the beneficial effect of countering the artificial selective effects of intense fishing pressure.

Eventually, it is cultural changes that may be at the same time most effective and most difficult to achieve to prevent artificial selection in cases where it is undesirable. Unfortunately, there seems to be a perception that the size of a harvested animal's horns or antlers somehow reflects the ability of the hunter. Trophy hunting can be evolutionarily sustainable if directed to age classes old enough to ensure that males can contribute to breeding before being harvested, and limited to a small proportion of mature males. In practical terms this can best be ensured by harvest of animals beyond a certain age threshold, which will vary across species but is likely to be about nine years for wild sheep and many deer, possibly more for species like ibex and possibly less for smaller cervids. Research on red deer, for instance, suggests that males aged twelve years and older grow very large antlers but have limited reproductive success (Nussey et al. 2009). Inevitably, redirection of the harvest to older males will involve a substantial reduction in the harvest, because adult males typically suffer a high rate of natural mortality (Festa-Bianchet 2007). For example, the median age of bighorn rams currently harvested in Alberta and British Columbia, Canada, is about seven years. Data on age-specific survival for bighorn rams suggest that about 35 percent of yearling males will survive to seven years of age (excluding hunting mortality), but only 19 percent will survive to age nine. In some cases it may simply be impractical to direct the harvest to the oldest age classes because most hunters will be unable to identify them.

Harvesting policies are driven by science but also by societal demands. Evolutionary ecologists have now clearly raised the point that selective harvests will, indeed, select, and not necessarily for traits favored by harvesters. So far, there is little evidence that evolutionary considerations have affected harvesting policies for fish or wildlife (Allendorf and Hard 2009). Part of the reluctance to embrace evolutionarily enlightened management may stem from the limited amount of scientific information available, particularly for species other than fishes (Darimont et al. 2009). In some cases, the reduced income that would result from less intensive trophy hunting may also motivate people to deny the role of selective human-caused mortality in the evolution of harvested species. In several parts of the world, the commercial interests of those providing "canned hunt" for animals that have been artificially fed or otherwise "managed" to grow large ornaments (Mysterud 2010) may be partly behind the apparent increased emphasis on "trophy size" among hunting promoters. That emphasis is likely to further stimulate the antihunting attitude that is increasingly prevalent in urbanized societies and may ultimately decrease the perception of sport hunting as part of a conservation strategy.

Evolutionarily (as well as ecologically) enlightened management may seek to mimic natural mortality patterns, directing harvest as far as possible toward senescent individuals and young of the year. For example, redirection of the harvest toward juveniles has allowed substantial increases in ungulate harvests in Scandinavian countries. Conservationists may then face a dilemma because the revenues generated by hunters interested in meat will likely be lower than those possible when exploiting the willingness to pay of hunters seeking trophies. I suggest that sustainable compromises are possible with a knowledge of the mating ecology of each species and a realization that long-term sustainability includes evolution as well as ecology.

References

Allendorf, F. W., P. R. England, G. Luikart, P. A. Ritchie, and N. Ryman. 2008. "Genetic Effects of Harvest on Wild Animal Populations." *Trends in Ecology and Evolution* 23: 327–37.

Allendorf, F. W., and J. J. Hard. 2009. "Human-Induced Evolution Caused by Unnatural Selection through Harvest of Wild Animals." *Proceedings of the National Academy of Sciences* 106: 9987–94.

Carlson, S. M., E. Edeline, L. A. Vøllestad, T. O. Haugen, I. J. Winfield, J. M. Fletcher, J. B. James, and N. C. Stenseth. 2007. "Four Decades of Opposing Natural and Human-Induced Artificial Selection Acting on Windermere Pike (*Esox lucius*)." *Ecology Letters* 10: 512–21.

Coltman, D. W. 2008. "Molecular Ecological Approaches to Studying the Evolutionary Impact of Selective Harvesting in Wildlife." *Molecular Ecology* 17: 221–35.

Coltman, D. W., M. Festa-Bianchet, J. T. Jorgenson, and C. Strobeck. 2002. "Age-Dependent Sexual Selection in Bighorn Rams." *Proceedings of the Royal Society: Biological Sciences* 269: 165–72.

Coltman, D. W., P. O'Donoghue, J. T. Jorgenson, J. T. Hogg, and M. Festa-Bianchet. 2005. "Selection and Genetic (Co)variance in Bighorn Sheep." *Evolution* 59: 1372–82.

Darimont, C. T., S. M. Carlson, M. T. Kinnison, P. C. Paquet, T. E. Reimchen, and C. C. Wilmers. 2009. "Human Predators Outpace Other Agents of Trait Change in the Wild." *Proceedings of the National Academy of Sciences* 106: 952–54.

Festa-Bianchet, M. 2003. "Exploitative Wildlife Management as a Selective Pressure for the Life-History Evolution of Large Mammals." In *Animal Behavior and Wildlife Conservation*, edited by M. Festa-Bianchet and M. Apollonio, 191–207. Washington, DC: Island Press.

———. 2007. "Ecology, Evolution, Economics, and Ungulate Management." In *Wildlife Science: Linking Ecological Theory and Management Applications*, edited by T. E. Fulbright and D. G. Hewitt, 183–202. Boca Raton, FL: CRC Press.

Fryxell, J. M., J. B. Falls, E. A. Falls, R. J. Brooks, L. Dix, and M. A. Strickland. 1999. "Density Dependence, Prey Dependence, and Population Dynamics of Martens in Ontario." *Ecology* 80: 1311–21.

Gaillard, J.-M., M. Festa-Bianchet, N. G. Yoccoz, A. Loison, and C. Toïgo. 2000. "Temporal Variation in Fitness Components and Population Dynamics of Large Herbivores." *Annual Review of Ecology and Systematics* 31: 367–93.

Harris, R. B., and D. H. Pletscher. 2002. "Incentives toward Conservation of Argali *Ovis ammon*: A Case Study of Trophy Hunting in Western China." *Oryx* 36: 373–81.

Hogg, J. T. 2000. "Mating Systems and Conservation at Large Spatial Scales." In *Vertebrate Mating Systems*, edited by M. Apollonio, M. Festa-Bianchet, and D. Mainardi, 214–52. Singapore: World Scientific.

Hutchings, J. A., and M. Festa-Bianchet. 2009. "Canadian Species at Risk (2006–2008), with Particular Emphasis on Fishes." *Environmental Reviews* 17: 53–65.

Kruuk, L. E. B., J. Slate, J. M. Pemberton, S. Brotherstone, F. Guinness, and T. Clutton-Brock. 2002. "Antler Size in Red Deer: Heritability and Selection but No Evolution." *Evolution* 56: 1683–95.

Lundrigan, B. 1996. Morphology of Horns and Fighting Behavior in the Family Bovidae. *Journal of Mammology* 77 (2): 462–75.

Mainguy, J., S. D. Côté, M. Festa-Bianchet, and D. W. Coltman. 2009. "Father-Offspring Phenotypic Correlations Suggest Intralocus Sexual Conflict for a Fitness-Linked Trait in a Wild Sexually Dimorphic Mammal." *Proceedings of the Royal Society of London B* 276: 4067–75.

Martin, J. G. A., M. and Festa-Bianchet. 2010. "Bighorn Ewes Transfer the Costs of Reproduction to Their Lambs." *American Naturalist* 176 (4): 414–23.

McElligott, A. G., T. J. and Hayden. 2000. "Lifetime Mating Success, Sexual Selection and Life History of Fallow Bucks (*Dama dama*)." *Behavioral Ecology and Sociobiology* 48: 203–10.

Milner, J. M., C. Bonenfant, A. Mysterud, J. M. Gaillard, S. Csany, and N. C. Stenseth. 2006. "Temporal and Spatial Development of Red Deer Harvesting in Europe: Biological and Cultural Factors." *Journal of Applied Ecology* 43: 721–34.

Mysterud, A. 2010. "Still Walking on the Wild Side? Management Actions as Steps towards 'Semi-Domestication' of Hunted Ungulates." *Journal of Applied Ecology* 47: 920–25.

Mysterud, A., E. J. Solberg, and N. G. Yoccoz. 2005. "Ageing and Reproductive Effort in Male Moose under Variable Levels of Intrasexual Competition." *Journal of Animal Ecology* 74: 742–54.

Mysterud, A., N. G. Yoccoz, and R. Langvatn. 2009. "Maturation Trends in Red Deer Females over 39 Years in Harvested Populations." *Journal of Animal Ecology* 78: 595–99.

Nussey, D. H., L. E. B. Kruuk, A. Morris, M. N. Clements, J. Pemberton, and T. H. Clutton-Brock. 2009. "Inter- and Intrsexual Variation in Aging Patterns across Reproductive Traits in a Wild Red Deer Population." *American Naturalist* 174: 342–57.

Peterson, R. O. 1999. "Wolf-Moose Interaction on Isle Royale: The End of Natural Regulation?" *Ecological Applications* 9 (1): 10–16.

Poissant, J., A. J. Wilson, and D. W Coltman. 2010. "Sex-Specific Genetic Variance and the Evolution of Sexual Dimorphism: A Systematic Review of Cross-Sex Genetic Correlations." *Evolution* 64: 97–107.

Postma, E., J. Visser, and A. J. van Noordwijk. 2007. "Strong Artificial Selection in the Wild Results in Predicted Small Evolutionary Change." *Journal of Evolutionary Biology* 20: 1823–32.

Rughetti, M., and M. Festa-Bianchet. 2010. "Compensatory Growth Limits Opportunities for Artificial Selection in Alpine Chamois." *Journal of Wildlife Management* 74: 1024–29.

Sharpe, D. M. T., and A. P. Hendry. 2009. "Life History Change in Commercially Exploited Fish Stocks: An Analysis of Trends across Studies." *Evolutionary Applications* 2: 260–75.

Swain, D. P., A. F. Sinclair, and J. M. Hanson. 2007. "Evolutionary Response to Size-Selective Mortality in an Exploited Fish Population." *Proceedings of the Royal Society: Biological Sciences* 274: 1015–22.

Tenhumberg, B., A. J. Tyre, A. R. Pople, and H. P. Possingham. 2004. "Do Harvest Refuges Buffer Kangaroos against Evolutionary Responses to Selective Harvesting?" *Ecology* 85: 2003–17.

Toïgo, C., S. Servanty, J. M. Gaillard, S. Brandt, and É. Baubet. 2008. "Disentangling Natural from Hunting Mortality in an Intensively Hunted Wild Boar Population." *Journal of Wildlife Management* 72: 1532–39.

Walsh, M. R., S. B. Munch, S. Chiba, and D. O. Conover. 2006. "Maladaptive Changes in Multiple Traits Caused by Fishing: Impediments to Population Recovery." *Ecology Letters* 9: 142–48. doi:10.1111/j.1461-0248.2005.00858.x.

Whitman, K., A. M. Starfield, H. S. Quadling, and C. Packer. 2004. "Sustainable Trophy Hunting of African Lions." *Nature* 428: 175–78.

Willisch, C. S., and P. Neuhaus. 2009. "Alternative Mating Tactics and Their Impact on Survival in Adult Male Alpine Ibex (*Capra ibex ibex*)." *Journal of Mammalogy* 90: 1421–30.

Wright, G. J., R. O. Peterson, D. W. Smith, and T. O. Lemke. 2006. "Selection of Northern Yellowstone Elk by Gray Wolves and Hunters." *Journal of Wildlife Management* 70: 1070–78.

10

Reintroductions to "Ratchet Up" Public Perceptions of Biodiversity

Reversing the Extinction of Experience through Animal Restorations

Philip J. Seddon and Yolanda van Heezik

People who care conserve; people who don't know don't care. [Robert M. Pyle, 1993]

Introduction

ROBERT PYLE coined the phrase "extinction of experience" in 1978 in recognition of a widening gap between people and the natural world (Miller 2005). Pyle believed that "one of the greatest causes of the ecological crisis is the state of personal alienation from nature in which most people live" (Pyle 1993), with collective ignorance leading to collective indifference. In an increasingly urbanized world the majority of people in many countries encounter fewer wild species in their day-to-day lives than the generation before them (Turner, Nakamura, and Dinetti 2004). We now realize that the environment encountered in childhood becomes a baseline against which future degradation is assessed (Miller 2005). This "shifting baseline syndrome" (Pauly 1995) leads to a ratcheting down of expectations as people don't realize what has been lost but accept the highly modified and depauperate environment that surrounds them as normal. This estrangement from nature is at the heart of the failure to gain public support for conservation initiatives (Miller 2005). Loss of local wildlife reduces opportunities to experience the natural world, leads to a failure to appreciate the wonder and importance of biodiversity, and produces a general public who sees little point in changing either behavior or attitudes in order to conserve nature.

How can we break this "cycle of impoverishment" (Miller 2005)? Calls to reawaken an interest in the natural world and to encourage people to reengage with nature (Samways 2007) can achieve only limited success in environments that have lost much of their original biota. We need to reset public baselines through the restoration of biodiverse natural environments in which people can reconnect with nature and renew their expectations of what is normal. We need actively to put native species back into areas from which they have been lost.

By some measures the release of fifteen American bison (*Bison bison*) into a reserve in Oklahoma in 1907 (Kleiman 1989) marked the start of a modern era of species restorations (Beck 2001). Over the intervening one hundred or so years both the science and the practice of animal reintroductions has grown exponentially (Seddon, Armstrong, and Maloney 2007b). A reintroduction is formally defined as "the intentional movement of an organism into a part of its native range from which it has disappeared in historic times" (IUCN 1987). Despite a record of high failure rates in early reintroduction programs (Fischer and Lindenmayer 2000; Wolf et al. 1996; Wolf, Garland, and Griffith 1998), reintroduction programs are increasingly achieving their stated goals and leading to the resurrection of viable populations of animal species long missing from large parts of their natural distribution. Some recent examples range from large blue butterflies (*Phengaris arion*) (Barkham 2009) and red kites (*Milvus milvus*) in the United Kingdom (Tregaskis 2009) to lynx (*Lynx canadensis*) in the United States (Banda 2010) and (*L. lynx*) in Europe (Kramer-Schadt, Revilla, and Wiegand 2005). Wildlife reintroductions provide a means to reunite people with the natural heritage of the lands they occupy, to expose them to biodiversity writ large, to increase interest in conservation (Parker 2008), and to rekindle a sense of awe and reawaken a sense of responsibility for environmental stewardship. Most importantly, restoration of wildlife populations can reset baselines and "ratchet up" expectations of what the natural world around us should look like.

This halcyon vision of a reintroduction-recreated world is tempered by the plethora of biological, social, economic, technical, and political challenges faced in the restoration of wild animal populations. We have identified three specific challenges facing reintroduction practitioners in dealing with public perceptions of, and engagement with species restorations: (1) a biased focus on large-bodied charismatic species, (2) potential human-wildlife conflict in the restoration of keystone species, and (3) the need to acknowledge and adapt restoration targets to a human-modified world. Below we consider each of these challenges and offer some potential solutions.

Challenge 1: Taxonomic Bias and Single Species Focus in Animal Reintroductions

Ask a member of the general public to name three threatened species and we can be sure that this list will probably contain giant panda and perhaps tigers, and will certainly be dominated by large mammals and maybe some birds. Necessarily, conservation needs "poster children," iconic species that can capture public interest and sympathy and gain tangible public support for biodiversity conservation actions. But the rallying of public support around a handful of charismatic vertebrates has worked almost too well. By parading giant pandas, polar bears, dolphins, gorillas, and the like variously as species in peril or triumphs of conservation management, we seem to ignore or downplay the importance of other taxa that may actually be overwhelmingly more fundamental to the resilience of natural ecosystems. Public interest becomes public focus, and as a consequence conservation interventions, for which public support is essential, similarly focus on a handful of high-profile vertebrates.

This bias runs deeper than just an uninformed public preferring cuddly mammals. Studies have shown that research in taxonomy, ecology, and conservation is not proportional to the frequency of organisms in nature (Gaston and May 1992; Amori and Gippoliti 2000; Bonnet, Shine, and Lourdais 2002). This may not be a major concern if there is a difference between what scientists choose to study and that which is the focus of conservation action, but the bias not only pervades the conservation literature (Clark and May 2002a, b) but is also reflected in where we place the emphasis of conservation activities. A review of reintroduction projects for 489 animal species compared the numbers of projects observed with the number expected according the distribution of described taxa in nature and found a bias toward the reintroduction of vertebrates and a dearth of projects focusing on invertebrates (Seddon, Soorae, and Launay 2005). The bias continued within vertebrate taxa, with much more attention being paid to mammals and birds than to groups such as fishes. Even among the favored taxa there was a bias toward large mammals and recreationally or economically important species. This bias was not driven by any measure of conservation concern, with many reintroduction projects aimed at species listed as Least Concern by the IUCN (Seddon, Soorae, and Launay 2005). A specific assessment of 697 threatened mammals showed that 75 percent of them receive little or no conservation attention, with priority being given to the large-bodied, charismatic species (Sitas, Baillie, and Isaac 2009). This bias has been further reinforced through the publicity given to the success of a handful of high-profile reintroductions for species such as Arabian oryx (*Oryx leucoryx*), golden lion tamarin (*Leontopithecus rosalia*), and peregrine falcon (*Falco peregrinus*) (Seddon, Armstrong, and Maloney 2007b).

Clearly, the choice of conservation focus is subjective and self-reinforcing; funding for research or conservation action depends in large part on public support, and the general public inherently favors certain species. With fund raising focused on the larger-bodied and more charismatic species the impression that such species are most in need of conservation action is reinforced in the public mind. As a result, what we choose to expend conservation efforts on is not determined by considerations of rarity, endemism, conservation status, or ecological importance, but rather by what people perceive to be attractive or inspiring species.

Solution 1: Focus on Keystone Species and Ecosystem Function

What is the solution? The challenge of setting conservation priorities has spawned a multitude of potentially competing and conflicting approaches (Sitas, Baillie, and Isaac 2009), such as directing management toward a few surrogate species to benefit the wider ecosystem of which they are a part (Simberloff 1998; Fleishman, Murphy, and Brussard 2000; Poiani, Merrill, and Chapman 2001; Noon and Dale 2002). One conservation shortcut is the selection of an umbrella species, defined as a wide-ranging species whose "conservation confers protection to a large number of naturally co-occurring species" (Roberge and Anglestam 2004, 77). The umbrella species concept has been used to protect specific taxa at the local level (Suter, Graf, and Hess 2002; Caro 2003; Jones, McLeish, and Robertson 2004), and as a basis for reserve design (Wilcove 1994; Wallis de Vries 1995). It has, however, been subject to criticism for providing only limited protection to co-occurring species (Berger 1997; Noss, O'Connell, and Murphy 1997; Hitt and Frissell 2004; Rowland et al. 2006), and because the existing criteria for rigorous assessment of candidate umbrella species requires such detailed knowledge of both candidate and co-occurring species that it is less a shortcut and more of a "long and winding road" (Seddon and Leech 2008). In reality charismatic vertebrates tend to be selected as umbrella species with little reference to ecological criteria (Betrus, Fleishman, and Blair 2005), bringing us back to where we started.

Historically, reintroductions have focused on single species, but increasingly the impacts and effects on multiple species are being considered as part of ecosystem-level restorations, where the aim is to restore key ecological processes through the release of keystone species (Armstrong and Seddon 2008). A keystone species is any species that exerts an effect, even indirect, on other species that is disproportionate to its abundance or biomass (Paine 1995). Keystone species may be predators that regulate the abundance of prey, or they may be ecosystem engineers, organisms that modify, maintain, create, or destroy structure in their physical environment (Lawton 1994; Jones, Lawton, and Shachak 1997). Ecosystem engineers can change energy flow and resource availability to other species, change water flows, and nutrient levels and habitat quality (Decaens et al. 2002), and thereby increase species richness, diversity, and productivity (Day, Laland, and Odling-Smee 2003). Ecosystem engineers can be a focus of restoration efforts (Seddon 1999; Hastings et al.

2007); for example, in the Australian desert the reintroduction of native fossorial mammals has been critical in the restoration of natural fertile patches in arid landscapes through their burrowing activities (James and Eldridge 2008). Reintroduction of ecosystem engineers can reduce the human effort necessary to achieve a target state, making restoration cheaper, faster, easier, and more sustainable (Byers et al. 2006).

There needs to be a shift in focus away from species chosen solely for their public appeal and toward restorations that include keystone species in an explicit attempt to restore ecosystem function. Clearly, we can't expect widespread public support if we shift our focus simply away from all charismatic species and concentrate on the reintroduction of ecosystem engineers such as microorganisms and invertebrates. Perhaps the best way to begin this shift in emphasis is to identify a few charismatic keystone species, the reintroduction of which would spearhead increased acceptance and understanding of the need to consider biodiversity at an ecosystem level, not only at a single species level. Restoration of American bison populations is one such example. As many as sixty million bison once roamed the tallgrass prairies of the Great Plains before their near extirpation by the late 1800s. Reintroduction of bison has resulted in over 150,000 animals in the Great Plains, the restoration of large ungulate herbivory, and the realization of the keystone role bison play in altering plant species composition, changing nutrient cycling and soil resource availability, and in creating a spatial heterogeneity that is essential for the functioning of the tallgrass prairie ecosystem (Knapp et al. 1999).

Challenge 2: Human-Wildlife Conflicts in Species Restorations

If we address the taxonomic bias in conservation efforts by directing attention to reintroduction of keystone species to restore critical ecological processes, this means that reintroduction practitioners are no longer asking the public to support conservation measures for a handful of attractive species that may have little or no impact on daily lives or livelihoods. But some keystone species, owing to the ecosystem effects that make them keystones, have the potential to affect humans also. People have been so long without the endemic elements of their lands that they have learned to do without them and too often neither miss nor value them. Many people see restoration of animal biodiversity in human-dominated landscapes as an imposition, a risk, and an economic cost. We have become accustomed to living in areas dominated by a few species of numerous exotics and may have even forgotten which species are introduced and which are native. Engagement with nature may amount to feeding pigeons or invasive squirrels in an urban park or scaring away exotic crop-feeding wild birds, and while it might be nice to think of native wild animals running free somewhere, that somewhere needs to be somewhere else if it seems to threaten our way of life. Wolves and beavers are high priority reintroduction candidates through their respective roles as top predators and ecosystem engineers, and both exemplify the conflict that can arise when ecosystem restoration goals clash with public perceptions of risk and impact.

The European beaver (*Castor fiber*) was reduced to relict populations by the beginning of the twentieth century in France, Norway, Poland, Russia, and Sweden but has been absent from Scotland for much longer, having been hunted to extinction by the 1500s (O'Connell et al. 2008). Beavers are considered true ecosystem engineers, able to change the geomorphology, hydrology, and biotic properties of a landscape and thereby increase both habitat and species diversity (Rosell et al. 2005). Since the second half of the twentieth century, reintroduction of beaver across Europe has resulted in a global population of greater than 300,000 animals, but restoration of beaver in the United Kingdom has run up against negative perceptions about the environmental impacts arising from the beavers' keystone role in modifying riparian habitats that have reached a new functional, economic, and cultural equilibrium in the centuries since beaver extirpation (South, Ruston, and Macdonald 2000).

There is evidence that top predators can promote species richness by acting as structuring agents in some ecosystems (Sergio et al. 2008). The loss of large predators has led to the increased abundance of their larger herbivore prey and consequent impacts due to overgrazing and associated changes to community structure and composition (Nilsen et al. 2007). The wolf (*Canis lupus*) is a top predator that has been historically extirpated from much of its former range due to human persecution but is now the focus of a number of planned and ongoing restoration efforts throughout Europe and North America (Mech 1995), including wolf recovery programs in Montana, Idaho, Arizona, and New Mexico over the last fifteen or so years (Ripple and Beschta 2004). In 1995, after a seventy-year absence, wolves were reintroduced to Yellowstone National Park in the United States (Fritts et al. 1997), where loss of predation pressure on elk (*Cervus elaphus*) had resulted in overgrazing and a cessation of recruitment of woody plants, eventual loss of beaver, and a decline in food for other species (Ripple and Beschta 2004). Restoration of wolves not only resumed natural regulation of elk populations and reduced ungulate-caused landscape simplification (White and Garrott 2005), but also resulted in a behaviorally mediated trophic cascade whereby elk preferentially forage in sites that allow for early detection of wolves, leading to the selective release from grazing pressure for some woody species (Ripple and Beschta 2003). But public attitude to wolf reintroduction is not always supportive, with rural communities expressing concern over the possibility of wolf attacks on livestock, financial losses to ranchers, and potential risks to humans (Pate et al. 1996). Urban residents tend to be more supportive of wolf restoration, but people with the most positive attitudes toward wolves tend to be those with the least experience of them, and evidence suggests that the presence of wolves may lead to more balanced experience and a reduction in positive attitudes (Williams, Ericsson, and Heberlein 2002; Ericsson and Heberlein 2003). Wolves were eradicated from the Scottish Highlands by 1769, and there are proposals to reintroduce them to control high deer densities that are hampering reforestation efforts, reducing bird densities, and competing with livestock (Nilsen et al. 2007). But although the Scottish public is generally supportive of the proposal, farmers have more negative attitudes (Wilson 2004). It is critical to ensure there is local community support and benefit from the reintroduction of potentially disruptive species to reduce the likelihood of attempts to disrupt or sabotage a project (Nilsen et al. 2007).

Solution 2: Increasing Public Support for Keystone Species Restoration Projects

More than ever the restoration of keystone species that have at least the perceived potential to have impacts on human lives and livelihoods must be undertaken in a bottom-up manner, with real, fundamental local community support. To address the specific concerns of land and livestock owners and the general public it is insufficient to cite generic research from other times and sites; rather, it is necessary to provide new, locally gathered evidence of the potential for impacts by released keystone species and the mitigation of any such impacts. In the case of European beaver restoration in Scotland, public consultation is seen as essential (Macdonald et al. 1995), and progress has been made through small-scale pilot reintroductions of beaver specifically to quantify environmental impacts; for instance, these have succeeded in demonstrating that beavers do not damage "human timbers" such as wooden fence posts, as landowners believe (O'Connell et al. 2008).

The situation with predators can be more fraught, but there has been a shift in the focus of wolf reintroductions, from the preservation of the species toward the restoration ecosystem function as the importance of wolves as keystone species has been realized (Nilsen et al. 2007). As research from existing restored wolf populations is compiled and the critical role of top predators is unraveled, public understanding and support for the restored presence of wolves has grown. In the United States, preservation of wolves was only one of the reasons given for supporting wolf reintroduction, with positive attitudes toward wolves being driven also by the belief that wolves will keep deer and elk and rodent populations in balance, will restore the natural environment, and will help people understand the importance of wilderness (Pate et al. 1996). However, there are some important patterns in public attitudes to wolves that have emerged. While attitudes to wolves are generally more positive in the United States than, for example, in Scandinavia or Western Europe, in the United States there is real concern that the pro-wolf interests of the dominant urban society could be forced on the rural communities that will bear most of the economic costs (Williams et al. 2002). Critically therefore, attitudes towards the reintroduction of wolves are tied to level of awareness and experience, economic interests, but also to broader social identity and ideological conflicts that exist over and above the specifics of wolf restoration (Williams, Ericsson, and Heberlein 2002; Naughton-Treves, Grossberg, and Treves 2003). Managers must therefore consider two things: meaningful engagement with the general public to fully understand the basis for negative attitudes and to try to use the best available information from ongoing wolf reintroductions to address concerns and mitigate impacts, and to be aware that attitudes to wolves will change in the presence of wolves. In Norway, for example, it has been proposed that innate fear of wolves can best be addressed through educational programs where the public can learn more about wolf biology and gain direct experience of wolves in the outdoors (Roskaft et al. 2003). In this way a wolf reintroduction can itself provide the means to increase public exposure to and knowledge of the ecosystem role of these top predators and gain further support for such ecosystem-based restoration approaches.

Challenge 3: Flawed Dictates of Historical Species Distributions as Restoration Targets

We've proposed that we could address the existing taxonomic bias in wildlife restorations by shifting our focus away from charismatic species to those that are critical to the restoration of ecosystem processes, the keystone species such as top predators, and ecosystem engineers. However, this raises the possibility of increased human wildlife conflict with local communities who have rapidly grown used to the absence of wildlife in the surroundings and who understandably focus on concerns over economic and other impacts on lifestyles. Part of reengaging the public is through the promotion of community-engaged restoration projects, allowing a reconnection with nature to derive from bottom-up action rather than be dictated by top-down decree, and thereby increase public interest, participation, and investment in biodiversity conservation (Parker 2008). But with an increase in the public awareness of and support for restoration projects a third challenge will be evident: restore to what? What is the target state for an ecosystem restoration? What is the appropriate habitat for the restoration of critically endangered species?

Reintroduction practitioners have tended to assume that because a given species has been lost within historic times, then releases should take place within the recent range of that species, but the assumption that historical range is a reliable guide to future habitat suitability is weakened by unreliable historical records, arbitrary reference points, and accelerating habitat change (Seddon 2010). The use of documented species distributions to determine release sites also assumes static distributions, stable environmental conditions, and an absence of historical human influences. Setting targets is a challenge that is being debated also by restoration ecologists who increasingly recognize the inherent problems in trying to replicate some arbitrary past condition (Temperton 2007) given the lack of accurate historical records (Hobbs 2007), the dynamic nature of ecological systems (Choi et al. 2008), and the occurrence of irreversible losses or change (Hobbs and Harris 2001; Jackson and Hobbs 2009).

Solution 3: Plan for Conservation Introductions and the Creation of Novel Ecosystems

Restoration ecologists are starting to address the problem of setting of restoration goals with new directions that do not seek to construct replicas of the past (Temperton 2007) but that acknowledge the dynamic nature of ecosystems and human-induced change (Hobbs 2007; Hobbs and Cramer 2008; Jackson and Hobbs 2009). There is recognition that global climate change, in particular, reduces the usefulness of historical ecosystem conditions as restoration reference points (Harris et al. 2006). Instead of using historical reference points, restoration ecologists are moving toward management for ecosystem function (Harris et al. 2006) and persistence in future environments (Choi 2007).

Reintroduction biologists term any mediated movement of organisms outside their native range as a species introduction (IUCN 1987). If the intent for such releases is the establishment of a new population explicitly for conservation, then it is a conservation introduction (IUCN 1998). There are three rationales for conservation introductions: when no other habitat exists, ecological replacement, and assisted colonization. Ecological replacement is the release of a species outside its historic range in order to fill an ecological niche left vacant by the extinction of a native species. Extinction removes the option of reintroduction and may mean the loss of critical ecological functions. One option is to restore this lost function through the establishment of an ecologically similar species (Atkinson 2001), perhaps through the release of a closely related taxon (Seddon and Soorae 1999). For example, Aldabra giant tortoises (*Aldabrachelys* sp.) have been used to restore selective grazing and seed dispersal functions once performed by the now extinct giant *Cylindraspis* tortoises on islands in the Indian Ocean (Griffiths et al. 2010). The best definition of assisted colonization is that of Ricciardi and Simberloff (2009b, 248): "translocation of a species to favourable habitat beyond their native range to protect them from human induced threats, such as climate change." So while climate change may loom as perhaps the most significant future threat, assisted colonization could be and has been used to mitigate a variety of threats.

The debate around assisted colonization sensibly focuses on uncertainty and the risk posed by introduced species (Ricciardi and Simberloff 2009a, b; Sax, Smith, and Thompson 2009; Seddon et al. 2009; Vitt, Havens, and Hoegh-Guldberg 2009). The deliberate moving of species is already positioned on the translocation spectrum (Seddon 2010). For example, in New Zealand, extinction threats to endemic birds by introduced mammalian predators have been addressed through translocations to predator-free offshore islands (Saunders and Norton 2001) that in many cases are not documented parts of the species' historic range. These translocations are effectively assisted colonizations, resulting in viable new populations in new areas (Atkinson 2001).

Restoration ecologists have recognized that anthropogenic drivers of environmental change may result in the development of emerging ecosystems, defined as "ecosystem[s] whose species composition and relative abundance have not previously occurred within a given biome" (Milton 2003). Also termed "novel ecosystems" (Chapin and Starfield 1997), or "no-analog communities" (Jackson and Hobbs 2009), these new assemblages of species have challenged the prevailing paradigm that holds that by managing human impacts it is possible to return nature to some stable, pristine state (Hobbs and Cramer 2008). Rather than attempting to force changed ecosystems back to some likely unsustainable or unattainable preexisting conditions, the development of novel ecosystems could be guided to maximize benefits (Hobbs et al. 2006) and to promote ecosystems that are feasible and resilient (Seastedt, Hobbs, and Suding 2008). Radically, this may include the active creation of "novel systems using species not native to the region" (Hobbs and Harris 2001), to "maximise genetic, species and functional diversity" (Seastedt, Hobbs, and Suding 2008), thus shifting from "historic" to "futuristic restoration" (Choi 2004), and the creation of "designer" (Temperton 2007) or "engineered" ecosystems (Jackson and Hobbs 2009), in which ecosystem function has been rehabilitated for future environments (Choi et al. 2008).

Restoration ecologists and reintroduction biologists need to join forces (Seddon et al. 2007a) to use conservation introductions to establish species into suitable habitat outside their historic distribution range in order to contribute to both species conservation objectives and ecosystem restoration goals in the face of climate-driven habitat change. The most provocative recent proposal for the creation a novel ecosystem through species translocations is that of "Pleistocene rewilding" (Donlan et al. 2006), whereby the multitude of ecological functions once performed by now-extinct North American megafauna could be replaced through the translocation of a suite of ecological replacements, some of which may be threatened by habitat loss or change. Although considered extreme (Rubenstein et al. 2006; Caro 2007), the notion of Pleistocene rewilding has engaged public interest and stimulated a rethink of the future of restoration strategies.

Last Thoughts

There remains a misconception that a desirable goal of conservation is preservation of, or restoration to, some pristine state. This perception has been challenged by archaeological evidence that the human impact on landscapes has been substantial, complex, and very old (Grayson 2001). For example, humans have lived in Amazonia for at least 12,000 years, and there is evidence that what may seem to be untouched Amazonian forest has in fact been substantially altered by humans (Guix 2009). It is now considered extremely unlikely that there are any habitable places on earth where the terrestrial biota has not been changed by prehistoric human activities, and thus it is the activities of humans that have shaped the current biodiversity that conservationist are attempting to conserve (Grayson 2001). Accepting the human-changed state of the world we live in frees us from an obligation to try to preserve isolated patches in a pristine state free from human influence, since the recreation of reserves that exclude people can never result in the recreation of biotic communities that once existed in the absence of human influence (Grayson 2001). We need to acknowledge that we are already making de facto decisions about the biodiversity we have around us, but these are decisions based on ignorance, economic imperatives, and inaction. Instead, we have the opportunity to foster a sense of place, to decide what kind of biodiversity we wish to live among, have access to, and benefit from. And while the reintroduction of iconic species back into part of their historical range will necessarily be a part of our conservation efforts, the selected restoration of keystone species to recreate critical ecosystem processes, perhaps within novel species assemblages, can facilitate the vital reconnection of humans with nature.

References

Amori, G., and S. Gippoliti. 2000. "The Problem of Subspecies and Biased Taxonomy in Conservation Lists: The Case of Mammals." *Folia Zoologica* 56: 113–17.

Armstrong, D. P., and P. J. Seddon. 2008. "Directions in Reintroduction Biology." *Trends in Ecology and Evolution* 23: 20–25.

Atkinson, I. A. E. 2001. "Introduced Mammals and Models for Restoration." *Biological Conservation* 99: 81–96.

Banda, S. 2010. "Colorado Declares Lynx Reintroduction a Success." http://www.denverpost.com/search/ci_16107178. Accessed December 10, 2010.

Barkham, P. 2009. "Four Decades of Hard Labour and Battenburg Cake: What It Takes to Bring the Large Blue Butterfly Back to Britain." http://www.guardian.co.uk/environment/2009/jun/16/large-blue-butterfly. Accessed December 10, 2010.

Beck, B. B. 2001. "A Vision for Reintroduction." Communiqué September 2001: 20–21. American Zoo and Aquarium Association, Silver Spring, Maryland, USA.

Berger, J. 1997. "Population Constraints Associated with the Use of Black Rhinos as an Umbrella Species for Desert Herbivores." *Conservation Biology* 11: 69–78.

Betrus, C. J., E. Fleishman, and R. B. Blair. 2005. "Cross-Taxonomic Potential and Spatial Transferability of an Umbrella Species Index." *Journal of Environmental Management* 74: 79–87.

Bonnet, X., R. Shine, and O. Lourdais. 2002. "Taxonomic Chauvinism." *Trends in Ecology and Evolution* 17: 1–3.

Byers, J. E., K. Cuddington, C. G. Jones, T. S. Talley, A. Hastings, and J. G. Lambrinos. 2006. "Using Ecosystem Engineers to Restore Ecological Systems." *Trends in Ecology and Evolution* 21: 493–500.

Caro, T. M. 2003. "Umbrella Species: Critique and Lessons from East Africa." *Animal Conservation* 6: 171–81.

———. 2007. "The Pleistocene Re-Wilding Gambit." *Trends in Ecology and Evolution* 22: 281–83.

Chapin, F. S., III, and A. M. Starfield. 1997. "Time Lags and Novel Ecosystems in Response to Transient Climatic Change in Arctic Alaska." *Climatic Change* 35: 449–61.

Choi, Y. D. 2004. "Theories for Ecological Restoration in Changing Environment: Toward 'Futuristic' Restoration." *Ecological Research* 19: 75–81.

———. 2007. "Restoration Ecology to the Future: A Call for New Paradigm." *Restoration Ecology* 15: 351–53.

Choi, Y. D., V. M. Temperton, E. B. Allen, A. P. Grootjans, M. Halassy, R. J. Hobbs, M. A. Naeth, and K. Torok. 2008. "Ecological Restoration for Future Sustainability in a Changing Environment." *Ecoscience* 15: 53–64.

Clark, J. A., and R. May. 2002a. "How Biased Are We?" *Conservation in Practice* 3: 28–29.

———. 2002b. "Taxonomic Bias in Conservation Research." *Science* 297: 191.

Davis, M. A. 2000. "'Restoration'—A Misnomer?" *Science* 287: 1203.

Day, R. L., K. N. Laland, and J. Odling-Smee. 2003. "Rethinking Adaptation: The Niche Construction Perspective." *Perspectives and Biology and Medicine* 46: 80–95.

Decaens, T., N. Asakawa, J. H. Galvis, R. J. Thomas, and E. Amezquita. 2002. "Surface Activity of Soil Ecosystem Engineers and Soil Structure in Contrasting Land Use Systems in Colombia." *European Journal of Soil Biology* 38: 267–71.

Donlan, C. J., J. Berger, C. E. Bock, J. H. Bock, D. A. Burney, J. A. Estes, D. Foreman, P. S. Martin, G. W. Roemer, F. A. Smith, M. E. Soule, and H. W. Greene. 2006. "Pleistocene Rewilding: An Optimistic Agenda for Twenty-First Century Conservation." *American Naturalist* 168: 660–81.

Ericsson, G., and T. A. Heberlein. 2003. "Attitudes of Hunters, Locals, and the General Public in Sweden Now That the Wolves Are Back." *Biological Conservation* 111: 149–59.

Fischer, J., and D. B. Lindenmayer. 2000. "An Assessment of the Published Results of Animal Relocations." *Biological Conservation* 96: 1–11.

Fleishman, E., D. D. Murphy, and P. F. Brussard. 2000. "A New Method for Selection of Umbrella Species for Conservation Planning." *Ecological Applications*, 10: 569–79.

Fritts, S. H., E. E. Bangs, J. A. Fontane, M. R. Johnson, K. E. Phillips, E. D. Koch, and J. R. Gunson. 1997. "Planning and Implementing a Reintroduction of Wolves to Yellowstone National Park and Central Idaho." *Restoration Ecology* 5: 7–27.

Gaston, K., and R. May. 1992. "Taxonomy of Taxonomists." *Nature* 356: 281–82.

Grayson, D. K. 2001. "The Archaeological Record of Human Impacts on Animal Populations." *Journal of World Prehistory* 15: 1–68.

Griffiths, C. J., C. G. Jones, D. M. Hansen, M. Puttoo, R. V. Tatayah, C. B. Muller, and S. Harris. 2010. "The Use of Extant Non-Indigenous Tortoises as a Restoration Tool to Replace Extinct Ecosystem Engineers." *Restoration Ecology* 18: 1–7.

Guix, J. C. 2009. "Amazonian Forests Need Indians and Caboclos." *Orsis* 24: 33–40.

Harris, J. A., R. J. Hobbs, E. Higgs, and J. Aronson. 2006. "Ecological Restoration and Global Climate Change." *Restoration Ecology* 14: 170–76.

Hastings, A., J. E. Byers, J. A. Crooks, K. Cuddington, C. G. Jones, J. G. Lambrinos, T. S. Talley, and W. G. Wilson. 2007. "Ecosystem Engineering in Space and Time." *Ecology Letters* 10: 153–64.

Hitt, N. P., and C. A. Frissell. 2004. "A Case Study of Surrogate Species in Aquatic Conservation Planning." *Aquatic Conservation: Marine and Freshwater Ecosystems* 14: 625–33.

Hobbs, R. J. 2007. "Setting Effective and Realistic Restoration Goals: Key Directions for Research." *Restoration Ecology* 15: 354–57.

Hobbs, R. J., and V. A. Cramer. 2008. "Restoration Ecology: Interventionist Approaches for Restoring and Maintaining Ecosystem Function in the Face of Rapid Environmental Change." *Annual Review of Environment and Resources* 33: 39–61.

Hobbs, R. J., and J. A. Harris. 2001. "Restoration Ecology: Repairing the Earth's Ecosystems in the New Millennium." *Restoration Ecology* 9: 239–46.

Hobbs, R. J., S. Arico, J. Aronson, J. S. Baron, P. Bridgewater, V. A. Cramer, P. R. Epstein, J. J. Ewel, C. A. Klink, A. E. Lugo, D. Norton, D. Ojima, D. M. Richardson, E. W. Sanderson, F. Valladares, M. Vila, R. Zamora, and M. Zobel. 2006. "Novel Ecosystems: Theoretical and Management Aspects of the New Ecological World Order." *Global Ecology and Biogeography* 15: 1–7.

IUCN (International Union for the Conservation of Nature). 1987. "IUCN Position Statement on the Translocation of Living Organisms: Introductions, Re-Introductions, and Re-Stocking." IUCN, Gland, Switzerland.

———. 1998. "Guidelines for Re-Introductions." IUCN/SSC Re-Introduction Specialist Group, IUCN, Gland, Switzerland and Cambridge, United Kingdom.

Jackson, S. T., and R. J. Hobbs. 2009. "Ecological Restoration in the Light of Ecological History." *Science* 325: 567–69.

James, A. I., and D. J. Eldridge. 2008. "Reintroduction of Fossorial Native Mammals and Potential Impacts on Ecosystem Processes in an Australian Desert Landscape." *Biological Conservation* 138: 351–59.

Jones, C. G., J. H. Lawton, and M. Shachak. 1997. "Positive and Negative Effects of Organisms as Physical Ecosystem Engineers." *Ecology* 78: 1946–57.

Jones, J., W. J. McLeish, and R. J. Robertson. 2004. "Predicting the Effects of Cerulean Warbler, *Dendroica cerulea* Management on Eastern Ontario Bird Species." *Canadian Field Naturalist* 118: 229–34.

Kleiman, D. G. 1989. "Reintroduction of Captive Mammals for Conservation." *BioScience* 39: 152–61.

Knapp, A. K., J. M. Blair, J. M. Briggs, S. L. Collins, D. C. Hartnett, L. C. Johnson, and E. G. Towne. 1999. "The Keystone Role of Bison in North American Tallgrass Prairie." *Bioscience* 49: 39–50.

Kramer-Schadt, S., E. Revilla, and T. Wiegand. 2005. "Lynx Reintroduction in Fragmented Landscapes of Germany: Projects with a Future or Misunderstood Wildlife Conservation?" *Biological Conservation* 125: 169–82.

Lawton, J. H. 1994. "What Do Species Do in Ecosystems?" *Oikos* 71: 367–74.

Macdonald, D. W., F. H. Tattersal, E. D. Brown, and D. Balharray. 1995. "Reintroducing the European Beaver to Britain: Nostalgic Meddling or Restoring Biodiversity?" *Mammal Review* 25: 161–200.

Mech, L. D. 1995. "The Challenge and Opportunity of Recovering Wolf Populations." *Conservation Biology* 9: 270–78.

Miller, J. R. 2005. "Biodiversity Conservation and the Extinction of Experience." *Trends in Ecology and Evolution* 20: 430–34.

Milton, S. J. 2003. "'Emerging Ecosystems'—A Washing-Stone for Ecologists, Economists and Sociologists?" *South African Journal of Science* 99: 404–6.

Naughton-Treves, L., R. Grossberg, and A. Treves. 2003. "Paying for Tolerance: Rural Citizens' Attitudes toward Wolf Depredation and Compensation." *Conservation Biology* 17: 1500–1511.

Nilsen, E. B., E. J. Milner-Gulland, L. Schofield, A. Mysterud, N. C. Stenseth, and T. Coulson. 2007. "Wolf Reintroduction to Scotland: Public Attitudes and Consequences for Red Deer Management." *Proceedings of the Royal Society B* 274: 995–1002.

Noon, B. R., and V. H. Dale. 2002. "Broad-Scale Ecological Science and Its Application." In *Applying Landscape Ecology in Biological Conservation*, edited by K. J. Gutzwiller, 34–52. New York: Springer-Verlag.

Noss, R. F., M. A. O'Connell, and D. D. Murphy. 1997. *The Science of Conservation Planning*. Washington, DC: Island Press.

O'Connell, M. J., Atkinson, S. R., Gamez, K., Pickering, S. P., and Dutton, J. S. 2008. "Forage Preferences of the European Beaver *Castor fiber*: Implications for Re-Introduction." *Conservation and Society* 62: 190–94.

Paine, R. T. 1995. "A Conversation on Refining the Concept of Keystone Species." *Conservation Biology* 9: 962–64.

Parker, K. A. 2008. "Translocations: Providing Outcomes for Wildlife, Resource Managers, Scientists, and the Human Community." *Restoration Ecology* 16: 204–9.

Pate, J., M. J. Manfredo, A. D. Bright, and G. Tischbein, G. 1996. "Coloradans' Attitudes toward Reintroducing the Gray Wolf into Colorado." *Wildlife Society Bulletin* 24: 421–28.

Pauly, D. 1995. "Anecdotes and the Shifting Baseline Syndrome of Fisheries." *Trends in Ecology and Evolution* 10: 430.

Poiani, K. A., M. D. Merrill, and K. A. Chapman. 2001. "Identifying Conservation-Priority Areas in a Fragmented Minnesota Landscape Based on the Umbrella Species Concept and Selection of Large Patches of Natural Vegetation." *Conservation Biology* 15: 513–22.

Pyle, R. M. 1993. *The Thunder Tree: Lessons from an Urban Wildland*. Boston: Houghton Mifflin. http://www.morning-earth.org/DE6103/Read%20DE/Extinction%20of%20Experience.pdf.

Ricciardi, A., and D. Simberloff. 2009a. "Assisted Colonization: Good Intentions and Dubious Risk Assessment." *Trends in Ecology and Evolution* 24: 476–77.

———. 2009b. "Assisted Colonization Is Not a Viable Conservation Strategy." *Trends in Ecology and Evolution* 24: 248–53.

Ripple, W. J., and R. L. Beschta. 2003. "Wolf Reintroduction, Predation Risk, and Cottonwood Recovery in Yellowstone National Park." *Forest Ecology and Management* 184: 299–313.

———. 2004. "Wolves and the Ecology of Fear: Can Predation Risk Structure Ecosystems?" *BioScience* 54: 755–66.

Roberge, J. M., and P. Angelstam. 2004. "Usefulness of the Umbrella Species Concept as a Conservation Tool." *Conservation Biology* 18: 76–85.

Rosell, F., O. Bozser, P. Collen, and H. Parker. 2005. "Ecological Impact of Beavers *Castor fiber* and *Castor canadensis* and Their Ability to Modify Ecosystems." *Mammal Review* 35: 248–76.

Roskaft, E., T. Bjerke, B. J. Kalternborn, D. C. Linnell, and R. Andersen. 2003. "Patterns of Self-Reported Fear towards Large Carnivores among Norwegian Public." *Evolution of Human Behaviour* 24: 184–98.

Rowland, M. M., M. J. Wisdom, L. H. Suring, and C. W. Meinke. 2006. "Greater Sage-Grouse as an Umbrella Species for Sagebrush-Associated Vertebrates." *Biological Conservation*, 129: 323–35.

Rubenstein, D. R., D. I. Rubenstein, P. W. Sherman, and T. A. Gavin. 2006. "Pleistocene Park: Does Re-Wildling North America Represent Sound Conservation for the 21st Century?" *Biological Conservation* 132: 232–38.

Samways, M. J. 2007. "Rescuing the Extinction of Experience." *Biodiversity and Conservation* 16: 1995–97.

Saunders, A., and D. A. Norton. 2001. "Ecological Restoration at Mainland Islands in New Zealand." *Biological Conservation* 99: 109–19.

Sax, D. F., K. F. Smith, and A. R. Thompson. 2009. "Managed Relocation: A Nuanced Evaluation Is Needed." *Trends in Ecology and Evolution* 24: 472–73.

Seastedt, T. R., R. J. Hobbs, and K. N. Suding. 2008. "Management of Novel Ecosystems: Are Novel Approaches Required?" *Frontiers in Ecology and Environment* 6: 547–53.

Seddon, P. J. 1999. "Persistence without Intervention: Assessing Success in Wildlife Re-Introductions." *Trends in Ecology and Evolution* 14: 503.

———. 2010. "From Reintroduction to Assisted Colonization: Moving along the Conservation Translocation Spectrum." *Restoration Ecology* 18: 796–802.

Seddon, P. J., and T. Leech. 2008. "Conservation Short-Cut, or Long and Winding Road? A Critique of Umbrella Species Criteria." *Oryx* 42: 240–45.

Seddon, P. J., and P. Soorae. 1999. "Guidelines for Subspecific Substitutions in Wildlife Restoration Projects." *Conservation Biology* 13: 177–84.

Seddon, P. J., P. S. Soorae, and F. Launay. 2005. "Taxonomic Bias in Reintroduction Projects." *Animal Conservation* 8: 51–58.

Seddon, P. J., D. P. Armstrong, and R. F. Maloney. 2007a. "Combining the Fields of Reintroduction Biology and Restoration Ecology." *Conservation Biology* 21: 1388–90.

———. 2007b. "Developing the Science of Reintroduction Biology." *Conservation Biology* 21: 303–12.

Seddon, P. J., D. P. Armstrong, P. Soorae, F. Launay, S. Walker, C. R. Ruiz-Miranda, S. Molur, H. Koldewey, and D. G. Kleiman. 2009. "The Risks of Assisted Colonization." *Conservation Biology* 23: 788–89.

Sergio, F., T. Caro, D. Brown, B. Clucas, J. Hunter, J. Ketchum, K. McHugh, and F. Hiraldo. 2008. "Top Predators as Conservation Tools: Ecological Rationale, Assumptions, and Efficacy." *Annual Review of Ecology Evolution and Systematics* 39: 1–19.

Simberloff, D. 1998. "Flagships, Umbrellas, and Keystones: Is Single-Species Management Passé in the Landscape Era?" *Biological Conservation* 83: 247–57.

Sitas, N., J. E. M. Baillie, and N. J. B. Isaac. 2009. "What Are We Saving? Developing a Standardized Approach for Conservation Action." *Animal Conservation* 12: 231–37.

South, A., S. Ruston, and D. Macdonald. 2000. "Simulating the Proposed Reintroduction of the European Beaver (*Castor fiber*) to Scotland." *Biological Conservation* 93: 103–16.

Suter, W., R. F. Graf, and R. Hess. 2002 "Capercaillie (*Tetrao urogallus*) and Avian Biodiversity: Testing the Umbrella-Species Concept." *Conservation Biology* 16: 778–88.

Temperton, V. M. 2007. "The Recent Double Paradigm Shift in Restoration Ecology." *Restoration Ecology* 15: 344–47.

Tregaskis, S. 2009. "Red Kites Exported after Success of Reintroduction Programme in Britain." *Guardian*, June 8, 2009. http://www.guardian.co.uk/environment/2009/jun/08/red-kites -reintroduction. Accessed December 10, 2010.

Turner, W. R., T. Nakamura, and M. Dinetti. 2004. "Global Urbanization and the Separation of Humans from Nature." *BioScience* 54: 585–90.

Vitt, P., K. Havens, and O. Hoegh-Guldberg. 2009. "Assisted Migration: Part of an Integrated Conservation Strategy." *Trends in Ecology and Evolution* 24: 473–74.

Wallis de Vries, M. F. 1995. "Large Herbivores and the Design of Large-Scale Nature Reserves in Western Europe." *Conservation Biology* 9: 25–33.

White, P. J., and R. A. Garrott. 2005. "Yellowstone's Ungulates after Wolves—Expectations, Realizations, and Predictions." *Biological Conservation* 125: 141–52.

Wilcove, D. S. 1994. "Turning Conservation Goals into Tangible Results: The Case of the Spotted Owl and Old-Growth Forests." In *Large-Scale Ecology and Conservation Biology*, edited by P. J. Edwards, R. M. May, and N. R. Webb, 313–29. Oxford: Blackwell Scientific Publications.

Williams, C. K., G. Ericsson, and T. A. Heberlein. 2002. "A Quantitative Summary of Attitudes toward Wolves and Their Reintroduction." *Wildlife Society Bulletin* 30: 575–84.

Wilson, C. J. 2004. "Could We Live with Reintroduced Large Carnivores in the UK?" *Mammal Review* 34: 211–32.

Wolf, C. M., T. Garland, and B. Griffith. 1998. "Predictors of Avian and Mammalian Translocation Success: Reanalysis with Phylogenetically Independent Contrasts." *Biological Conservation* 86: 243–55.

Wolf, C. M., B. Griffith, C. Reed, and S. A. Temple. 1996. "Avian and Mammalian Translocations: Update and Reanalysis of 1987 Survey Data." *Conservation Biology* 10: 1142–54.

11

Przewalski's Horses and Red Wolves

Importance of Behavioral Research for Species Brought Back from the Brink of Extinction

Sarah R. B. King

AT THE BEGINNING of the twenty-first century behavioral research conducted by direct observations on wild animals appears to have fallen out of fashion. This could be due partly to new technologies but also to the new generation of scientists, and it means that nature is increasingly being ignored. The employment of technologies such as trail cameras and telemetry collars means that biologists no longer need to be in the field watching their study animal all the time. Trail cameras show when an animal passes and give an idea of what species are in a particular place. Collars give information on the geographical position of an animal, thus its habitat use, and can even be set to provide an indication of behaviors such as feeding and sleeping. Although both of these technologies are important tools for research on elusive animals, they provide very few fine details of an animal's life and little information of the relevance of geographical position in relation to behavior or social structure. The current generation of young scientists has been referred to as "afraid of nature" (Hafner 2007). Due to their being brought up in an environment where most play was indoors, they have had little exposure to wilderness, and so perhaps they have a magnified view of its inherent risks. Concurrent with this attitude is the modern expectation that technology is always better, and things should be as automated as possible. However, observing animals only from a computer screen is tantamount to ignoring nature: it is necessary to actively watch animals outside and be receptive to subtle environmental variations to understand their behavior. Behavioral research requires a great deal of patience; this often means spending long hours watching animals do very little. But what it can tell us about a species' habitat use at a fine scale—social interactions, social system, activity patterns, and individual behavior—is invaluable.

Behavioral research is particularly critical for endangered species, where we need to know all we can to best conserve them. This is often difficult, and with extremely rare or elusive species sometimes can only be accomplished through collars or trail cameras. However, when direct observations are possible they can provide important insights into aspects such as feeding behavior, so that we know what resources to conserve, and wild mating behavior, so that we can maximize the success of captive breeding programs. Such research is also important for reintroduction programs, both before and after release. Knowledge of the animals' social behavior prior to release can inform what sex and age of animals to release together, and following the animals' postrelease allows us to assess how well they are adapting to the new environment and using the habitat.

Mammal reintroductions are lengthy and expensive, and frequently fail (Fischer and Lindenmayer 2000; Kleiman 1989; Morell 2008). Beyond aesthetic value and ideas of restoring what was anthropogenically removed, there are few reasons for their necessity. However, they do play a role in helping restore ecosystems, for example, wolves (*Canis lupus*) in Yellowstone (Beschta and Ripple 2009). They can also help local economies, especially in developing nations, by drawing in tourists eager to view the animals or by attracting additional development funds. Most reintroductions are to expand the range of an endangered species, or to restock areas where it was extirpated (Griffith et al. 1989). In a few cases the species has become extinct in the wild and is only kept extant in zoos. Reintroduction efforts should be based on the most precise knowledge of the species available, and that requires detailed observations of their daily behavior. This is difficult when it comes to species that were extinct in the wild before reintroduction, but observations of the animals either in large enclosures while captive or intensive observations immediately after their release can help provide information on behavior pre- and postreintroduction.

Two species that were reintroduced from a status of being extinct in the wild are the red wolf (*C. rufus*) and Przewalski's horse (*Equus ferus przewalskii*). Despite the differences between the animals, their stories have many similarities. Not much is known about the history of either species before they were "discovered" by Western scientists. Red wolves are thought to have once been found throughout the southeastern United States, but were extirpated from all but southeastern Texas by the 1970s (Paradiso and Nowak 1972). Recognizing the species' decline, scientists brought fourteen individuals thought to be pure red wolves into captivity for a breeding program. By 1980 the wolves were considered extinct in the wild due to extermination and habitat alteration (US Fish and Wildlife Service 2007). Przewalski's horses were first recorded from the

Chinese-Mongolian border in 1878, although from their similarity to horses depicted in European cave paintings it is likely that their range once covered Eurasia. Przewalski's horses were considered common at the time of their discovery, and there was a rush to capture individuals for European collections. Of the animals brought into captivity, only thirteen produced offspring. As no more individuals were captured before the species became extinct in the wild, these thirteen horses are the founders of all current Przewalski's horses (Bouman and Bouman 1994). The wild population dwindled to extinction by the 1970s due to a variety of factors, including severe winters, proliferation of weapons, competition with livestock for water sources, and possibly hybridization.

Both red wolves and Przewalski's horses are currently sympatric with closely related species with which they can hybridize and produce fertile offspring. In the case of red wolves this is with coyotes (*C. latrans*) and with Przewalski's horses it is domestic horses (*E. f. caballus*). Hybridization has affected these species even in captivity. When the first red wolves were brought into captivity it is likely that they had already hybridized with coyotes, so the first effort in the captive breeding was to remove individuals with coyote-like physical traits. Similarly, the thirteen Przewalski's horse founders brought into captivity included one domestic horse and possibly one hybrid. The studbook was then managed to remove individuals showing domestic horse characteristics, such as a nonerect mane or a lack of black pigment in the muzzle and coat (the "fox gene").

There was no research on either species before they were taken into captivity; we therefore know almost nothing about how they behaved under natural conditions. There is no published literature on captive red wolf behavior prior to their release. Przewalski's horses were studied in zoos and in larger grassy reserves but were in small human-formed groups, and research tended to focus on time budgets (Boyd, Carbonaro, and Houpt 1988) and aggression (Keiper 1988), with only a small amount of research on other behavior.

The reintroductions of the two species were handled differently, due to differences in their ecology and cultural differences where they were being reintroduced. The site planned for the release of red wolves was Alligator River National Wildlife Refuge (NWR) in North Carolina, and to facilitate support for the reintroduction they were classified as "nonessential experimental." This meant that the US Fish and Wildlife Service had to remove them from any land where the owner didn't wish their presence. The wolves were bred in St. Louis Zoo and initially were released from enclosures in the middle of Alligator River NWR. The first wolves released tended to wander widely and initially had high mortality. Subsequently, wolves were bred on barrier islands in South Carolina, Tennessee, and Florida and then released at Alligator River

NWR. These wolves were much more wary of people and roads and suffered less mortality after release.

Przewalski's horse reintroduction to Mongolia began in the early 1990s, with two organizations releasing horses at the same time but at different sites: one in desert habitat in Great Gobi B Special Protected Area, the other into a reserve (which subsequently became Hustai National Park) in steppe habitat in central Mongolia; neither area was fenced. The two reintroductions had different approaches. The Gobi reintroduction brought horses almost directly from zoos, allowed them to acclimatize in enclosures at a base camp, and then released them. The Hustai reintroduction moved horses from zoos to large grassy enclosures ("semireserves") in Holland where they lived with minimal human contact, before being transported to acclimatization enclosures at Hustai National Park. A third, later reintroduction of Przewalski's horses in 2004 was based more on observations of horse behavior. Horses were brought from zoos in Europe to a large semireserve in France, where over ten years they were allowed to choose breeding partners and form social groups without human interference. The naturally formed and established social groups of second- or third-generation horses were then transported to a 14,000-hectare enclosure in western Mongolia. The populations at all three sites had slow growth rates in the first years after release, followed by higher reproduction, resulting in the species being downlisted by the IUCN from Extinct in the Wild to Critically Endangered in 2008.

Although there are now breeding populations of red wolves and Przewalski's horses in the wild, both remain at risk. There are currently a total of about one hundred red wolves and three hundred Przewalski's horses in the wild; such small populations are at great risk from stochastic events, and both populations will continue to need management for the foreseeable future. Although the issues are different, the two reintroduction sites face problems with land use. In North Carolina this is due to private landowners wanting to be able to control wolves on their land, and in Mongolia this is due to the pressure of local herdsmen wanting access to the pasture. However, the biggest long-term risk to both populations is hybridization. Both Przewalski's horses and red wolves have undergone intensive breeding programs to ensure the species is pure, as they were brought back from the brink of extinction. The species with which hybridization would occur (domestic horses or coyotes) are abundant. So should it continue unchecked, at best hybrids will be mixed in with purebred animals, thereby complicating conservation actions for the species, and at worst Przewalski's horse and red wolf genes could become diluted until they effectively become extinct. This is possibly what happened to the last wild Przewalski's horses. Behavioral research is necessary to understand how and when hybridization happens, so that management can be targeted at preventing it.

Although useful for many ecological and management reasons, home range or habitat use studies based on periodically locating an animal do not provide the level of detail necessary to understand subtle behaviors such as mate choice, dispersal, or social structure. To understand this we need detailed life history information of individuals, which can only be gained through direct observation. It is often difficult to gather observational behavior data, especially on elusive animals like wolves in forests, or when horses have a long flight distance. It is also not always desirable to habituate animals to the presence of people. Despite the difficulties, the benefits of observational behavior of individuals could ultimately aid the management of the entire population, helping conserve the species in the wild.

There has been little published research on red wolf behavior, meaning that it is unclear what mechanisms cause red wolves to mate with coyotes. As these wolves form strong pair bonds, the consequence of a red wolf partnering with a coyote has implications not just for one litter but also for future litters produced by that individual. It appears that red wolves mate with coyotes when they cannot find an available conspecific. Work by Fredrickson and Hedrick (2006) has led to an adaptive management plan of coyote–red wolf pairs and hybrids, whereby nonpure red wolves are sterilized and left to be "placeholders" of a territory until displaced by pure individuals. This currently appears to be reducing introgression of coyote genes, but more research is needed into wolf-coyote interactions and mate choice behavior to elucidate what they select in a mate so that management can help avoid red wolf–coyote pairs.

Behavioral research has provided insights into horse social systems. The typical group is composed of a stallion living with three to five mares and their offspring. Unusually for social mammals, both male and female offspring disperse around the time of puberty, which in the case of females acts as an incest avoidance mechanism: they will not mate with familiar individuals (Monard and Duncan 1996). On dispersal the males form into bachelor groups, and females either directly join another harem group or spend some time with a bachelor group before forming a new group with a bachelor. Except for stallions that no longer have a group of mares, it is rare for horses to be solitary at any stage in their life. This suggests that they actively seek the company of other horses. More detailed research on how dispersing individuals form groups and choose social partners is needed so that we can determine if certain individuals are more at risk of finding and joining domestic horses, and therefore potentially producing hybrid foals.

Previous behavioral research based on long-term direct observations has helped inform the social structure of reintroduced groups, and then it provided an indication of how the animals were adjusting to life in the wild. Current behavioral research is helping management for hybrids and can show which habitat areas are most necessary for conservation. In concert with veterinary and ecological research, studies of the behavior of red wolves and Przewalski's horses should help them continue to persist in the wild. We cannot expect management plans of reintroduced species to work when nature is ignored, and there are inadequate observations of animals in the wild. Behavioral research involving knowledge of the species' biology and behavior is therefore vital for their conservation over the long term.

References

Beschta, R. L., and W. J. Ripple. 2009. "Large Predators and Trophic Cascades in Terrestrial Eco-
    systems of the Western United States." *Biological Conservation* 142: 2401–14.
Bouman, I., and J. Bouman. 1994. "The History of Przewalski's Horse." In *Przewalski's Horse: The
    History and Biology of an Endangered Species*, edited by L. Boyd and K. Houpt, 5–38. Albany:
    State University of New York Press.
Boyd, L., D. Carbonaro, and K. Houpt. 1988. "The 24-Hour Time Budget of Przewalski Horses."
    Applied Animal Behaviour Science 21: 5–17.
Fischer, J., and D. B. Lindenmayer. 2000. "An Assessment of the Published Results of Animal Re-
    locations." *Biological Conservation* 96: 1–11.
Fredrickson, R. J., and P. W. Hedrick. 2006. "Dynamics of Hybridization and Introgression in Red
    Wolves and Coyotes." *Conservation Biology* 20: 1272–83.
Griffith, B., M. Scott, J. W. Carpenter, and C. Reed. 1989. "Translocation as a Species Conservation
    Tool: Status and Strategy." *Science* 245: 477–80.
Hafner, M. S. 2007. "Field Research in Mammalogy: An Enterprise in Peril." *Journal of Mammal-
    ogy* 88: 1119–28.
Keiper, R. R. 1988. "Social Interactions of the Przewalski Horse (*Equus przewalskii* Poliakov, 1881)
    Herd at the Munich Zoo." *Applied Animal Behaviour Science* 21: 89–97.
Kleiman, D. G. 1989. "Reintroduction of Captive Mammals for Conservation." *BioScience* 39: 152–61.
Monard, A., and P. Duncan. 1996. "Consequences of Natal Dispersal in Female Horses." *Animal
    Behaviour* 52: 565–79.
Morell, V. 2008. "Into the Wild: Reintroduced Animals Face Daunting Odds." *Science* 320: 742–43.
Paradiso, J. L., and R. M. Nowak. 1972. "*Canis rufus.*" *Mammalian Species* 22: 1–4.
US Fish and Wildlife Service. 2007. "Red Wolf (*Canis rufus*) 5-Year Status Review: Summary and
    Evaluation." Red Wolf Recovery Program Office, Alligator River National Wildlife Refuge: 58.

12

Why Individuals Matter

Lessons in Animal Welfare and Conservation

Liv Baker

UNDERSTANDING INDIVIDUAL DIFFERENCES among animals is essential to our understanding of nature. Certainly, the importance of natural variation was not lost on Charles Darwin as he developed his evolutionary theory (Gosling and John 1999). Indeed, consideration of individual variation has continued to be critical to animal-related disciplines. We see this consideration, for example, in Ivan Pavlov's work on nervous system function, Robert Yerkes's studies of the animal mind, and Konrad Lorenz's behavioral research (Pavlov 1906, 1928, 1941; Yerkes and Yerkes 1917; Yerkes 1939; Lorenz 1954). Attention to the unique characteristics of individuals was especially notable in the work of primatologists such as Jane Goodall, who recorded complex narratives for individual animals (Goodall 1986). This approach offered an alternative paradigm that shifted the focus from species norms to individuality and arguably contributed to scientists' acceptance of the role of affective states in animal behavior (Fraser 2009).

Despite the contribution of these and other scientists, a great deal of research on animals continues to focus on norms for species rather than differences among individuals. In fact, differences among individuals have often been viewed as a problem—undesirable variation to be minimized—rather than interesting and biologically meaningful information (Sapolsky 1994; Carere and Locurto 2011), especially in nonprimate species.

However, when we ignore individuals and the variation among them, we run the risk of limiting our understanding of animals and natural processes. In behavioral ecology, for example, the general neglect of individual variation has led evolutionary theorists to predict that when individual differences are heritable and linked to fitness, natural selection should select for optimal behavioral suites and minimize behavioral variation (Wilson 1998), whereas recent studies suggest that individual differences are maintained by frequency-dependent natural selection (Carere, Caramaschi, and Fawcett 2010).

With this change of thinking (Sih, Bell, and Johnson 2004), behavioral ecologists have begun to explore the fitness consequences of different personalities within a population. They have shown that mortality, breeding, dispersal, and partner preference, as well as disease risk and vulnerability to parasites, are correlated with individual personality differences (Réale and Festa-Bianchet 2003; Dingemanse et al. 2004; Both et al. 2005; Korte et al. 2005).

Beyond a theoretical understanding of individual differences, the awareness of individuals and their variation also has applied and ethical implications, which can be seen in efforts to conserve threatened species. Conservation biology, like other animal-related sciences, has often disregarded individual differences in ways that reinforce the conservation movement's traditional emphasis on "ecological collectives" (populations, species, ecosystems) rather than individuals (Soulé 1985; Norton 1995; Rawles 1997; Vucetich and Nelson 2007; Fraser 2010). However, some scientists now recognize this blindness to individuals as counterproductive.

Wildlife translocations (including reintroductions) offer a case in point. Translocations are likely the most common conservation practice used to combat loss of species and habitat and are an integral part of the recovery plans for many at-risk species. Nevertheless, the success of translocations has traditionally been dismal. Some reviews put the success rate as low as 10–25 percent (Beck et al. 1994; Wolf, Garland, and Griffith 1998). And it is not uncommon for the mortality of animals in these translocations to be as high as 50–95 percent (Teixeira et al. 2007). These low success rates have likely been shaped by a lack of attention to individuals.

The ultimate measure of translocation success is a self-sustaining population, and this depends on the animals' surviving, settling, and reproducing at the release site (Gosling and Sutherland 2000; Letty et al. 2003). Establishing a viable population may be thwarted if there is an inadequate number of reproducing survivors, if the animals disperse without reproducing, or if they reproduce briefly and then languish as nonbreeding adults (Leopold 1933; Fischer and Lindenmayer 2000; Armstrong and Seddon 2008). More often than not, however, translocated animals simply die during the initial days or weeks after release (Kleiman 1989; Short et al. 1992; Beck et al. 1994; Teixeira et al. 2007).

Because of the translocation process, an animal may meet multifarious hardships that have direct and indirect effects on its welfare and survival. For example, animals can suffer from trapping injuries, capture and transport myopathy, excessive weight loss, stereotypy, anomalous predation, disease, suboptimal foraging and navigation, reproduction depression, and isolation (Teixeira et al. 2007). Moreover, different animals will handle such problems in different ways; therefore, we need attention to individuals for the sake of their welfare and survival.

By the mid-1990s, researchers were proposing that translocations were unsuccessful because of the limited understanding of the basic behavior and ecology of the species involved. To address this, behavioral ecologists looked into a range of factors including dispersal, foraging, social and antipredator behaviors, habitat requirements, social structure, and mating habits (Letty et al. 2000; Gerber et al. 2003; Shier 2006). Although such work has heightened our awareness of the complexity of animals' lives, and in some cases improved translocation success, the application of behavioral ecology remains infrequent and continues to focus on species norms rather than individuals (Caro 2007; Swaisgood 2010).

The impact of the translocation process on individual animals has only recently received attention, notably by Letty and colleagues (2007) and Teixeira and colleagues (2007), who call attention to potential translocation stressors that affect animals' coping abilities and thus influence the survival of the animals.

Typical translocations involve trapping, handling, marking, captivity, and transport, in addition to drastic environmental and social disturbance. These events, particularly in combination, may interfere with an animal's adaptive stress response, whereby its behavioral, physiological, and cognitive repertoire of coping strategies is affected. Ultimately, this may compromise the mechanisms that manage vital events.

Several studies have shown that translocation events do affect the levels of stress hormones in animals, with behavioral and physical consequences (e.g., Franceschini et al. 2008; Dickens, Delehanty, and Romero 2009, 2010). For example, Dickens, Delehanty, and Romero (2009) showed a correlation between stress hormone levels and body weight among translocated chukar partridge (*Alectoris chukar*). Specifically, partridges exhibited a depressed stress response and dramatic weight loss when exposed to the translocation process. In an attempt to determine if the method of soft release was beneficial to Grevy's zebra (*Equus grevyi*), Franceschini and colleagues (2008) monitored fecal glucocorticoid metabolites (FGMs). FGM levels suggested that the soft release did not reduce stress hormone levels for the animals; in fact, it seemed to be quite stressful to them.

Pinter-Wollman, Isbell, and Hart (2009) also looked at FGMs as well as body condition in translocated African elephants (*Loxodonta africana*). Although metabolite levels were not different compared to a local resident group, translocated individuals did have poorer body condition, and adults had a higher death rate.

While these studies looked for overall treatment effects, and did not report the variation between individual animals, current research indicates that stressors differentially affect individuals because of personality types (Koolhaas et al. 1999). For instance, different strategies of dispersal, migration, foraging, predation, and breeding cause individuals to be more or less vulnerable in a challenging environment. This is, in part, because of variation in animals' aggression levels, willingness to explore and approach novel objects, and their underlying physiological mediators (Verbeek, Drent, and Wiepkema 1994; Verbeek, Boon, and Drent 1996; Dingemanse et al. 2002; Bell 2004). As an example, Bremner-Harrison, Prodohl, and Elwood (2004) observed that among swift foxes (*Vulpes velox*), bolder individuals, characterized by their lower levels of neophobia, had higher death rates than more cautious individuals after release.

Researchers now believe that different personality types are preserved within a population because different types are favored under naturally shifting environmental pressures. Thus the persistence of a wide complement of personality types is important for maintaining a population (Dingemanse et al. 2004; Wolf et al. 2007). However, the direct and indirect effects of human activity may disrupt the selective forces that preserve personality composition. For example, in addition to being artificially relocated from one habitat to another, translocated animals are also opportunistically assembled. In particular, those that are trapped for relocation are likely to include fewer trap-shy individuals, and hence the translocated group may be skewed toward a particular personality type.

This presents a problem because translocation events are intense and varied enough that it is unlikely that any single personality type would be favored in all events and inadvertent selection for one type may be counterproductive if establishing a resilient population requires a mix of personality types.

Furthermore, knowledge of individual differences likely involves examining social dynamic and structure, habitat familiarity, feeding habits, foraging skills, and predator recognition and avoidance (Stamps and Swaisgood 2007). As with personality, an animal's life events shape its needs, preferences, and expectations. All of which have consequences for how individuals may respond to the stages of translocation.

Apart from those captive breeding programs that have developed successful antipredator training, there are scant examples from the translocation literature where such consideration at the individual level has been put into practice. Shier (2006) is a notable exception. By observing individuals and identifying family groups she showed that family support was integral to the translocation success of black-tailed prairie dogs (*Cynomys ludovicianus*). Those translocated in intact groups had greater survival than mixed groups, and they also exhibited greater reproductive success.

Thus not only are populations thought to need individuals with a mix of personalities to thrive under varying conditions, individuals with certain personality types and certain life experience may handle the stress of translocation better than others. Consequently, translocation practitioners need to be mindful of the individuals involved in order to have short-term animal welfare and translocation success and a population that will be viable in the long term.

When we attend to animals as individuals we should see our way to resolving both animal welfare and conservation implications.

References

Armstrong, D. P., and P. J. Seddon. 2008. "Directions in Reintroduction Biology." *Trends in Ecology and Evolution* 23: 20–25.

Beck, B. B., L. G. Rapaport, M. R. Stanley Price, and A. C. Wilson. 1994. "Reintroduction of Captive-Born Animals." In *Creative Conservation: Interactive Management of Wild and Captive Animals*, edited by P. Olney, 265–86. London: Chapman and Hall.

Bell, A. M. 2004. "Behavioural Differences between Individuals and Two Populations of Stickleback (*Gasterosteus aculeatus*)." *Journal of Evolutionary Biology* 18: 463–74.

Both, C., N. J. Dingemanse, P. J. Drent, and J. M. Tinbergen. 2005. "Pairs of Extreme Avian Personalities Have Highest Reproductive Success." *Journal of Animal Ecology* 74: 667–74.

Bremner-Harrison, S., P. A. Prodohl, and R. W. Elwood. 2004. "Behavioural Trait Assessment as a Release Criterion: Boldness Predicts Early Death in a Reintroduction Programme of Captive-Bred Swift Fox (*Vulpes vulpes*). *Animal Conservation* 7: 313–20.

Carere, C., D. Caramaschi, and T. W. Fawcett. 2010. "Covariation between Personalities and Individual Differences in Coping with Stress: Converging Evidence and Hypotheses." *Current Zoology* 56: 728–40.

Carere, C., and C. Locurto. 2011. "Interaction between Animal Personality and Animal Cognition." *Current Zoology* 57: 491–98.

Caro, T. 2007. "Behavior and Conservation: A Bridge Too Far." *Trends in Ecology and Evolution* 22: 394–400.

Dickens, M. J., D. J. Delehanty, and L. M. Romero. 2009. "Stress and Translocation: Alterations in the Stress Physiology of Translocated Birds." *Proceedings of Royal Society Biological Sciences* 276: 2051–56.

———. 2010. Stress: "An Inevitable Component of Animal Translocation." *Biological Conservation* 143: 1329–41.

Dingemanse N. J., C. Both, P. J. Drent, K. van Oers, and A. J. van Noordwijk. 2002. "Repeatability and Heritability of Exploratory Behaviour in Great Tits from the Wild." *Animal Behaviour* 64: 929–38.

Dingemanse, N. J., C. Both, P. J. Drent, and J. M. Tinbergen. 2004. "Fitness Consequences of Avian Personalities in a Fluctuating Environment." *Proceedings of Royal Society Biological Sciences* 271: 847–52.

Fischer, J., and D. B. Lindenmayer. 2000. "An Assessment of the Published Results of Animal Relocations." *Biological Conservation* 96: 1–11.

Franceschini, M.D., D. I. Rubenstein, B. Low, and L. M. Romero. 2008. "Fecal Glucocorticoid Metabolite Analysis as an Indicator of Stress during Translocation and Acclimation in an Endangered Large Mammal, the Grevy's Zebra." *Animal Conservation* 11: 263–69.

Fraser, D. 2009. "Animal Behaviour, Animal Welfare, and the Scientific Study of Affect." *Applied Animal Behaviour Science* 118: 108–17.

———. 2010. "Toward a Synthesis of Conservation and Animal Welfare Science." *Animal Welfare* 19: 121–24.

Gerber, L. R., E. W. Seabloom, R. S. Burton, and O. J. Reichman. 2003. "Translocation of an Imperiled Woodrat Population: Integrating Spatial and Habitat Patterns." *Animal Conservation* 6: 309–16.

Goodall, J. 1986. *The Chimpanzees of Gombe: Patterns of Behavior.* Cambridge, MA: Belknap Press of Harvard University Press.

Gosling, S. D., and O. P. John. 1999. "Personality Dimension in Nonhuman Animals: A Cross-Species Review." *Current Directions in Psychological Science* 8: 69–75.

Gosling, L. M., and W. J. Sutherland. 2000. *Behaviour and Conservation.* Cambridge: Cambridge University Press.

Kleiman, D. G. 1989. "Reintroduction of Captive Mammals for Conservation." *BioScience* 39: 152–61.

Koolhaas, J. M., S. M. Korte, S. F. De Boer, B. J. van der Vegt, C. G. van Reenen, H. Hopster, I. C. De Jong, M. A. W. Ruis, and H. J. Blokhuis. 1999. "Coping Styles in Animals: Current Status in Behaviour and Stress-Physiology." *Neuroscience and Biobehavioral Reviews* 23: 925–35.

Korte, S. M., J. M. Koolhaas, J. C. Wingfield, and B. S. McEwen. 2005. "The Darwinian Concept of Stress: Benefits of Allostasis and Costs of Allostatic Load and the Trade-Offs in Health and Disease." *Neuroscience and Biobehavioral Reviews* 29: 3–38.

Leopold, A. 1933. *Game Management.* New York: Charles Scribner's Sons, 481.

Letty, J., J. Aubineau, S. Marchandeau, and J. Clobert. 2003. "Effect of Translocation on Survival in Wild Rabbit (*Oryctolagus cuniculus*)." *Mammalian Biology* 68: 250–55.

Letty, J., S. Marchandeau, and J. Aubineau. 2007. "Problems Encountered by Individuals in Animal Translocations: Lessons from Field Studies." *Ecoscience* 14: 420–31.

Letty, J., S. Marchandeau, J. Clobert, and J. Aubineau. 2000. "Improving Translocation Success: An Experimental Study of Anti-Stress Treatment and Release Method for Wild Rabbits." *Animal Conservation* 3: 211–19.

Lorenz, K. 1954. *Man Meets Dog.* London: Methuen.

Norton, B. G. 1995. "Caring for Nature: A Broader Look at Animal Stewardship" In *Ethics on the Ark: Zoos, Animal Welfare, and Wildlife Conservation*, edited by B. G. Norton, M. Hutchins, E. F. Stevens, and T. L. Maple, 375–95. Washington, DC: Smithsonian Institution Press.

Pavlov, I. P. 1906. "The Scientific Investigation of the Psychical Faculties or Processes in the Higher Animals." *Science* 24: 613–19.

———. 1928. "The Inhibitory Type of Nervous Systems in the Dog." In *Lectures on Conditioned Reflexes: Twenty-Five Years of Objective Study of the Higher Nervous Activity (Behaviour) of Animals*, edited by I. P. Pavlov and translated by W. H. Gantt, 2:363–69. New York: International Publishers.

———. 1941. *Conditioned Reflexes and Psychiatry.* Translated by W. Horsley Gantt. New York: International Publishers.

Pinter-Wollman, N., L. A. Isbell, and L. A. Hart. 2009. "Translocation Outcome: Comparing Behavioral and Physiological Aspects of Translocated and Resident African Elephants (*Loxodonta africana*)." *Biological Conservation* 142: 1116–24.

Rawles, K. 1997. "Conservation and Animal Welfare." In *The Philosophy of the Environment*, edited by T. D. J. Chappell, 135–55. Edinburgh: Edinburgh University Press.

Réale, D., and M. Festa-Bianchet. 2003. "Predator-Induced Natural Selection on Temperament in Bighorn Ewes." *Animal Behaviour* 65: 463–70.

Sapolsky, R. M. 1994. "Individual-Differences and the Stress-Response." *Seminars in the Neurosciences* 6: 261–69.

Shier, D. M. 2006. "Effect of Family Support on the Success of Translocated Black-Tailed Prairie Dogs." *Conservation Biology* 20: 1780–90.

Short, J., S. Bradshaw, J. Giles, R. Prince, and G. Wilson. 1992. "Reintroduction of Macropods (Marsupialia: Macropodoidea) in Australia: A Review." *Biological Conservation* 62: 189–204.

Sih, A., A. M. Bell, and J. C. Johnson. 2004. "Behavioural Syndromes: An Ecological and Evolutionary Overview." *Trends in Ecology and Evolution* 19: 372–78.

Soulé, M. E. 1985. "What Is Conservation Biology?" *BioScience* 35: 727–34.

Stamps, J. A., and R. R. Swaisgood. 2007. "Someplace like Home: Experience, Habitat Selection, and Conservation Biology." In "Animal Behaviour, Conservation, and Enrichment," edited by R. R. Swaisgood. Special issue of *Applied Animal Behaviour Science* 102: 392–409.

Swaisgood, R. R. 2010. "The Conservation-Welfare Nexus in Reintroduction Programmes: A Role for Sensory Ecology." *Animal Welfare* 19: 125–37.

Teixeira, C. P., C. S. Azevedo, M. Mendl, C. F. Cipreste, and R. J. Young. 2007. "Revisiting Translocation and Reintroduction Programmes: The Importance of Considering Stress." *Animal Behaviour* 73: 1–13.

Verbeek, M. E. M., A. Boon, and P. J. Drent. 1996. "Exploration, Aggressive Behaviour, and Dominance in Pairwise Confrontations of Juvenile Male Great Tits." *Behaviour* 133: 945–63.

Verbeek, M. E. M., P. J. Drent, and P. R. Wiepkema. 1994. "Consistent Individual Differences in Early Exploratory Behaviour of Male Great Tits." *Animal Behaviour* 48: 1113–21.

Vucetich, J. A., and M. P. Nelson. 2007. "What Are 60 Warblers Worth? Killing in the Name of Conservation." *Oikos* 11: 1267–78.

Wilson, D. S. 1998. "Adaptive Individual Differences within Single Populations." *Philosophical Transactions of the Royal Society of London: Biological Sciences* 353: 199–205.

Wolf, C. M., T. Garland, and B. Griffith. 1998. "Predictors of Avian and Mammalian Translocation Success: Reanalysis with Phylogenetically Independent Contrasts." *Biological Conservation* 86: 243–55.

Wolf, M., G. S. van Doorn, O. Leimar, and F. J. Weissing. 2007. "Life-History Trade-Offs Favour the Evolution of Animal Personalities." *Nature* 447: 581–85.

Yerkes, R. M. 1939. "The Life History and Personality of the Chimpanzee." *American Naturalist* 73: 97–112.

Yerkes, R. M., and A. W. Yerkes. 1917. "Individuality, Temperament, and Genius in Animals." *American Museum Journal*. http://www.naturalhistorymag.com.

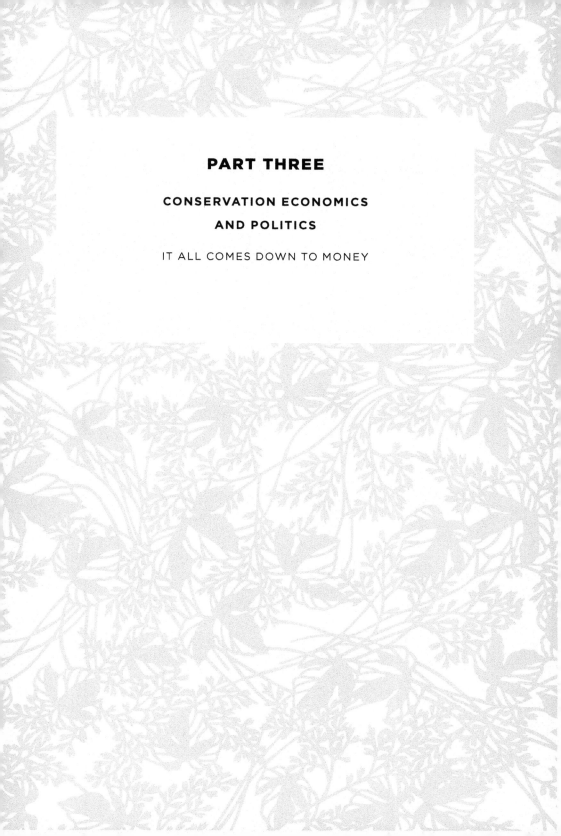

PART THREE

CONSERVATION ECONOMICS AND POLITICS

IT ALL COMES DOWN TO MONEY

LIKE MANY OTHER aspects of life, much of what we are able to do comes down to money, an extremely limited resource that severely restricts our efforts to make things better. Brian Czech points out, as have others, that the animal welfare movement tends to ignore wild animals, including the environment that maximizes wild animal welfare. He stresses that steady state economics is essential for wild animal welfare and that people who think about macroeconomics tend to ignore the environment and animal welfare. His basic arguments center on this presumption: "The process of economic growth simply entails more economic activity and, therefore, more habitat destruction and more inhumane treatment of wild animals. Economic growth is not intended to kill, torture, or harass animals, and in that respect is not as detestable as various other forms of inhumanity. *Yet economic growth is surely the greatest of all forms of inhumanity in terms of the gross amount of wild animal suffering that results*" (my emphasis). Furthermore, when considering endangered species, Czech writes, "These economic activities imperil species because they remove or degrade the food, water, cover, and space required to sustain wild animals. To put the scale of the problem into perspective, consider how many individual animals suffer when these economic activities imperil an entire species. Yet this is precisely what has occurred when a species is listed as threatened or endangered." The sequence of logic pertaining to the humane treatment of wild animals is: "(1) Wild animal welfare requires wildlife habitats. (2) Economic growth occurs at the expense of wildlife habitats. (3) Stabilization of wildlife habitats, and, therefore, the humane treatment of wild animals, requires the establishment of a steady state economy."

Eric Shelton's essay focuses on conservation, biodiversity, tourism, and the conservation economy in New Zealand and how different conceptions of nature add complexity to the notion of "ignoring nature." For example, the United States' idea of wilderness as standing in for nature is not replicated in New Zealand, where the current neoliberal move to a conservation economy emphasizes the role of the natural environment as being a provider of ecosystem services and as a resource for tourism. Shelton discusses the interplay among politics, economics, and activism. He argues that Kate Soper's (1995) philosophical framework provides a basis for clarifying ideas on the status and role of nature, as they are espoused, and may lead to better-informed community and professional involvement in the production of habitat and in species reintroduction. Soper's framework facilitates informed discussion about which concepts of nature inform which management preferences and decisions. Shelton discusses conservation, biodiversity, and tourism in New Zealand and highlights a project at Long Point intended to recreate and restore sea birds' colonies. The Long Point Project illustrates a social movement from environmental quietism to urgent ecosystem activism, a development that reflects an increased sense of individual agency, a key component of the neoliberal agenda.

An important question that's not considered in any detail in this book is, "How can money be raised?" It's been very difficult for animal protection and conservation organizations to raise the money needed to have widespread benefits. While tourism can raise some funds, Marco Festa-Bianchet (part 2, chapter 9) offers that trophy hunting "is the only form of economic activity that can protect some habitats and ecosystems as an alternative to more destructive human uses" because, unlike ecotourists who go to safe places where they are sure to see animals, trophy and sport hunters will travel to remote areas and are willing to spend large sums for the opportunity to kill a prize animal. Surely, this will not sit well with those interested in animal protection, but others have also made the same argument (e.g., Whitman et al. 2004). Once again, we're faced with difficult choices about who lives and who dies.

References

Soper, Kate. 1995. *What Is Nature?* Oxford: Blackwell.
Whitman, K., A. M. Starfield, H. S. Quadling, and C. Packer. 2004. "Sustainable Trophy Hunting of African Lions." *Nature* 428: 175–78.

13

The Imperative of Steady State Economics
for Wild Animal Welfare

Brian Czech

AS A LONGTIME WILDLIFE BIOLOGIST, in my estimation the most important source of wild animal suffering is habitat destruction. Habitat includes food, water, cover, and space. When any of these components are eliminated or degraded, wild animals suffer and many die, often in more insidious, protracted, and torturous ways than if killed or crippled by a hunter or natural predator.

Many wild animals survive an initial onslaught of habitat destruction only to be stranded in a foreign, inhospitable environment. When a food or water source is eliminated or degraded, wild animals may starve, die of thirst, or suffer agonizing debilities associated with malnutrition. When thermal cover is destroyed, wild animals must expend precious time and energy to regulate body temperatures, decreasing or eliminating other activities such as feeding, playing, or reproducing. When hiding cover is lost, wild animals enter a constant state of fear and stress, instinctively seeking cover, in vain, from predators that may or may not be present.

Wild animals who are able to escape to nearby suitable habitats (assuming such habitats exist) face the difficulty of competing with already-established individuals of their own species. The problems these animals face are very similar to the problems faced by those who remain in an area where habitat has contracted. In general, populations within an ecosystem tend to fluctuate near carrying capacity, so the immigration of displaced animals results in a stressful attempt for survival by all animals, including the original inhabitants and the migrating refugees.

Habitat destruction, meanwhile, occurs in the normal course of human affairs, and we often hear of "human activity" being identified as the cause of many environmental problems. However, it behooves the environmental and animal protection communities to specify what type of human activity is problematic. The habitat destruction human beings cause is virtually always a result of economic activity.

The process of economic growth simply entails more economic activity and, therefore, more habitat destruction and more inhumane treatment of wild animals. Economic growth is not intended to kill, torture, or harass animals, and in that respect is not as detestable as various other forms of inhumanity. Yet economic growth is surely the greatest of all forms of inhumanity in terms of the gross amount of wild animal suffering that results.

Economic Growth and Habitat Destruction

Economic growth is an increase in the production and consumption of goods and services. It entails increasing human populations, per capita consumption, or both. The size of an economy is generally indicated by gross domestic product (GDP).

The relationship between economic growth and habitat destruction is readily apparent when we consider the causes of species endangerment. For example, in the United States these causes include agriculture, domestic livestock production, mining, logging, and other extractive sectors (Czech, Krausman, and Devers 2000). These economic activities imperil species because they remove or degrade the food, water, cover, and space required to sustain wild animals. To put the scale of the problem into perspective, consider how many individual animals suffer when these economic activities imperil an entire species. Yet this is precisely what has occurred when a species is listed as threatened or endangered.

Another primary cause of species endangerment is urbanization. *Urbanization*, used here in the simplest sense of expanding urban area, reflects the growth of the national labor force and the consumer population as well as a variety of industrial and service sectors. Few types of habitat destruction are as thorough and permanent as urbanization. While the logging of a forest, for example, is a traumatic experience for its wild denizens, some of them are able to carve a niche out of what is left after the harvest. When a city expands, it usually does so by adding pavement, buildings, and infrastructure, all of which are absolutely inhospitable to most of the area's original species. Economic infrastructure extends far into the countryside, too, providing the matrix of a national economy. Roads, reservoirs, pipelines, power lines, telecommunications facilities, and wind farms are examples and constitute another major cause of species endangerment.

It is hard to imagine a more omnipresent danger than roads, upon which countless animals are mangled and left, during their final hours, to be slowly, opportunistically picked apart by vertebrate scavengers and insects. As The Humane Society of the United States (2006) noted, "Millions upon millions of wild animals are killed on our nation's highways every year. Some scientists estimate that humans kill more wild animals with their cars than with any other instrument, including guns." The damage that highways inflict on wildlife is not limited to direct mortality. It starts with the destruction of habitat and continues with the construction of the road itself, which causes more wildlife mortality. Chemical and physical alteration of the surrounding environment and introduction of potentially invasive species accompany the construction and use of roads.

Power lines present the menace of electrocution, the outcome of which may be death or permanent crippling. Harness and Wilson (2001) documented the electrocutions of 1,450 raptors representing sixteen species between 1986 and 1996. Golden eagles accounted for the largest percentage of fatalities. Data on power line electrocution are not easy to acquire, and it is logical to assume that a large number of birds, especially, are electrocuted each year on power lines, electric fences, and other electric infrastructure.

Power line collisions are also a significant source of bird crippling and death. As with electrocution, most instances of power line collision go undocumented, and often documentation occurs only for the most studied species. For example, power line collisions have been documented as a significant source of mortality for waterfowl species in many areas (Erickson, Johnson, and Young 2005).

This is an opportune time to mention an inevitable trade-off that occurs any time a habitat is transformed, lest we be charged with bias. Power lines and power poles, as anyone who has driven a country road can testify, do not only electrocute birds. They also provide perching habitats, as do grain elevators, skyscrapers, and even nuclear plants. All is relative, however, and what concerns us here is the net effect for wild animal welfare. To understand net effects, we must keep in mind what our economic infrastructure has replaced. When a forest, for example, is cleared of its trees, plowed, and fragmented by roads and power lines to feed the local economy, it is inane to conclude that economic growth was good for birds because power lines provide perches. The effects of economic growth on wild animal welfare must be considered in the aggregate.

Wind farms, seen as a great hope for "green" economic growth, are the newest gauntlets in the routes of migratory birds. Wind farms are often situated in areas where winds are favorable not only to harvesting for energy but also to birds for migrating. Substantial bird death and injury is inevitable. For example, wind turbines at Altamont Pass, California, kill approximately one thousand birds of prey per year, including hundreds of red-tailed hawks, burrowing owls,

American kestrels, great horned owls, ferruginous hawks, and barn owls. Birds of more than forty species have been killed at this single wind farm (Center for Biological Diversity 2006).

Outdoor recreation is another threat to species and may be classified as a distinct economic sector with many subsectors, including hunting, fishing, hiking, biking, four-wheeling, boating, and bird watching. Americans spent $108 billion in 2001 on wildlife-related outdoor recreation (US Fish and Wildlife Service 2002). Clearly, these various forms of outdoor recreation vary dramatically in their impact on wild animals, but most typically, the direct threat of outdoor recreation to wild animals is trampling, killing, or disturbance. Outdoor recreation constitutes the fourth most prominent cause of species endangerment in the United States (Czech, Krausman, and Devers 2000).

Pollution is an inevitable by-product of economic production. Along with the goods and services produced in an economy, pollution may be classified in economic terms as "coproduction." Pollution is an insidious, ubiquitous, and constant threat to wild animals, who are mostly helpless to understand when a pollutant has permeated their environment, what the pollutant may do to them, and how to avoid the pollutant, if indeed avoidance is possible. Whether it be respiratory failure stemming from pesticides, bone loss from lead poisoning, or ataxia (loss of coordination) from organic chemicals, or any symptom from a long, harrowing list, pollutants ensure some of the most torturous deaths in the animal kingdom (Sheffield, Sullivan, and Hill 2005). All else equal ("ceteris paribus," as the economist would say) economic growth means more such torture, more such death.

Nonnative invasive species, which disperse largely as a function of international trade and interstate commerce (Erickson, Johnson, and Young 2005), constitute one of the biggest and most rapidly growing threats to ecological integrity and animal welfare. Most wild animals, including native species in pristine environments, live lives of frequent or even constant danger. However, adaptation and evolution have equipped them to deal with other species in their natural ecosystems, and the very existence of a species is an indication of evolutionary success. However, when a totally foreign species is introduced via ship ballast, cargo plane, or railway car, native species may suddenly find themselves in a nightmarish ecosystem, occupied by one or more species before which they are defenseless. Sea lampreys slowly sucking the life out of lake trout, mice eating seabird chicks alive, and, most recently, giant pythons in Florida, constricting unsuspecting, slow-reacting animals—the fisherman's hook and the hunter's bullet are merciful in comparison.

Climate change is another threat to species (Malcolm et al. 2006), although its mechanisms are less direct. Temperature is a key variable in ecological functioning and species composition. Global warming is "pushing" polar species (such as polar bears) off the ends of the earth and creating unprecedented niches near the equator that will only be filled through the slow process of evolution. It has also been implicated in increased incidences of human and wildlife diseases (Harvell et al. 2002). Climate change is largely a function of greenhouse gas emissions from the burning of fossil fuels. The large, industrialized economies are primarily fossil fueled; therefore, climate change is also a function of economic growth. This is the real "inconvenient truth," the eight-hundred-pound gorilla in the room when climate change is discussed.

The threats to wild animals are essentially a who's who of the human economy. This is readily explained using basic principles of ecology. The principle of competitive exclusion, for example, states that no species succeeds except at the expense of other species with overlapping niches. Due to the tremendous breadth of the human niche, which expands via new technology, the human economy grows at the competitive exclusion of wild animals in the aggregate.

Another relevant aspect of ecology is trophic theory. The entire "economy of nature" (the production and consumption activities of nonhuman species) is founded upon the producers, or plants, that produce their own food via photosynthesis. Primary consumers, or animals that eat plants, constitute the next trophic level. Secondary consumers prey on primary consumers, and so forth. Mixed throughout this trophic system are "service providers" that are not readily categorized in trophic levels. These include decomposers, scavengers, and parasites. In addition, many species that do fit neatly into a particular trophic level also provide incidental services such as pollination, soil aeration, and nutrient cycling.

For our purposes, perhaps the most important thing to be gleaned from trophic theory is that the size of the entire enterprise, the whole economy of nature, depends on the size of the producer trophic level. Growth in the economy of nature requires growth of the producer trophic level. It requires an increase in primary production (i.e., photosynthesis). There is a limit to the size of the economy of nature imposed by primary production, which in turn is limited by solar energy and the availability of resources such as soils, minerals, and water.

The human economy, too, has a trophic structure, with the entire enterprise founded on agricultural and extractive surplus. As Adam Smith pointed out in *The Wealth of Nations*, the origins of money are in agricultural surplus. "No food, no stock market," we might say, along with no video games, no outdoor recreation, no sports, and so on. The economy is an integrated whole consisting of many and diverse sectors, but none of them grows without concomitant growth in some or all of the others. Most important, more agricultural and extractive surplus is required for the growth of the economy at large.

Philosophically, some prefer to classify humans within the economy of nature, in which case they clearly constitute the very highest trophic level. They are the supercarnivores. They can acquire for consumption virtually anything edible to them and are rarely threatened themselves by predators. As the trophic level comprising humans expands in biomass, it exerts "trophic compression" on the lower trophic levels that comprise the rest of the economy of nature. In other words, the growing human economy puts the squeeze on most other species.

Economic Growth as National Policy

Economic growth is a high priority in the domestic policy arena of virtually every nation, indeed the highest priority in most. In the United States, economic growth has been an explicit bipartisan goal since the Great Depression. There is still a significant populace in the United States living in poverty, and instead of instituting progressive reforms for redistributing wealth, federal and state governments have more often adopted supply-side economics and the logic that "a rising tide lifts all boats." Unfortunately, supply-siders have failed to acknowledge a limit to the supply of "water" or the number of "boats" in the "tide."

Critics of economic growth as a national goal in the United States tend to be suppressed, censored, and censured, and their arguments get very little media attention. Therefore, the American public seldom hears about environmental threats in terms of economic growth. Nor does it hear about the inhumane treatment of wild animals that accompanies economic growth.

Ever since President Eisenhower warned Americans of the "military-industrial complex" in his famous farewell address (Eisenhower 1961), political scientists have talked about "iron triangles." An iron triangle consists of a special interest group, a political faction, and a profession or professional society that is well represented in one or more government agencies. Iron triangles dominate policy arenas and fend off all comers. They are not necessarily conspiratorial, and probably seldom are, but they are extremely effective in charting the course of public policy.

In the United States, the iron triangle most relevant to the conflict between economic growth and the humane treatment of wild animals is a juggernaut. The "special interest" is the corporate community at large, and the political "faction" is the political community at large. The corporate community is concerned primarily with profits and is served by a policy of aggressive growth, while the campaign-financing system ensures political fealty to the corporate community (Korten 2001). The third side of the macroeconomic iron triangle is conventional or "neoclassical" economics, which feeds the politicians the expedient theory of unlimited economic growth and the corollary that there is no conflict between economic growth and environmental protection. The neoclassical theory of unlimited growth also helps maintain "consumer confidence."

Professional natural resource societies are also beginning to scrutinize neoclassical economics and the implications of economic growth for conservation. For example, the Wildlife Society (2003, 2) published a technical review on economic growth that described a "fundamental conflict between economic growth and wildlife conservation" and adopted a position on economic growth. The US Society for Ecological Economics, American Society of Mammalogists, and the North America Section of the Society for Conservation Biology have each taken strong positions on economic growth. The American Fisheries Society, Ecological Society of America, and American Society of Mammalogists have all been considering related positions. Hopefully the organizations concerned with animal welfare will supplement or participate in these efforts.

The Center for the Advancement of the Steady State Economy (CASSE, http://www.steadystate.org) plays a leading role in this movement. For example, the CASSE position on economic growth is often used as a template from which economic growth positions are developed. As of late 2010, over six thousand individuals (including many top scientists) and 140 organizations have endorsed the CASSE position on economic growth.

The Steady State Economy as an Alternative to Economic Growth

With economic growth as a primary policy goal and the greatest of threats to wild animal welfare, it behooves us to consider alternatives to growth. The basic alternatives are not so difficult to identify, for there are but two: economic degrowth and the steady state economy. Degrowth is also known as "recession" and is hardly a viable policy goal. (It is, however, inevitable if growth is pursued too long.) The steady state economy, on the other hand, has sociopolitical potential, especially in a world of climate change, resource shortages, and financial crises that threaten the credibility of neoclassical economics.

A steady state economy is an economy with stabilized (mildly fluctuating) population and per capita consumption. The result is a steady rate of production and consumption of goods and services in the whole. All else equal, a steady state economy is indicated by stabilized GDP. A steady state economy provides for a stable, secure, base of habitats that are required for the humane treatment of wild animals. Just as consumers and economic policy makers can steer an economy toward economic growth, they can do likewise toward a steady state economy.

Czech and Daly (2004) addressed frequently asked questions about the steady state economy. Perhaps the most relevant question for our purposes is, "How big should a steady state economy be?" This question naturally generates discussion about the capacity of the ecosystem, as if we would want to grow the economy to the limit. However, for our purposes, we can ask a different question: how much wild animal welfare do we want? Presumably many animal protection advocates would answer, "As much as possible of what is left." This gives us the answer to the original question, because maintaining as much wild animal welfare as possible requires the establishment of a steady state economy as soon as possible and as close to the current size as possible.

Ceteris paribus, then, there is an optimum size of the economy for society as a whole. There is also an optimal size, and certainly a smaller size, from the perspective of the humane treatment of wild animals. Humane treatment was not a pressing concern in primitive economies emerging from the wilderness. As the economy grows, however, natural capital is liquidated, wildlife habitats are lost, and wild animal welfare declines. Society begins devoting fiscal resources to conserving wildlife habitats and tending to wild animal welfare, and humane societies thrive. As vast areas become devoid of wildlife, however, there is less wild animal welfare to protect. For those concerned with the humane treatment of wild animals, the time for advocating a steady state economy is now.

Economic Growth and Animal Protection

Readers are now familiar with a sequence of logic pertaining to the humane treatment of wild animals. (1) Wild animal welfare requires wildlife habitats. (2) Economic growth occurs at the expense of wildlife habitats. (3) Stabilization of wildlife habitats, and, therefore, the humane treatment of wild animals, requires the establishment of a steady state economy. It remains only to consider some of the means available to animal protection advocates for pursuing the establishment of a steady state economy.

Fortunately, animal protection advocates do not have to start from scratch. Wildlife ecologists, conservation biologists, and ecological economists have been developing solidarity on this issue, informally for many years, and formally with policy statements on the conflict between economic growth and environmental protection. In other words, animal protection advocates have a foundation of professional, scientific findings and positions to stand on in educating the public and policy makers on the threat of economic growth to wild animal welfare. This is a crucial distinction from, for example, the efforts of Friends of the Earth in the 1970s. Friends of the Earth did a remarkable job of raising Americans' awareness of the perils of economic growth to the environment and wildlife, garnering coverage in such mainstream media as *U.S. News and World Report*, yet the effort seemed not to resonate in the American psyche and certainly made even less of an impact in the public policy arena. Why?

One major reason was that Friends of the Earth had no backing from the professional, scientific organizations that have established credibility over the decades with the public and politicians. That situation has changed, and we can hope that Friends of the Earth retrenches and once again confronts the eight-hundred-pound gorilla of economic growth, along with other key conservation organizations such as the National Wildlife Federation, Defenders of Wildlife, and the World Wildlife Fund.

Yet none of those organizations will bring to the table in prominent, urgent fashion the plight of individual, innocent wild animals who are crushed under the plow, poisoned by pollution, or summarily displaced by the roads, factories, and commercial metropolises that comprise our economies. It is left to animal protection organizations such as, for example, The Humane Society of the United States, The Fund for Animals, the International Fund for Animal Welfare, and the Animal Welfare Institute, to occupy this unique niche. There are many reasons beyond animal welfare for developed nations, beginning with the wealthiest, to adopt steady state economies, but there are just as many commercial and political barriers. It will take solidarity on the part of those advocating a steady state economy, and the animal welfare community's involvement is paramount in developing public support. Aside from the prospects of their own children and grandchildren (prospects that are likewise threatened in the long run by economic growth), many Americans genuinely care about the humane treatment of wild animals. They just need to see how this concern conflicts with the goal and process of economic growth.

One may ask, "But what, specifically, can animal protection advocates do to help in the establishment of a steady state economy?" A thorough answer requires a book of its own, but a short answer is easy and in order. First, animal protection organizations can educate their members on the conflict between economic growth and the humane treatment of wild animals. Use CASSE resources for this purpose. Once their members are sufficiently conversant with the subject, animal protection advocates can begin to educate the general public, beginning with the civic groups and organizations with which they already partner on other issues. A slightly more advanced step is to develop educational campaigns in cooperation with other animal welfare groups and conservation organizations.

We can expect the public to "get it" because, when we really think about it, this is an issue of common sense. Nothing grows forever. We can't have our cake and eat it, too. We can't kill the goose that lays the golden eggs. The American lexicon is laden with pithy proverbs and apt anecdotes about the fallacies of perpetual economic growth and the perils of pursuing it. The iron triangle of economic growth will defend itself with a plethora of propaganda, but one dollar's worth of solid common sense can defeat thousands of dollars of propaganda.

When we have engaged the public's common sense, there will remain a whole world of political work toward the establishment of a steady state economy through consumer behavior and public policy. This will entail macroeconomic policy reform. Fiscal and monetary policy levers will have to be ratcheted down gradually, from the current expansionary settings to the steady state economy.

Macroeconomic policy reform is off in the future. We can't get there without widespread public education and outreach. Yet that future is something to cherish and strive for. After all, it is the only future conducive to the humane treatment of wild animals.

References

Beder, S. 2002. *Global Spin: The Corporate Assault on Environmentalism*. Rev. ed. White River Junction, VT: Chelsea Green.

Center for Biological Diversity. 2006. "Altamont Pass Wind Resource Area." http://www.biologicaldiversity.org/swcbd/programs/bdes/altamont/altamont.html.

Czech, B. 2002. "A Transdisciplinary Approach to Conservation Land Acquisition." *Conservation Biology* 16 (6): 1488–97.

Czech, B., and H. Daly. 2004. "The Steady State Economy: What It Is, Entails, and Connotes." *Wildlife Society Bulletin* 32 (2): 598–605.

Czech, B., P. R. Krausman, and P. K. Devers. 2000. "Economic Associations among Causes of Species Endangerment in the United States." *Bioscience* 50 (7): 593–601.

Czech, B., D. L. Trauger, J. Farley, R. Costanza, H. E. Daly, C. A. S. Hall, R. F. Noss, L. Krall, and P. R. Krausman. 2005. "Establishing Indicators for Biodiversity." *Science* 308: 791–92.

Daly, H., and J. Cobb. 1989. *For the Common Good: Redirecting the Economy towards Community, the Environment, and Sustainable Development*. Boston: Beacon Press.

Daly, H. E., and J. Farley. 2003. *Ecological Economics: Principles and Applications*. Washington, DC: Island Press.

Deffeyes, K. S. 2001. *Hubbert's Peak: The Impending World Oil Shortage*. Princeton, NJ: Princeton University Press.

Eisenhower, D. D. 1961. "Public Papers of the Presidents." http://coursesa.matrix.msu.edu/~hst306/documents/indust.html.

Erickson, W. P., G. D. Johnson, and D. P. Young Jr. 2005. "A Summary and Comparison of Bird Mortality from Anthropogenic Causes with an Emphasis on Collisions." USDA Forest Service General Technical Report. PSW-GTR-191.

Gaffney, M., and F. Harrison. 1994. *The Corruption of Economics*. London: Shepheard-Walwyn.

George, H. 1929. *Progress and Poverty*. New York: Vanguard Press.

Harness, R. E., and K. R. Wilson. 2001. "Electric-Utility Structures Associated with Raptor Electrocutions in Rural Areas." *Wildlife Society Bulletin* 29: 612–23.

Harvell, C. D., C. E. Mitchell, J. R. Ward, S. Altizer, A. Dobson, R. S. Ostfeld, and M. D. Samuel. 2002. "Climate Warming and Disease Risks for Terrestrial and Marine Biota." *Science* 296: 2158–62.

Heilbroner, R. L. 1992. *The Worldly Philosophers: The Lives, Times, and Ideas of the Great Economic Thinkers*. 6th ed. New York: Simon and Schuster.

Humane Society of the United States, The. 2006. "Wildlife Crossings—Wild Animals and Roads." http://www.hsus.org/wildlife/issues_facing_wildlife/wildlife_crossings_wild_animals_and_roads/.

Jones, C. I. 1998. *Introduction to Economic Growth*. New York: W. W. Norton.

Korten, D. 2001. *When Corporations Rule the World.* 2nd ed. Bloomfield, CT: Kumarian Press.

Malcolm, J. R., C. Liu, R. P. Neilson, L. Hansen, and L. Hannah. 2006. "Global Warming and Extinctions of Endemic Species from Biodiversity Hotspots." *Conservation Biology* 20: 538–48.

Mill, J. S. 1900. *Principles of Political Economy, with Some of Their Applications to Social Philosophy.* Rev. ed. New York: Colonial Press.

Nørgård, J. S. 2006. "Consumer Efficiency in Conflict with GDP Growth." *Ecological Economics* 57 (1): 15–29.

Ormerod, P. 1997. *The Death of Economics.* New York: John Wiley and Sons.

Romer, P. M. 1990. "Endogenous Technological Change." *Journal of Political Economy* 98 (October): S71–S102.

Sardar, Z., and M. W. Davies. 2003. *Why Do People Hate America?* New York: Disinformation Company.

Sheffield, S. R., J. P. Sullivan, and E. F. Hill. 2005. "Identifying and Handling Contaminant-Related Wildlife Mortality/Morbidity." In *Techniques for Wildlife Investigations and Management.* 6th ed., edited by C. Braun, 213–38. Bethesda, MD: The Wildlife Society.

Simon, J. L. 1981. *The Ultimate Resource.* Princeton, NJ: Princeton University Press.

US Fish and Wildlife Service. 2002. "2001 National Survey of Fishing, Hunting, and Wildlife Associated Recreation." Washington, DC.

Venetoulis, J., and C. Cobb. 2004. *Genuine Progress Indicator (GPI) 1950–2002* (2004 update). San Francisco: Redefining Progress.

Wackernagel, M., and W. Rees. 1996. *Our Ecological Footprint: Reducing Human Impact on the Earth.* Gabriola Island, BC: New Society.

Wildlife Society, The. 2003. "The Relationship of Economic Growth to Wildlife Conservation." Wildlife Society Technical Review 03-1. Bethesda, MD: The Wildlife Society.

14

Conservation, Biodiversity, and Tourism in New Zealand

Engaging with the Conservation Economy

Eric J. Shelton

The Conservation Economy: Where We All Are Now

IN NEW ZEALAND, there has been a radical reordering of how things are. Since 1984, this country has embraced neoliberal economic reform, including the recent launching, by the minister of conservation, of the Conservation Economy:

> When I talk of the "conservation economy," the danger here is that some will incorrectly read into that phrase a lack of appreciation of the traditional and intrinsic conservation values—running the whole gamut from the preservationist view (and there must be a place in this wonderful country for the preservationist view to hold sway) to more mainstream public views. But in trying to highlight the very considerable economic stake New Zealand has in conservation, my real purpose is to broaden the long-term level of public support for conservation. I want to see New Zealanders encompass the conservation economy, so that they come to appreciate that conservation is not only in our hearts and minds, but is also the lungs of our economy. (Groser 2009, 2)

At the heart of the minister's comments is a call for a change in perception of the conservation estate, which comprises about one-third of the country's area. The speech had been foreshadowed by a report on "the value of conservation: what does conservation contribute to the economy" (New Zealand Department of Conservation 2006). While acknowledging a place for preservationist values to continue to be represented, most likely through parts of some existing national parks, the minister signals a move from intrinsic valuation of the estate to extrinsic valuation: what are the ecosystem services delivered and how is tourism serviced? These questions fit within a larger debate on "what is land for?" (Winter and Lobley 2009), which can be approached as a question centered in ethics, made all the more difficult "because the core purpose of land is no longer self-evident" (Carruthers 2009, 293).

> The most easily understood connection between conservation and the economy is through tourism. . . . Our environment is a huge part of our brand. Climate change awareness, resource shortages, and intolerance of environmental degradation are playing a part in the choices that tourists make—especially those from wealthier markets. . . . The government will work to protect the resources that tourism providers rely on—clean air, clean water, and unique landscape. . . . The logic is simple enough. Healthy natural biodiversity means healthy ecosystems, and healthy ecosystems deliver well-functioning ecosystem services. Together these things form natural capital." (Groser 2009, 2)

The text of this speech on "the conservation economy," yet another discursive form "in a world suffused in discursive forms" (Mels 2009, 385) could have been delivered to any number of audiences in any number of countries; it has the hallmarks of an approach to conservation consistent with neoliberal reform and easily sits alongside other features of *conservation for a new generation* (Knight and White 2009), specifically:

> The creation of capitalist markets for natural resource exchange and consumption . . . privatisation of resource control within these markets . . . commodification of resources so they can be traded within markets . . . withdrawal of direct government intervention from direct market transactions; and . . . decentralization of resource governance to local authorities and non-state actors such as non-governmental organizations (NGOs). (Fletcher 2010, 172)

The idea of capitalist markets for natural resource exchange has been well described:

> A common response to the misuse, abuse, or misdirection of market forces is to call for a retreat from capitalism and a return to heavy-handed regulation. But in addressing these problems, natural capitalism does not aim to discard market economics, nor reject its valid and important principles or its powerful mechanisms. It does suggest that we should vigorously employ markets for their proper purpose as a tool for solving the problems we face, while better understanding markets' boundaries and limitations. (Hawken, Lovins, and Lovins 1999, 260)

The stated goal of "better understanding markets' boundaries and limitations" is one of those neoliberal ambitions that prove difficult to critique; on the face of it they seem sensible, logical and uncontentious, but nonetheless arouse suspicion when applied to situations involving ecosystem services

(Kumar and Kumar 2008) or the possible exhaustion of irreplaceable assets (Henry 1974; Sukhdev 2008). Market approaches may be framed within the question "when does it pay to be green?" (Boons 2009, 7). By the time market understanding is achieved there may well have been avoidable and irreparable ecosystem damage caused; for example, widespread species extinction events. Just as "the emergence of life changed the Earth to make it more favourable to the survival of life" (Pretty 2007, 5), so may major changes to Earth make it inimical to life forms that we have come to take for granted (Eldredge 2010). Morton, in his promotion of *the ecological thought*, is suspicious of *natural capitalism*:

> Environmentalism, inheriting economic ideas from the long eighteenth century, runs the risk of being a rebranded version of regular economics. Paul Hawken's "natural capitalism" takes account of a wider view, without changing the basic model. . . . Since it looks like capitalism is about to use an ecological rhetoric of scarcity to justify future developments, it is vital that we recognize that there are serious problems with imagining an ecological view based on limits. . . . What if the problem were in fact one of a badly distributed and reified *surplus*. (Morton 2007, 109)

These concepts of scarcity and surplus currently are being examined more closely:

> scarcity is not merely a natural phenomenon that can be isolated from planning models, allocation politics, policy choices, market forces and local power, social and gender dynamics. . . . *Homo economicus* is neither universal nor desirable. The scarcity postulate (in other words, that needs, wants and desires are unlimited and the means to achieve these are scarce and limited) that unpins modern economics need not be universal. Needs, wants and desires do not have to be endless and unlimited . . . "scarcity" has emerged as a totalizing discourse in both the north and south with science and technology often expected to provide solutions, but such expectations embody a multitude of unexamined assumptions about the nature of the "problem," about the technologies and about the so-called institutional fixes that are put forward as the "solutions." (Mehta 2010, 2)

Mels (2009), in *Analysing Environmental Discourses and Representations*, notes that "environmental discourses draw attention to how the production, circulation and justification of meaning within particular constellations of power permeate all social practices and thereby always enter into the constitution of the biogeophysical environment" (387). This comment is important in that the Conservation Economy may be treated as one such strategic form of representation, one with its own set of implicit relationships with scarcity and surplus (Monk 2007).

Nature and the Conservation Economy

What does it mean to ignore nature? How have we modern humans ignored nature? Specifically, does the Conservation Economy ignore nature? This essay is about conservation, biodiversity, and tourism in New Zealand, eventually mentioning a project at Long Point intended to recreate and restore sea birds' colonies. Nature is central to each of these fields, but I argue that it is not the same nature in each case. Nature can be different depending on what purpose it serves (Clayton and Opotow 2003). Mine is not a new argument; rather, it is an application of a well-articulated ongoing debate, set out in part in *The Great New Wilderness Debate* (Callicott and Nelson 1998) and *The Wilderness Debate Rages On* (Nelson and Callicott 2008). My material is intended to be accessible to a wide readership, but of necessity it must engage with philosophical issues, since, despite Roderick Nash's (2001) claim that "what was new as the end of the twentieth century approached was the realization that the future of wilderness on earth depended as never before on the promulgation of a convincing philosophy" (257), "the wilderness debate," as presented in the two volumes mentioned, has failed to overcome a philosophical impasse. One philosophical position fits within poststructuralist thinking and "one aspect of poststructuralism is its power to resist and work against settled truths and oppositions" (Williams 2005, 3–4). For example, Morton (2007) believes that nature is no more than "an arbitrary rhetorical construct, empty of independent, genuine existence behind or beyond the texts we create about it" (21–22). In a contrasting important contribution to the "debate books," Eileen Crist presented the "immanence of nature" position (500–525) and J. Baird Callicott explored ecological flux (571–600). Crist's concerns about the immanence of nature not needing to be part of any social constructivist position relate to how social constructivism allows an endless process of formulation and reformulation: "Nature becomes narrated, theorized, inventoried, and comprehended—birthed into significant existence—by human activity" (503). Callicott's concern is with how "fields of endeavor that have been informed by ecology will have to take account of the paradigm shift in ecology (from a 'balance of nature' focus [e.g., Suzuki and McConnell's 1997 *The Sacred Balance*] to a 'flux of nature' paradigm) that is now virtually complete" (571). Taken together, these two concerns underline a fear that a combination of social constructivism and a lack of a scientific model of fixity removes authoritative support for environmental protection. The concern is well founded. Mels (2009), for example, concludes "the environment is no more" since "environmental discourses are power systems, which seek to systemise, capture and fix what is constantly mediated, in process and getting away. As soon as it looks as if all the shapes are in place and audiences convinced, the environment has somehow always already made its escape, only to return in a different guise" (397). Morton (2007) is more blunt: "There is no such 'thing' as the environment, since, being involved in it already, we are not separate from it" (163).

Regardless of concerns about social construction and environmental fixity, the passion and bitterness of the United States–based debate over wilderness has not been exported to New Zealand. Turner (2007) suggested that

> the great new wilderness debate was framed as a broad critique of the American wilderness ideal. Yet, for all of the debate's merits in raising questions about the assumptions that underlie the modern wilderness ideal, many scholars over-looked the historical contingency of wilderness in American environmental thought. In many regards, however, the great new wilderness debate marked a specific reaction to the new wilderness fundamentalism of the late 1980s, not a general reaction to the more pragmatic concerns that guided wilderness advocacy during the rest of the twentieth century. . . . The wilderness politics of the last forty years . . . emerged out of unique circumstances that reflected the changing place of wilderness in American environmental thought and politics. (258–59)

The contingencies and discourse affecting environmental protection in New Zealand have elements in common with those operating in the United States and have also significant points of difference. The North American preoccupation with wilderness is not replicated here. There is a discussion document, *The State of Wilderness in New Zealand* (New Zealand Department of Conservation 2001 [Lucas 2001]), and a protected areas category of Wilderness Area, but discussions and policy making lack the passion that resulted in, and have emanated from, the passage of the United States' Wilderness Act of 1964.

> Wilderness studies have come a long way in the short history of wilderness management in this country. Indeed, wilderness studies have come further than the formal establishment of new Wilderness Areas. . . . [I]t may be time to expand the focus of wilderness . . . to consider "wild lands" of New Zealand. (Lucas 2001, xiii)

Just as, in North America, wilderness has operated metonymically for "nature," so has "the bush" functioned in this way in New Zealand.

> What do New Zealanders mean when they talk about "the bush"? What does it evoke in their imaginations? Is it a particular type of vegetation, a place to visit, something to save, a source of identity or a refuge from modern life? . . . "Going bush," the deliberate act of escaping modern life or leaving unannounced, is also a quintessential New Zealand myth and reality. . . . The bush is a bountiful starting point for a history of nature's meaning and significance in New Zealand's culture. (Ross 2008, 1)

The Conservation Economy is enmeshed with these ideas and meanings of nature, and nature and its conservation is a problematic concept economically and socially (Scandrett 2010), philosophically (Soper 1995; Jamieson 2008), linguistically (Morton 2007; 2010a, b), and as a basis for environmental analysis (Crastree 2009; Mels 2009). Jamieson's (2008) approach to the relationship between philosophy, specifically ethics, and environmental problems is that "I will assume that among their many dimensions, environmental goods involve morally relevant values, and that environmental problems involve moral failings of some sort . . . environmental problems challenge our ethical and value systems. If I am right about this . . . our moral and political conceptions will themselves become more sophisticated as a result of their confrontations with real environmental problems" (25).

Also, our political conceptions may become more sophisticated if we examine how the different ways in which the term *nature* is used. Different writers approach nature sometimes from very different theoretical positions (Reis and Shelton 2011). One helpful way of structuring how we think about the idea of nature and how various people write and talk about it is provided within environmental philosophy. Kate Soper (1995) explains how humans situate nature within one or more of three different approaches. The first approach is to view nature as a metaphysical concept, "through which humanity thinks its difference and specificity." The second approach is of *nature* as a realist concept referring to "the structures, processes and causal powers that are constantly operative within the physical world. . . . It is the nature to whose laws we are always subject, even as we harness them to human purposes, and whose processes we can neither escape nor destroy." The third and most common approach to thinking, talking, and writing about nature is as "a 'lay' or 'surface' concept . . . in reference to ordinarily observable features of the world: the 'natural' as opposed to the urban or industrial . . . animals . . . and raw materials. This is the nature of immediate experience and aesthetic appreciation; the nature we have destroyed and polluted and are asked to conserve and preserve" (125).

Soper's framework promotes analysis of the contested discourse around nature and facilitates close reading of the texts involved. Nevertheless, Soper notes that it is pointless to "attempt to expound these concepts any further in isolation from each other, since they are interlocking" (126). In particular:

> when the Green Movement speaks of nature, it is most commonly in this third "lay" or "surface" sense: it is referring to nature as wildlife, raw materials, the non-urban environment . . . but when it appeals to humanity to preserve nature or make use of it in sustainable ways it is . . . employing the idea in a metaphysical sense to designate an object in relation to a subject (humanity) . . . [D]rawing

attention to human transformation (destruction, wastage, pollution, manipulation, instrumental use) of nature . . . is, at least implicitly, invoking the realist idea of nature and referring us to structures and processes that are common to all organic and inorganic entities, human beings included. (Soper 1995, 126)

Morton (2010b), citing Darwin, elaborates on the difficulty in separating organic and inorganic forms and how "the further scholarship investigates life forms . . . the less those forms can be said to have a single, independent and lasting identity" (1).

Projects intended to slow or reverse biodiversity decline also are informed by ideas of nature, and one implication of Morton's disavowal of reified nature, that is, as a thing independent of us, is the possibility of escaping from being trapped in the dystopian present, where many "environmentalists" are trapped, in a despoliation narrative of scarcity, and moving instead toward a utopian future (Hayden and El-Ojeili 2009) where ecocities accommodate the world's burgeoning urban population, in a properly ecological manner, and the current protected area/unprotected area dichotomy gives way to a world where various kinds of protection are afforded all locations, features, flora and fauna, *on the grounds of their interconnectedness.* Morton (2007) labels this process as moving to "the ecological thought" (81). Whether or not neoliberal economics persists, in order to avoid dystopia protection will still need to be incorporated within production in a way that avoids the creation of the ecological and economic "winners and losers" that both neoliberal economic policy and "fortress conservation" create (Brockington, Duffy, and Igoe 2009). A challenge for individuals and groups engaged in slowing biodiversity decline is that "ecological criticism must politicize the aesthetic" (Morton 2007, 205), and this process must include critiquing discourses of nature and environment.

> What I suggest is that the myriad things, processes, and relations we call environment, how they work, and how we should act towards them, are inherently discursive problems. . . . The very awareness of discursive forms awakens a feeling that the environment presented "as-it-really-is" may not be all that natural, but the exact expression of manipulable, authoritative discourses. . . . The term discourse has frequently been associated with a broad range of more or less strategic forms of representation (maps, imagery, narratives) mobilized within the ongoing struggle over spaces and places. (Mels 2009, 385–87)

Is Critique Ever a Bad Idea?

Is this a good time in history to interrogate the "natures that shouldn't be ignored" of our title? This contemplation is what I call Cronon's dilemma. As Cronon (1996) discovered, offering a critique of a core nature-culture concept at a time of political crisis was as unpopular with his environmental fellow travelers as he felt it to be necessary. At this distance in time it is difficult to conceive of how challenging a notion was his claim that wilderness, a special kind of nature, was entirely of our own making. Cronon was acutely aware of the issues involved in his critiquing and was indeed, as expected, criticized for questioning the value of a shibboleth, in that instance wilderness, at a time when "a Republican-dominated Congress . . . quickly distinguished itself as the most hostile toward environmental protection in all of U.S. history" (Cronon 1996, 19). A similar dilemma exists now. Should the popular models of conservation, biodiversity and nature-based tourism as they are widely understood to operate in New Zealand, be critiqued at the very time that a neoliberal political agenda is being so forcefully pursued? Cronon's answer is "that we should be willing to question some of our own moral certainty in an effort to understand why we ourselves think of nature as we do, and why others do not always agree with us" (Cronon 1996, 21). Such times of crisis are exactly when critique is most needed in order for well-informed debate to occur.

Any critique of nature in contemporary Aotearoa/New Zealand must acknowledge the ongoing struggle over spaces and places involving essentialist notions of ethnicity that are enshrined in the Treaty of Waitangi, "whereby all activities of the Crown, and by implication the Department of Conservation, must take into account Maori aspirations" (Shelton and Tucker 2008, 202). Thus these activities are part of the structure within which academic activity, conservation effort, and tourism development are situated. The postcolonial tension created, for example, between scientific knowledge and traditional knowledge, need not prevent a fruitful attempt being made at developing a nuanced understanding of the complexities involved in providing a bicultural critical reception to the notion of a conservation economy, necessarily centered on tourism (Amoamo and Thompson 2010). Such a belief, that a change from a colonial to a postcolonial worldview will actually change the world, "is deeply rooted in the Romantic period, as is the notion of worldview itself (*Weltanschauung*)" (Morton 2007, 2). For New Zealand, where "conservation and tourism are inextricably linked" (Shelton and Tucker 2008, 198), how "nature's meaning and significance" have been constructed is central to the "tourism product" that is offered. Tourists in New Zealand are better understood as performing themselves as tourists rather than as simply gazing at sights and sites (Larsen 2010; Perkins and Thorns 2001). The same landforms, for example mountains, presented to be performed as sublime nature as part of a European tourism product may equally be presented as ancestors demanding, in part as an act of indigenous resistance (Hokowhitu 2010), a performance of respect as part of a Maori product.

Does the Conservation Economy Ignore Nature?

Implicit in Hawken, Lovins, and Lovins's (1999) promotion of *natural* capital, albeit hidden and unacknowledged, is the target of Timothy Morton's concern: the concept of nature. Discussions of the conservation economy that ignore nature, that is, that fail to acknowledge that *nature* is a problematic term that has been a source of debate and confusion, fail also to consider that different ideas of nature inform conservation, biodiversity, and tourism — key elements of the conservation economy. *Conservation* threatens to join *environment* and many other words as elements of Morton's "metonymic list" of things that comprise *nature* or stand in for *nature* but which become so numerous that the list becomes long and all-encompassing to the point of being useless.

The excerpt above, from the minister's speech, so rich in metaphors of embodiment, may be read in various ways. Those who are of a preservationist bent view the term *conservation economy* as an example of oxymoron; the landforms, flora, and fauna of the conservation estate, approximately one-third of the country's area, must be seen as public goods with intrinsic value that cannot be treated as an economic resource or be subject to economic valuation or cost-benefit analysis. This approach fits well within Soper's (1995) metaphysical concept of nature: the ways in which humans separate themselves from the rest of the planet. The minister deftly acknowledges the preservationist position, and simultaneously diminishes it with — "running the whole gamut from the preservationist view (and there must be a place in this wonderful country for the preservationist view to hold sway) to more mainstream public views" (Groser 2009). The implication here is that "more mainstream public views" will draw either on Soper's "realist" concept of nature — the "structures, processes and causal powers," or the "lay" concept of nature — wildlife, raw materials, the nonurban. However, when the minister states that "our environment is a huge part of our brand. . . . The government will work to protect the resources that tourism providers rely on — clean air, clean water, and unique landscape. . . . Healthy natural biodiversity means healthy ecosystems, and healthy ecosystems deliver well-functioning ecosystem services," he is invoking the object / subject notion of metaphysical nature. What may characterize "healthy" in this case?

> [A] healthy and sustainable environment might well be defined as *an ecosystem that persists over time; remains productive, exhibits biodiversity, resilience, and adaptivity to change; and that can be expected to continue doing so into the foreseeable future* (emphasis in original). (Wimberley 2009, 128)

The minister's speech launching the Conservation Economy doesn't so much ignore nature as fail to make explicit the dynamic mixture of theoretical positions that inform his vision. For traditional conservation organizations like the New Zealand Forest and Bird Protection Society, and groups who recreate within conservation areas, including the Federated Mountain Clubs of New Zealand, an emphasis on economic considerations in the processes of the creation, management, retention, or disestablishment of protected areas, and in particular the application of neoliberal market models to these processes, was perceived as putting the whole notion of conservation at risk (Barnett 2009; Mitchell 2009; Wards 2010). These organizations too meld the metaphysical, realist, and lay concepts of nature in their discourse, and so it is little wonder that public and private debate on the merits of the Conservation Economy all too often generate more heat than light. If we engage with Cronon's dilemma and "question some of our own moral certainty in an effort to understand why we ourselves think of nature as we do, and why others do not always agree with us," then it seems that a good first step is to apply Soper's framework when preparing to debate contentious issues. Different concepts of nature inform different positions; on the role and function of national parks and other protected areas for example (Maffi and Woodley 2010; Stolton and Dudley 2010), and in practice the concepts "are interlocking, and getting clear about one will necessarily depend on establishing the meaning of the other" (Soper 1995, 126). In real-life applications these contested concepts of nature are situated within a dynamic set of power relations, often oppositional. The biodiversity issues (Maclaurin and Sterelny 2008) involved in the representation of native versus exotic flora and fauna is one such debate (O'Brien 2006; Warren 2007), situated. Williams's (2005, 3–4) comment that "one aspect of poststructuralism is its power to resist and work against settled truths and oppositions" is apposite. Attempting to engage with Soper's concepts is inherently unsettling and is consistent with Morton's (2007, 1) notion of "properly ecological forms of culture, philosophy, politics, and art" rather than ones based on some aspect of reified nature.

Application and Conclusion: The Long Point Project

Long Point is a significant headland on the Catlins coast of New Zealand. The existing forest was converted into poor-quality pasture over a period of 130 years, until 1984 often under direct or indirect central government subsidy. As is common in New Zealand, the cleared area included public land designated *road reserve*, *scenic reserve*, and *coastal strip*, although in practice these areas were all grazed as part of the farm. Also, an area supporting a colony of yellow-eyed penguin was designated *scientific reserve*, which is the most stringent protection available and excludes public visitation. Recently, a substantial part of the point was purchased by the Yellow-Eyed Penguin Trust, an environmental NGO supported by a major corporate sponsor, and amalgamated with the existing public land to form a 50ha block. Governance and management of the site has been vested in the trust in a working partnership with the Department of Conservation. The department recently has worked to strengthen community capacity to undertake conservation work (Johnson and Wouters 2008), has described the process of developing effective partnerships between itself and community groups (Wilson 2005), and has calculated the monetary value of the time that community groups contribute to conservation (Hardie-Boys 2010). Internationally, reporting of restoration projects such as this one at Long Point is now encouraged (McDonald 2008). Ongoing public access to all but the scientific reserve is a condition of being granted governance. This process and structure fits with the "decentralization of resource governance to local authorities and non-state actors such as non-governmental organizations (NGOs)" that Fletcher (2010, 172) labels neoliberal environmentality and certainly embodies the Conservation Economy. Also, the change in land use raises the issue of whether land always is scarce or alternatively may at times constitute a poorly managed surplus. The site is to be developed to promote the production of seabird habitat and the colonization and recolonization by seabirds and will integrate tourist activity into site planning and management. Two broad development and management approaches are available: Restoration Ecology (Wilson 2004) and Reintroduction Biology (Armstrong and Seddon 2007; Seddon, Armstrong, and Mahoney 2007). These approaches are not entirely incompatible, but neither do they wholly overlap. The former emphasizes *letting nature take its course*, though not to the point of simple pasture abandonment (Cramer, Hobbs, and Standish 2008), while the latter promotes more vigorous environmental intervention. Deliberate translocation of target species is one such management technique. Debate within the professional and volunteer conservation community centers on the optimum degree of hands-on manipulation of revegetation and avian recolonization. Soper's framework facilitates informed discussion about which concepts of nature inform which management preferences and decisions.

The Long Point Project is only one of many such projects nationwide and illustrates a social movement from environmental quietism to urgent ecosystem activism (Bekoff 2010), a development that reflects an increased sense of individual agency—a key component of the neoliberal agenda. Robertson (2009) points out that opponents of the environmental application of this agenda will find "there is no clear inside to penetrate and there is no unambiguous outside from which to launch external critique" (10). Soper's three concepts of nature have the potential to engage with the Conservation Economy both from the opaque inside and the ambiguous outside.

References

Amoamo, M., and A. Thompson. 2010. "Re-Imaging Maori Tourism: Representation and Cultural Hybridity in Postcolonial New Zealand." *Tourism Studies* 10: 35–55.

Armstrong, D., and P. Seddon. 2007. "Directions in Reintroduction Biology." *Trends in Ecology and Evolution* 23: 20–25.

Barnett, S. 2009. "The Mines of Moria." *Federated Mountain Clubs of New Zealand Bulletin* 178: 4–6.

Bekoff, M. 2010. "Ignoring Nature: Why We Do It, the Dire Consequences, and the Need for a Paradigm Shift to Save Animals, Habitats, and Ourselves." *Human Ecology Review* 17: 70–74.

Boons, Frank. 2010. *Creating Ecological Value: An Evolutionary Approach to Business Strategies and the Natural Environment.* Cheltenham: Edward Elgar.

Brockington, D., R. Duffy, and J. Igoe. 2008. *Nature Unbound: Conservation, Capitalism, and the Future of Protected Areas.* London: Earthscan.

Callicott, J. 2008. "The Implication of the 'Shifting Paradigm' in Ecology for Paradigm Shifts in the Philosophy of Conservation." In *The Wilderness Debate Rages On: Continuing the Great New Wilderness Debate,* edited by M. P. Nelson and J. B. Callicott, 571–600. Athens: University of Georgia Press.

Callicott, J. B., and M. P. Nelson, eds. 1998. *The Great New Wilderness Debate: An Expansive Collection of Writings from John Muir to Gary Snyder.* Athens: University of Georgia Press.

Carruthers, P. 2009. "The Land Debate—'Doing the Right Thing': Ethical Approaches to Land-Use Decision-Making." In *What Is Land For? The Food, Fuel, and Climate Change Debate,* edited by M. Winter and M. Lobley, 293–316. London: Earthscan.

Clayton, Susan, and Susan Opotow, eds. 2003. *Identity and the Natural Environment: The Psychological Significance of Nature.* Cambridge, MA: MIT Press.

Cramer, V., R. Hobbs, and R. Standish. 2008. "What's New about Old Fields? Land Abandonment and Ecosystem Assembly." *Trends in Ecology and Evolution* 23: 104–12.

Crist, Eileen. 2008. "Against the Social Production of Nature and Wilderness." In *The Wilderness Debate Rages On: Continuing the Great Wilderness Debate,* edited by M. P. Nelson and J. B. Callicott, 500–525. Athens: University of Georgia Press.

Cronon, W. 1996. Foreword to the paperback edition. In *Uncommon Ground: Rethinking the Human Place in Nature,* edited by W. Cronon, 19–22. New York: W. W. Norton.

Eldredge, N. 2010. *The Sixth Extinction.* Retrieved December 1, 2010, from http://www.actionbioscience.org/newfrontiers/eldredge2.html.

Fletcher, R. 2010. "Neoliberal Environmentality: Towards a Poststructural Political Ecology of the Conservation Estate." *Conservation and Society* 8: 171–81.

Groser, T. 2009. "The Conservation Economy." *Research Cluster for Natural Resources Law Newsletter* 2: 2–3. Also at http://www.beehive.govt.nz/speech/address+conservation+estate+symposium.

Hardie-Boys, N. 2009. *Valuing Community Group Contributions to Conservation.* Wellington: New Zealand Department of Conservation.

Hawken, Paul, A. B. Lovins, and L. H. Lovins. 1999. *Natural Capitalism: The Next Industrial Revolution.* London: Earthscan.

Hayden, Patrick, and Chamsy el-Ojeili, eds. 2009. *Globalization and Utopia: Critical Essays.* Houndsmills and New York: Palgrave Macmillan.

Henry, C. 1974. "Option Values in the Economics of Irreplaceable Assets." *Review of Economic Studies* 41: 89–104.

Hokowhitu, Brendan. 2010. "Indigenous Studies: Research, Identity and Resistance." In *Indigenous Identity and Resistance: Researching the Diversity of Knowledge,* edited by B. Hokowhitu, N. Kermoal, C. Andersen, M. Reilly, A. Petersen, I. Altamirano-Jiménez, and P. Rewi, 9–22. Dunedin: Otago University Press.

Jamieson, D. 2008. *Ethics and Environment.* Cambridge: Cambridge University Press.

Johnson, A., and M. Wouters. 2008. *Strengthening Community Capacity to Undertake Conservation Work: Sharing Conservation Skills and Knowledge*. Wellington: New Zealand Department of Conservation.

Knight, R., and C. White, eds. *Conservation for a New Generation: Redefining Natural Resources Management*. Washington, DC: Island Press.

Kumar, M., and P. Kumar. 2008. "Valuation of the Ecosystem Services: A Psycho-Cultural Perspective." *Ecological Economics* 64: 808–19.

Larsen, J. 2010. "Goffman and the Tourist Gaze: A Performative Perspective on Tourism Mobilities." In *The Contemporary Goffman*, edited by M. H. Jacobsen, 313–51. New York: Routledge.

Lucas, P. 2001. Preface. In *The State of Wilderness in New Zealand*, edited by G. Cessford, xi–xiii. Wellington: New Zealand Department of Conservation.

Maclaurin, James, and Kim Sterelny. 2008. *What Is Biodiversity?* Chicago: University of Chicago Press.

Maffi, L., and E. Woodley. 2010. *Biocultural Diversity Conservation: A Global Sourcebook*. London: Earthscan.

McDonald, T. 2008. "Further Refining This Journal's Commitment to Restoration Practice." *Ecological Management and Restoration* 9: 1.

Mehta, Lyla, ed. 2010. *The Limits to Scarcity: Contesting the Politics of Allocation*. London: Earthscan.

Mels, T. 2009. "Analysing Environmental Discourses and Representations." In *Companion to Environmental Geography*, edited by N. Crastree, 385–39. Chichester: Wiley Blackwell.

Mitchell, R. 2009. President's Column. *Federated Mountain Clubs of New Zealand Bulletin* 178: 7–8.

Monk, Daniel. 2007. "Hives and Swarms: On the 'Nature' of Neoliberalism and the Rise of the Ecological Insurgent." In *Evil Paradises: Dreamworlds of Neoliberalism*, edited by M. Davis and D. Monk, 262–74. New York: New Press.

Morton, Timothy. 2007. *Ecology without Nature: Rethinking Environmental Aesthetics*. Cambridge, MA: Harvard University Press.

———. 2010a. *The Ecological Thought*. Cambridge, MA: Harvard University Press.

———. 2010b. "Ecology as Text: Text as Ecology." *Oxford Literary Review* 32: 1–17.

Nash, Roderick. 2001. *Wilderness and the American Mind*. 4th ed. New Haven, CT: Yale University Press.

Nelson, M. P., and J. B. Callicott, eds. 2008. *The Wilderness Debate Rages On: Continuing the Great Wilderness Debate*. Athens: University of Georgia Press.

New Zealand Department of Conservation. 2006. *The Value of Conservation: What Does Conservation Contribute to the Economy?* Wellington: New Zealand Department of Conservation.

O'Brien, W. 2006. "Exotic Invasions, Nativism, and Ecological Restoration: On the Persistence of a Contentious Debate." *Ethics, Place and Environment* 9: 63–77.

Perkins, H., and D. Thorns. 2001. "Gazing or Performing? Reflections on Urry's Tourist Gaze in the Context of Contemporary Experience in the Antipodes." *International Sociology* 16: 185–204.

Pretty, Jules. 2007. *The Earth Only Endures: On Reconnecting with Nature and Our Place in It*. London: Earthscan.

Reis, A., and E. Shelton. 2011. "The Nature of Tourism Studies." *Tourism Analysis* 16: 375–84.

Robertson, M. 2009. "Performing Environmental Governance." *Geoforum* 41: 7–10.

Ross, Kirstie. 2008. *Going Bush: New Zealanders and Nature in the Twentieth Century*. Auckland: Auckland University Press.

Scandrett, E. 2010. "Environmental Justice in Scotland: Incorporation and Conflict." In *Neoliberal Scotland: Class and Society in a Stateless Nation*, edited by N. Davidson, P. McCafferty, and D. Miller, 183–201. Cambridge: Cambridge Scholars Publishing.

Seddon, P., D. Armstrong, and R. Mahoney. 2007. "Developing the Science of Reintroduction Biology." *Conservation Biology* 21: 303–12.

Shelton, E., and H. Tucker. 2008. "Managed to Be Wild: Species Recovery, Island Restoration, and Nature-Based Tourism in New Zealand." *Tourism Review International* 11: 197–204.

Soper, Kate. 1995. *What Is Nature?* Oxford: Blackwell.

Stolton, S., and N. Dudley, eds. 2010. *Arguments for Protected Areas: Multiple Benefits for Conservation and Use*. London: Earthscan.

Sukhdev, Pavan. 2008. *The Economics of Ecosystems and Biodiversity: An Interim Report*. Brussels: European Communities.

Suzuki, David, and A. McConnell. 1997. *The Sacred Balance: Rediscovering Our Place in Nature*. Vancouver: Bantam Books.

Turner, J. 2007. "The Politics of Modern Wilderness." In *American Wilderness: A New History*, edited by M. Lewis, 243–62. Oxford: Oxford University Press.

Wards, B. 2010. Editorial. *Forest and Bird* 335: 2.

Warren, C. 2007. "Perspectives on the 'Alien' versus 'Native' Species Debate: A Critique of Concepts, Language, and Practice." *Progress in Human Geography* 31: 427–46.

Williams, J. 2005. *Understanding Poststructuralism*. Chesham, England: Acumen.

Wilson, Kerry-Jayne. 2004. *Flight of the Huia: Ecology and Conservation of New Zealand's Frogs, Reptiles, Birds, and Mammals*. Christchurch: Canterbury University Press.

Wimberley, Edward. 2009. *Nested Ecology: The Place of Humans in the Ecological Hierarchy*. Baltimore: Johns Hopkins University Press.

Winter, M., and M. Lobley, eds. 2009. *What Is Land For? The Food, Fuel, and Climate Change Debate*. London: Earthscan.

PART FOUR

HUMAN DIMENSIONS OF SOCIAL JUSTICE, EMPATHY, AND COMPASSION FOR ANIMALS AND OTHER NATURE

AMONG THE CLEAREST and most common messages throughout this book are those that stress the importance of looking at animal and conservation activism as social movements (also see, for example, Cooney 2011 and Ryan 2011) and the importance of education for getting people back in touch with nature. While some people ignore nature because they don't know they're doing so or don't care, others do so because they simply don't know the facts or don't feel part of an organized and coherent movement.

The essays in this section deal with important issues that center on "human dimensions." Among the topics considered are how we got to where we are and human perceptions and behavior regarding nature as studied in disciplines such as anthropology, conservation education, humane education, conservation psychology, and conservation social work. These, and other essays in this book, all point to the need to develop a strongly interdisciplinary field that can be called "conservation science" (Kareiva and Marvier 2010; for good examples of how this eclectic approach can be applied to a given group of animals see Musiani, Boitani, and Paquet [2009] and Brakes and Simmonds [2011]). Although globally there seem to be trends pointing to an increase in positive environmental values and beliefs, only rarely do these feelings translate into environmentally supportive actions. Apathy still prevails, and there's a serious gap between words and actually doing something (Lertzman 2011). A nationwide survey in Canada showed that 72 percent of the respondents reported "environmental values/ behavior gap" between their intentions and their actions (Kennedy et al. 2009). We need to reduce this gap, and the essays in this section point to ways to do this.

Anthropologist Barbara King begins this section by noting that early humans evolved in the context of thriving, multispecies, in-balance ecosystems, and that being with animals made us human. As we co-evolved with other species our senses, our minds, and our emotions developed in part because of our interactions with other species. We also began to think with animals, as animals became integrated in art. Animals were much more than food. But over time our relationships with other nature became unbalanced, moving from mutual and reciprocal interactions to human domination. King stresses that it's not our nature to dominate, and it has not always been this way. Now, neither we nor other animals are able to successfully adapt to rapidly changing and novel environments. And we need to do something about it now, not when it's more convenient to do so.

Next, conservation psychologist Susan Clayton is concerned with fostering environmental identity. She argues people ignore nature partly because of a perceived, and illusory, distinction between what is of relevance to humans and what pertains to nature. However, people also like nature and are predisposed to have positive emotional responses to it. Clayton stresses that we should encourage opportunities to make a connection to nature particularly within a social context because social interactions with nature make it less likely that we will artificially separate humans and nature. This connection increases the likelihood that people will care about protecting nature. She also discusses Peter Kahn's (1999) notion of environmental generational amnesia that points out that each generation is oblivious to the environmental degradation that has taken place since the time of the previous generation—the shifting baseline syndrome. It's sad to learn that a Gallup poll in 2010 showed only 31 percent of respondents said they were concerned with the extinction of plant and animal species. Clayton writes, "What explains the simultaneous concern and lack of concern for animals? In part, this reflects a long-standing disjunction between humans and nature that seems to be a dominant theme in Western and perhaps particularly American culture. Nature is defined as what is free from human influence, and when we decide to protect nature we do so by removing it from a human sphere of influence. What is out of our daily experience, however, is all too easily ignored. And when it is ignored it becomes unfamiliar, scary, and potentially disliked."

Clearly, there's a lot of work to be done. What can we do? Early childhood education is critical, and we should play off of the idea that many people are perhaps genetically inclined to form emotional connections with nature, a phenomenon renowned biologist E. O. Wilson calls *biophilia*. Conservation psychology can also help us along, as Clayton continues: "Conservation psychologists are trying to use the methodological tools, theoretical understandings, and research results from psychology to encourage a healthy relationship between humans and nature, one in which people obtain the benefits that nature can provide for physical and mental well-being and people, in turn, protect ecosystems in which animal species can flourish . . . [A]nimals bridge the gap between humans and nature and provide a powerful route for promoting both human and environmental health. By their charismatic attributes and behavior, animals attract our limited attention and remind us that nature deserves notice."

Philip Tedeschi and his colleagues focus on the role that social work can play in fostering better relationships among humans, animals, and other nature (see also Ryan 2011). They note that "the defining feature of social work is the profession's focus on human well-being in a social context, often referred to as *person-in environment,* and our mission is to enhance the health of individuals, families, and communities through action and advocacy." They also stress that being connected to the natural world is not only good for other animals but also centrally important to human health and well-being. Issues of social injustice also are apparent: "It is a grave global justice concern that those who suffer most from climate change have done the least to cause it. Developing countries bear over nine-tenths of the climate change burden: 98 percent of the seriously affected and 99 percent of all deaths from weather-related disasters, along with over 90 percent of the total economic losses (Global Humanitarian Forum 2009)." Conservation social work, in combination with conservation psychology, provides a powerful way to combat the numerous and serious problems that arise when individuals do not feel connected to the natural world in which they live.

David Johns's essay on conservation activism is a deeply thoughtful and challenging "how to do it" guide for developing a strong social movement that also is concerned with decisions about life and death, not only of individual animals but also organizations trying to make the world a better place. There is a strong social psychological bent to Johns's arguments because first and foremost, activists are human beings, often jostling for power and engaging in self-destructive behavior. Johns calls for people with varied interests and a *bold vision* to join forces: "Bold vision is also a hallmark of successful movements. Movements emerge in response to the failure of existing societal structures. Structurally rooted failures can only be addressed by bold solutions, not band-aids. Visions of the end of racism, economic and political domination, or the lethal exploitation of the natural world are visions of fundamental change. Boldness is also a tactical imperative—one can bargain down, but not up." Money can be saved when groups with common interests and visions work together. Johns's general rules of thumb for success include building a strong community, exploiting divisions among opponents, being persistent, having patience, remaining uncompromising on goals but being open to flexible means to achieve them, and understanding power. I hope that all potential and current activists will take to heart the incredibly important messages in his inspirational essay.

Mass media also plays significant roles in how animals and nature are perceived and for getting people to do something about the current state of affairs. In their essay on the cultural politics of animals and the environment in the media, Carrie Packwood Freeman and Jason Leigh Jarvis aptly claim: "The commercially driven mass media package human identity and all our surrounding environments for daily consumption in the public sphere. It is of critical importance whether they choose to ignore humanity's responsibility toward the natural world and simply have us consume it as a product, or whether they actively cultivate ecological responsibility and newfound respect toward animals as fellow sentient beings." They argue media has the strong potential to raise awareness about habitat and wildlife protection and "inspire the social change needed to reverse the destructive behaviors and beliefs that are contributing to our global ecological calamity." Media can also "change humanist worldviews and consumptive lifestyles to promote self-awareness of humanity's position as a fellow species in an ecological web in crisis." Nonetheless, to date, mass media hasn't been all that effective in getting people to be more concerned with major losses in biodiversity (Robbins 2011). It's on the lips of many people but not in their hearts, and this means there's little direct action to right the many wrongs.

Education, education, education. One doesn't have to search far or deep to recognize that education is key, for youngsters and adults. But many argue that we really need to focus on children to get the ball rolling in the right direction. Gene Myers does just this by discussing many different aspects of humane and conservation education for children. He begins most poignantly: "In a remarkable public address, the interests of children in real conservation action was articulated by twelve-year-old Severn Suzuki, David Suzuki's daughter, at the 1992 Rio Earth Summit conference: 'In my life, I have dreamt of seeing the great herds of wild animals, jungles and rain forests full of birds and butterflies, but now I wonder if they will even exist for my children to see. Did you have to worry about these little things when you were my age? All this is happening before our eyes and yet we act as if we have all the time we want and all the solutions. I'm only a child and I don't have all the solutions, but I want you to realize, neither do you! You don't know how to fix the holes in our ozone layer. You don't know how to bring salmon back up a dead stream. You don't know how to bring back an animal now extinct. And you can't bring back forests that once grew where there is now desert. If you don't know how to fix it, please stop breaking it!'" (Suzuki 1992).

Among the many other topics he considers, Myers explains the bases for children's basic predilection to acknowledge and care about animals, touching on empathy and recent neuroscientific research, including what we're learning about mirror neurons, that could include the less charismatic and the ecological extensions of the animal. He notes, "Children recognize and respond to animals in reliably patterned ways. Infants distinguish animals very early. Specific brain injuries eliminate the ability to identify or name animals but not artifacts, and vice versa . . . [and] PET scans of normal brains show areas specialized for categorizing animals versus artifacts." The question then becomes, why is that not what we always observe? Myers suggests several reasons, including the traits of the animal, situations when the animal is feared, getting caught in vying values, psychological defenses when animals are exploited, and cultural norms. He also discusses children's ability to cope with bad news and how adults can model, support, and enhance their coping abilities, participation, and resilience.

Myer's essay is bookended by some pointed moral framings about children's interest in animals and the question of "protecting" them versus involving them in a more robust and supportive social movement. David Sobel's (1996) notion of "ecophobia," that bad news can make children feel disempowered, leads educators to conclude that direct experiences with local nature and local community efforts are very helpful and necessary for building an affinity with nature. No environmental tragedies before age ten. The essays in this section lead nicely into the final five essays in which education, empathy, and compassion are highlighted as being of critical importance globally as we move on to deal with existing and future problems.

References

Brakes, Philippa, and Mark Peter Simmonds, eds. 2011. *Whales and Dolphins: Cognition, Culture, Conservation, and Human Perceptions*. London: Earthscan.

Cooney, Nick. 2011. *Change of Heart: What Psychology Can Teach Us about Spreading Social Change*. New York: Lantern Books.

Global Humanitarian Forum. 2009. *Human Impact Report: Climate Change—The Anatomy of a Silent Crisis*. Geneva: Global Humanitarian Forum.

Kahn, P. H., Jr. 1999. *The Human Relationship with Nature*. Cambridge, MA: MIT Press.

Kareiva, Peter, and Michelle Marvier. 2010. *Conservation Science: Balancing the Needs of People and Nature*. Greenwood Village, CO: Roberts and Company.

Kennedy, Emily H., Thomas M. Beckley, Bonita L. McFarlane, and Solange Nadeau. 2009. "Why We Don't 'Walk the Talk': Understanding the Environmental Values/Behaviour Gap in Canada." *Human Ecology Review* 16: 151–60. http://www.humanecologyreview.org/pastissues/her162/kennedyetal.pdf.

Lertzman, Renee. 2011. "Mind the 'Gap.'" http://www.sustainablelifemedia.com/content/column/brands/beyond_the_gap.

Musiani, Marco, Luigi Boitani, and Paul C. Paquet, eds. 2009. *A New Era for Wolves and People: Wolf Recovery, Human Attitudes, and Policy*. Calgary: University of Calgary Press.

Robbins, Jim. 2011. "For Many Species, Moving Day Has Added Stress." *New York Times*, December 19. http://www.nytimes.com/2011/12/20/science/moving-day-for-many-species-is-becoming-more -fraught.html?_r=1&scp=1&sq=robbins%20moving%20day&st=cse.

Ryan, Thomas. 2011. *Animals and Social Work: A Moral Introduction*. New York: Palgrave Macmillan.

Sobel, David. 1996. *Beyond Ecophobia*. Great Barrington, VT: Orion Society.

Suzuki, Severn. 1992. http://www.youtube.com/watch?v=uZsDliXzyAY.

15

Anthropological Perspectives on Ignoring Nature

Barbara J. King

THREE MILLION YEARS AGO, in a region now called Hadar in Ethiopia, a young female walked the grasslands. Together with others of her kind, *Australopithecus afarensis*, this female strode upright in a gait not so different from ours today. Named Lucy by the scientists who found her in 1974 (see Kimbel and Delezene 2009), she rocketed to worldwide fame because of the clues her skeleton held for how our human lineage evolved.

We know now that our lineage began millennia earlier than Lucy's day, but even so, for many of us Lucy remains a touchstone in bringing an understanding of the human family tree to life. And these australopithecines have a point to make in regard to ignoring nature, the theme of this volume. Lucy and her kind lived in a notable way, as equally deserving of notice as their bones: Without stone tools or spears, and with a brain the size of a modern ape, they survived in an ecosystem teeming with elephants, pigs, saber-toothed cats, hyenas, baboons, colobus monkeys, bison, gazelles, giraffes, and rabbits (see Efossils site).

Lucy points us to an important fact: We humans *became* human through our interaction with other animals. Knowing this fact gives us a new way to think about our responsibilities to help our hurt natural world today; it lends "deep time" support to the key realization that *as we save other animals, we'll be saving ourselves* (Bekoff, preface, this volume).

Thinking of our past in terms of the other mammals, birds, reptiles, and insects that surrounded us is common in paleoecological scholarship (e.g., Bobe and Behrensmeyer 2004; Foley 1987). Yet this perspective hasn't penetrated the popular imagination, and we don't get much help in that regard from museum exhibits or TV documentaries. With a focus on human domination of the landscape through discoveries of toolmaking, hunting, and fire keeping, the "conquering hero" myth (Landau 1984) still frames the visual presentation of our past. A deeper look into prehistory suggests that it was in significant part through our interaction with other animals that we developed our senses, our patterns of problem solving, and our emotions.

In my work I have suggested (e.g., King 2010) that the web of interspecies interactions has shaped our ancestors' lives since the very dawn of our lineage at around six million years ago. Paleoanthropologist Pat Shipman notes that from at least 2.6 million years onward, "a defining trait of the human species has been a connection with animals" (2011, 13). Shipman chose that date because it represents the onset of stone-tool-making. From this point forward, our ancestors routinely butchered carcasses with stone tools.

The date one chooses for the start point matters a lot less than the recognition that the trajectory of human evolution is fully bound up with animals. It's a powerful thing for those of us who feel a connection with animals today to know: our becoming a thinking, feeling species is in large part due to our coevolution with other species (some of them also thinking, feeling animals).

It's important to resist romanticizing our past, of course. We shouldn't imagine that our ancestors coexisted in idyllic harmony with the other animals around them. For many thousands of years, we were more often the prey rather than the predator (or even the scavenger). Sometimes the history of our physical vulnerability is written into bone; the "Taung child" skull discovered by famous fossil hunter Raymond Dart in 1924 bears marks of a predatory eagle (Berger 2006). I have often wondered what it must have been like for the australopithecines, perhaps including the child's mother, to witness the seizing of a young group member by a bird of prey.

Yet individual losses to predation may in the long run result in new strengths, because think of how evolution works: Danger presented by other animals around us would have honed our senses. It would have taught us to pay close attention to how other animals move, behave, and think.

Nor should we imagine that humans and our ancestors left no ecological footprint on the landscape until recent times. Human overhunting almost certainly contributed to the demise of large flightless birds in New Zealand, for instance, and also of the large megafauna in North America, such as mammoths, mastodons, and saber-toothed cats, at around 10,000 years ago. (The latter example is strenuously debated; see Brook and Bowman 2002 and references therein.)

It may be hard, I have found, to let go of the myth of an ecologically pristine past. Concerned with conservation of the Amazonian forest and its remarkable biodiversity, for a long time I had imagined an unbroken canopy in the preindustrial past. Anthropologists have soundly exposed the inaccuracy of such an image. Five hundred years ago the forested regions contained more people, in bigger settlements carrying out intensive farming, than the region does today (see Mann 2007).

Anthropogenic alteration of the landscape is natural. Yet as many scholars across the disciplines now emphasize (Vince 2011), damage to our ecosystems via human action in the past was to the *local* environment. What we're confronting now, in terms of animal extinctions and climate change, is on a *global* scale. We have entered the Anthropocene, "the age of Man," a term used by scientists who see humanity's "geophysical force on a par with supervolcanoes, asteroid impacts, or the kinds of tectonic shift that led to the massive glaciation" that characterized the distant past (Vince 2011, 33).

But here is the hopeful point: in seeking to right the environmental harm of the Anthropocene, we needn't seek to replicate some ecologically perfect Garden of Eden from the human past, because its existence was mythic in the first place!

The message from anthropology is different. We can learn from our prehistory, when humans and other animals coconstructed their lives via mutual interaction in accordance with conditions of the local environment. We evolving humans ate animals, thought with animals, and became emotionally attached to animals in our daily lives. Even when consumed, animals were central to our lives.

Two examples from prehistory support the points I have just made (for more, see King 2010). One of my favorite images in ancient art comes from Chauvet Cave in France. Once a bear cave, Chauvet's floor is littered with over 170 bear skulls and about 2,500 bear bones. Chauvet was used by more than bears, however; our *Homo sapiens* forebears turned it into a glorious artist's studio around 32,000 years ago.

The Chauvet paintings—the subject of Werner Herzog's visual feast of a film, *Cave of Forgotten Dreams*—focus overwhelmingly on animals. Here are rhinoceroses, bears, mammoths, lions, and horses painted on the walls. The favorite image I mentioned is different, though, because it depicts a figure that is half-woman and half-bison. The legs, hips, and pubic area are human, the head and chest, bison. From the bison portion flows an arm with human hand and fingers.

Why an ancient cave painter created this image, and what it meant to the Chauvet people, we cannot know. But across the years it telegraphs a message: Even 32,000 years ago, animals were more than just food to us. We thought about our relationship to them—a relationship that mattered to the group.

The village of Çatalhöyük in ancient Turkey is the setting for my second example. There, about 8,000 years ago, a man was buried together with a lamb (Russell and During 2006). The man's skull had been crushed; he was interred in a flexed position with three objects atop his body: a worked bird bone, a flint object, and a bone point. The lamb was right in with him, at an unusual contorted position, though separated from him by some kind of a mat or blanket.

The burial's location is telling: The man and the lamb were placed beneath a house floor, in a pit. In later years other individuals were buried there, even as people continued to live in the house above. Archaeologists interpret this choice of location as recognition by the community of some sort of kinship between the man and the lamb.

Once again, a message beams through time; here, at the early stages of village life and domestication of animals, was some type of emotional relationship between a human and a young animal. As the process of domestication proceeded across the world, involving more and more species, more and more such ties would have been forged.

We see then how far back in time go humans' thinking with other animals, and humans' feeling with other animals. We have been active copartners with other animals all along. The task now is to harness those flexible cultural adaptations, based on social learning, that define our lives as *Homo sapiens*.

The world is globalized now to an unprecedented degree, but think what would happen if each of us took responsibility for what is happening locally, and then did a little bit more. The distribution of resources in our world requires those of us with more resources to do a "little bit more" for the environment; for our evolutionary partners, other animals; and for our fellow human beings who may not have enough to eat to feed their families. One of our evolved gifts is cognitive empathy—shared with other species to be sure (Bekoff and Pierce 2009; King 2013; de Waal 2010), yet elaborated in a meaningful way, and we can put it to good use.

The key point is that an evolutionary perspective on the Anthropocene is only genuinely helpful when it is coupled with action. Here are some ways each of us may choose to help (see also Bekoff and Bexell 2010), a list crafted with an explicitly evolutionary framework in mind:

- Make a difference for animals in your local environment. Small acts matter! Move the turtle out of the busy traffic lanes, to safety. Spay or neuter the family cat or dog. Plant a tree, or protest the destruction of a patch of woodland for another mall.

- Build on our attunement, as social primates, to others in our group; teach accordingly about the Anthropocene and what we can do to change our harmful actions. Our children have evolved specifically to pay attention to what we do and what we say.

- Keep your senses open to animals. No matter where we live, each of us can observe what animals really do out in the world, and connect to animals more deeply as we recognize their ways of thinking and feeling. In writing a book about animal grief (King 2013), I have discovered what happens when I observe and read about not only our closest living relatives, the primates (a traditional task of anthropologists), but also other animals ranging from domestic goats to wild bison, from dog and cat animal companions to wild crows and ravens (see Crist this volume).

- Talk, write, and teach others about animals as individuals with distinct personalities, skills, moods, friends, allies, and loved ones. We have evolved to respond to *specific known* individuals, not abstraction concerns. Two books I discovered recently do just this in wonderful ways: *Among African Apes* (Robbins and Boesch 2011) and *Smiling Bears* (Poulsen 2009). Animal-oriented documentaries on the *Nature* series from American public television are equally fantastic in this way; full episodes are available online (Nature/PBS).

- Support conservation organizations that fund jobs for local people, especially in regions where people and wild animals suffer conflict over space and food. If people can't eat and feed their children, they are going to kill and eat wildlife, even endangered species. Wouldn't you?

- Make time for animals. Sometimes I get very caught up, as many of us do, in the demands of family and work. But the twenty-seven rescued cats under our care (mostly by Charles Hogg, my husband) here at home don't much care if I have an urgent writing deadline to meet. We are responsible for their food, shelter, and medical care, and we want to shower them with love. Their needs are important. I am inspired daily by the work of people who do so much more. *The Chimps of Fauna Sanctuary* (Westoll 2011) is a magnificent example.

- Above all, resist those who proclaim loudly and confidently that it is our human evolutionary birthright to be *the* dominant animal in the landscape. On the contrary, an anthropological perspective shows that the Anthropocene is not a natural outcome of our evolutionary trajectory.

References

Bekoff, Marc, and Sarah M. Bexell. 2010. "Ignoring Nature: Why We Do It, the Dire Consequences, and the Need for a Paradigm Shift to Save Animals, Habitats, and Ourselves." *Human Ecology Review* 17 (1): 70–74.

Bekoff, Marc, and Jessica Pierce. 2009. *Wild Justice: The Moral Lives of Animals.* Chicago: University of Chicago Press.

Berger, Lee R. 2006. "Predatory Bird Damage to the Taung Type-Skull of *Australopithecus africanus* Dart 1925." *American Journal of Physical Anthropology* 131: 166–68.

Bobe, Rene, and Anna K. Behrensmeyer. 2004. "The Expansion of Grassland Ecosystems in Africa in Relation to Mammalian Evolution and the Origin of the Genus *Homo.*" *Palaeogeography, Palaeoclimatology, Palaeoecology* 207 (3–4): 399–420.

Brook, Barry W., and David M. J. S. Bowman. 2002. "Explaining the Pleistocene Megafaunal Extinctions: Models, Chronologies, and Assumptions." *Proceedings of the National Academy of Sciences* 99 (23): 14624–27.

Efossils site on Lucy and surrounding fauna: http://www.efossils.org/site/hadar#.

Foley, Robert A. 1987. *Another Unique Species: Patterns in Human Evolutionary Ecology.* New York: Wiley.

Kimbel, William H., and Lucas K. Delezene. 2009 "'Lucy' Redux: A Review of Research on *Australopithecus afarensis.*" *Yearbook of Physical Anthropology* 52: 2–48.

King, Barbara J. 2010. *Being with Animals.* New York: Doubleday.

————. 2013. *How Animals Grieve.* Chicago: University of Chicago Press.

Landau, M. 1984. "Human Evolution as Narrative." *American Scientist* 72: 262–68.

Mann, Charles C. 2005. *1491: New Revelations of the Americas before Columbus.* New York: Knopf.

Nature/PBS. Full episodes available at: http://www.pbs.org/wnet/nature/category/video/watch-full-episodes/.

Poulsen, Else. 2009. *Smiling Bears: A Zookeeper Explores the Behavior and Emotional Life of Bears.* Vancouver: Greystone Books.

Robbins, Martha M., and Christophe Boesch. 2011. *Among African Apes: Stories and Photos from the Field.* Berkeley: University of California Press.

Russell, Nerissa, and Bleda S. During. 2006. "Worthy Is the Lamb: A Double Burial at Neolithic Çatalhöyük (Turkey)." *Paleorient* 32 (1): 73–84.

Shipman, Pat. 2011. *The Animal Connection.* New York: W. W. Norton.

Vince, Gaia. 2011. "An Epoch Debate." *Science* 334: 32–37.

de Waal, Frans. 2010. *The Age of Empathy: Nature's Lessons for a Kinder Society.* New York: Three Rivers Press.

Westoll, Andrew. 2011. *The Chimps of Fauna Sanctuary: A True Story of Resilience and Recovery.* Boston: Houghton Mifflin Harcourt.

16

Nature and Animals in Human Social Interactions

Fostering Environmental Identity

Susan Clayton

What We Ignore

AS THE PREMISE of this volume suggests, humans ignore much of what goes on in the world. This is a necessary consequence of our limited perceptual and attentional capacities. In recent years research has provided dramatic and sometimes humorous examples of what people fail to notice: gorillas walking across a basketball court, a clown on a unicycle, the fact that a person we were speaking with has become someone else entirely (e.g., Hyman et al. 2009; Simons and Chabris 1999). Wrapped up in our internal monologues and goal pursuits, we are particularly inattentive to our environmental context. Robert Gifford coined the phrase "environmental numbness" to describe this phenomenon (see Gifford 2007). More recently, Peter Kahn's (1999) notion of "environmental generational amnesia" comments on the way each generation is oblivious to the environmental degradation that has taken place since the time of the previous generation, something that is also captured in the more general concept of "shifting baseline syndrome" (e.g., Papworth et al. 2008).

This lack of attention and awareness is worrying and dangerous when it encompasses our most pressing environmental problems. Recent polls show a decline in attention to climate change, for example, among the US and British populations (e.g., Jones 2010; Rosenthal 2010). Even among those members of the public who give some thought to environmental issues, the loss of biodiversity tends to get less attention than topics such as pollution. In a recent Gallup poll, only 31 percent said they were concerned about the extinction of plant and animal species (Jones 2010).

Of course, the failure to attend to these issues reflects more than cognitive limits. Emotional defenses also come into play: denial, for example, is a common response to big problems when people aren't sure what to do about them or don't perceive themselves as being able to solve them. Another emotional response is based on group identification. People can be angered if they perceive that their ways of life and social norms are being attacked and are likely to respond by dismissing the message (about environmental problems) and/or derogating the messenger (those who suggest we need to rethink the status quo). This is one reason why environmental discussions have become so politicized in the United States. The result, however, is a public that is largely unaware, inattentive, and unconcerned about the loss of species.

What We Value

This lack of attention to animal conservation does not connote a general apathy. People do care about nature, and about animals. According to the US Fish and Wildlife Service, for example, more Americans are participating in wildlife-related recreation; while the number of hunters and anglers is falling, a growing number of people report watching wildlife (31 percent, 21 percent are birdwatchers; USFWS 2009). Even stronger are attitudes toward animals that are closer to home. Approximately two-thirds of American households include pets (Walsh 2009). As an indication of how highly pets are valued from a purely economic perspective, total US pet expenditures for 2010 are projected to reach $47.7 billion (APPA 2010).

In systematic investigations, both Kellert (1996) and Manfredo (2008) have documented various ways in which we value wildlife. These range from abstract and objective, like a scientific interest or aesthetic appreciation, through utilitarian or moralistic values, to more emotionally based values like affection for nature and affection for individual animals. Both authors describe the possibility of mutualistic relationships between humans and animals; these relationships may be more likely when the animal possesses traits that are similar to humans.

In the middle ground between the domesticated animal-as-family-member and the wild animal, the popularity of zoos suggests that people value the experience of observing animals. Zoos and aquariums are very popular destinations, attracting more yearly visitors than all sporting events combined and more than any other type of informal learning environment; visitors to the zoo describe their experience as enjoyable (Clayton and Myers 2009). Zoo visits reflect not just a value for animals but also, significantly, a value for social interaction. Some of the primary motives reported behind zoo visits include the opportunity to spend time with friends and family. Klenosky and Saunders (2008) examined the factors that were implicated in whether visitors had a good time at the zoo and found two to be predictive: enjoyment and family togetherness.

In other words social motivations underlie many of the interactions people have with animals, from thinking about their pets as friends (95 percent) and/or family members (87 percent; Walsh 2009) to bonding with family members in front of zoo exhibits or developing a sense of egalitarian relationship with wildlife.

Why the Disjunction?

What explains the simultaneous concern and lack of concern for animals? In part, this reflects a long-standing disjunction between humans and nature that seems to be a dominant theme in Western and perhaps particularly American culture. Nature is defined as what is free from human influence, and when we decide to protect nature we do so by removing it from a human sphere of influence. What is out of our daily experience, however, is all too easily ignored. And when it is ignored it becomes unfamiliar, scary, and potentially disliked (cf. Bixler and Floyd 1997).

Animals may be defined as the ultimate out-group: beings that are totally dissimilar, foreign, and unknowable. Crompton and Kasser (2009) review research showing that animal terms are often used to describe and disparage human out-groups and thus to indicate dissimilarity. Although pet owners form close attachments to "their" animals, it may be that in some cases animals who become part of the domestic sphere no longer represent wild nature. This is particularly evident to the extent that pets are treated more and more like people: wearing clothes, receiving medical care, eating from plates, and being house trained.

And yet even domesticated animals may provide an opportunity for people to be reminded of their love for the natural world. Several studies have found relationships between caring for pets and attitudes toward wild animals that attribute them with greater rights and moral standing (Kafer et al. 1992; Vining and Merrick 2006).

Why We Need to Connect Humans and Nature

People should be encouraged to make the connection between the things they value and the natural world as a whole, for three reasons. First, from an objective point of view, everything is connected: we can't draw a line between the human domain and the natural domain that will make a clear division between the domestic and the wild. Urban ecosystems are still ecosystems, and humans are still animals. Threats to one will affect the rest.

The objective truth, however, is the least important part of the issue. A second reason for making the connection is that human health is intricately dependent on the health of the natural environment. It is clear that our physical well-being requires healthy environments, including access to clean water and clean air, and food from organisms that have not become toxic due to contamination by manufactured chemicals. A growing body of evidence suggests that our mental well-being is also connected to the natural world (e.g., American Psychological Association 2009; Clayton and Myers 2009; Kuo and Faber Taylor 2004; Fuller et al. 2007; Wells 2000; Wells and Evans 2003): natural environments, and perhaps particularly healthy, diverse ecosystems, promote cognitive restoration and stress reduction. The potential benefits of relationships with animals has led to an increasing number of therapy programs in which children are given opportunities to interact with, or care for, a variety of animals.

A third reason to promote connections between the human realm and the rest of nature is in order to encourage people to care, actively, for their environment. Conserving nature and protecting species require people to act in ways that are not immediately, or obviously, connected to their own self-interest. Decades of research on prosocial behavior have shown that people are capable of profoundly altruistic behavior, sacrificing their own self-interest in order to promote the welfare of others. One of the best predictors of this type of altruism is the extent to which the person being helped is seen as similar to the helper, that, in fact, the helper is able to empathize with the one who needs help (e.g., Batson 1987).

Although most of the research has, unsurprisingly, studied humans helping other humans, several studies have shown that people are able to make this empathic connection to animals or even other elements of nature. My research has consistently found that a perception of animals as similar to humans is positively correlated with interest in protecting that animal or species (Clayton, Fraser, and Burgess 2008; Clayton, Fraser, and Saunders 2009). In an experimental manipulation, Schultz (2000) found that students who were explicitly told to take the perspective of an animal harmed by pollution showed a greater level of environmental concern than students who merely read about the problem without taking the animal's perspective. Berenguer (2007) showed a similar effect, not only when people were asked to take the perspective of a bird but also if they were asked to take the perspective of a tree!

Interestingly, a dispositional tendency toward empathy for other humans has been found to be associated with concern for animals, at least as reflected in vegetarianism (Filippi et al. 2010; Preylo and Arikawa 2008). Unfortunately, recent research also suggests that empathy as a whole may be declining (O'Brien, Hsing, and Konrath 2010). According to a meta-analysis, American college students are less likely to report that they try to take the perspective of another person than they were twenty or thirty years ago. If we are less inclined to empathize with other humans, it seems even more unlikely that we would try to take the perspective of a nonhuman animal.

What to Do

What is needed, then, is an emphasis on ways for people to make personal connections with nature. These connections can be found in opportunities to experience nature on a regular basis, through walks in the woods, bird watching, and other nonspectacular encounters with nonhuman nature. Research consistently shows that one of the strongest determinants of environmentalism as an adult is regular experience of the natural world as a child (Wells and Lekies 2006). Sheer familiarity with something tends to increase liking; in addition, repeated exposures allow us to learn more and feel more confident in our ability to navigate the wild or coexist with wildlife.

Humans are highly social animals, and their responses are correspondingly dependent on their social context. This is one reason why studies often find such disappointingly low correlations between individual attitudes or values and their behavior: conformity is a stronger determinant of behavior than principle. Connections to nature will be strongest when these connections are supported by, and have legitimacy within, a social context. Nature should be part of family outings and educational curricula. Neighborhoods should include pockets of public green space. City dwellers should encounter nature on their way to work.

Repeated encounters with nature and with animals, in social groups, allow people to recognize their connections with nature and their similarities with animals in a social context that validates those intersections, perhaps by providing experiences of shared emotional responses (e.g., Kals, Schumacher, and Montada 1999). The social context, in turn, can encourage people to link those emotional responses to a sense of responsibility and action.

Environmental Identity

My recent work has explored the concept of *environmental identity*: a sense of oneself as interdependent with the natural world, a self-concept that encourages cognitive and emotional connections between self and nature. Although everyone has the potential to have an environmental identity, just as everyone has the potential for an identity based on race or gender, people vary in the extent to which they consider it an important part of their self-definition. Using a questionnaire that was developed to measure this construct, I have found that people high in environmental identity are likely to have proenvironmental attitudes and to behave in more environmentally sustainable ways (Clayton 2003). They are also more sympathetic to the needs of animals and likely to support animal rights (Clayton 2008).

An environmental identity can be most effectively developed within a supportive social network. Early childhood experiences in nature are even more significant when they include a social component, such as the presence of a family member or other significant relationship (Chawla 1986; Kals, Schumacher, and Montada 1999); sharing nature with others provides a richer context for the interpretations of nature, emotional experiences, and self-understandings that constitute the environmental identity. Conversely, an unsupportive social context makes it harder to maintain a sense of oneself as a part of nature.

Fortunately, research suggests that many people value the opportunity to interact with others in a natural environment and experience nature in the presence of others—to develop, in other words, what Fraser has described as an *environmental social identity* (Fraser et al. 2009). An environmental social identity situates the sense of self as interdependent with nature within a social context that makes it a basis for a group identity as well: we are connected by our love of nature, care for animals, or work to promote conservation. Such an identity allows people to create or strengthen social ties based on shared values for nature, to gain social capital based on understanding of nature, to adopt a social role that emphasizes interactions with nature, and to enhance self-esteem and self-efficacy by working with others to take actions on behalf of nature.

In research on the environmental social identity of zoo volunteers, Fraser et al. (2009) found that both love of animals and the desire for social interaction were among the reasons to volunteer. Importantly, volunteers reported becoming more aware of, and committed to, conservation as a result of volunteering because of the ways in which their interactions with others solidified and amplified the initial support for animal conservation.

Social Interactions in Zoos

Zoo volunteers, as in the Fraser et al. (2009) study, are more committed to the protection of animals than the average person. But many average people are also motivated to cultivate an environmental social identity—that is, at a minimum, to have nature-based social experiences in which shared values are nurtured or expressed. Many families, for example, consider camping or hiking to be important to developing family bonds. A common way to create social experiences of nature can be found in zoos. As described above, zoos are a popular leisure destination and are particularly favored for family visits.

What do zoos do? Traditionally, they have been seen as providing educational information, but they do much more. They provide experiences that are emotionally engaging and socially significant. The animals themselves naturally arouse emotional responses of awe, amusement, even fear or disgust; these shared emotional experiences provide an important basis for strengthening bonds among the observers, and the observers can use the opportunity to discuss their relationship to the natural world.

Beyond strengthened social relationships, these experiences provide an opportunity to communicate shared values and conceptualizations of the human relationship with nature. Adult visitors often describe their motives for visiting as related to fostering wildlife appreciation or learning among the others in the group (Clayton and Myers 2009). In a recent qualitative study, Fraser (2009) interviewed parents about their motivation for taking children to the zoo. He reported four themes that emerged from the responses: encouraging altruism and empathy, promoting environmental values, enhancing self-esteem, and communicating cultural norms. Parents specifically valued the way in which the zoo facilitated bonding within the family and allowed them to talk to their children about respect, responsibility, care, and appreciation for nature. There is research suggesting that the development of empathy may be fostered by such experiences. For example, Kidd and Kidd (1996), in an observational study, found that the youngest children (three to five) took an egocentric perspective to animals in a petting museum, but the older children were increasingly likely to express concern (six to eight) and empathy (nine to twelve) for the animals.

Research I have conducted in zoos shows some fascinating and creative ways in which responses to the animal exhibits are used to reflect values and create shared social experiences. In order to explore the manifestations of environmental social identities, my colleagues and I observed people as they observed a variety of zoo exhibits. In a first set of studies (Clayton, Fraser, and Saunders 2009), we were most interested in emotional responses and in reactions that indicated awareness of similarity. We found that a significant proportion of comments reflected some sense of shared identity with the animals, as indicated by statements explicitly comparing humans and animals, taking the perspective of the animal, making inferences about what the animal was thinking, or imitating the animal. Importantly, survey data showed that a sense of connection to the animal was a strong predictor of wanting to know more about the animal and to help protect the species in the wild. A sense of connection was also significantly correlated with a happy response to the animals; similarly, Tunnicliffe (1996) found that more anthropomorphic comments about animals were made in a zoo than in a natural history museum, and that the number of anthropomorphic comments was related to the number of comments about liking the animals.

The tendency to imitate the animal may be significant, because some interesting recent work suggests that imitation might have a direct link to empathy. In a study of neurological processes, imitation of others led to greater activation of brain areas that are associated with action representation and that are in turn linked with the emotional response system (Carr et al. 2003). Thus the ways in which children hop like kangaroos, growl like tigers, and flap their arms like birds may be more important than we think in fostering emotional responses and an environmental identity.

In a second set of studies (Clayton, Fraser, and Burgess 2008), our focus was on the ways in which the animal exhibits prompted social interactions among visitors. We observed nonverbal and verbal interactions among 409 and 396 small (family or peer) groups, respectively. The most common responses were calling attention to the animal (seen in 84 percent of groups) and pointing to the animal (89 percent). We were struck by these seemingly banal behaviors: in the vast majority of groups, the response to the animal was to use it as a stimulus for a social interaction. People were not content to view the animal as solitary individuals; they wanted to share their awe, wonder, or amusement. Other common reactions included leaning together, found among almost 40 percent of the groups, and discussing the animal, found among almost 80 percent. (We distinguished between conveying actual information and merely discussing the animal in a descriptive way.) Thus the animal exhibits served to bring people together, physically and through conversation.

A further important type of response was a positive comment, which occurred among 47 percent of groups. This suggests that the exhibit was not just prompting a social interaction, but allowing the groups to affirm a shared value, that of appreciation for the animals. Some of the comments specifically articulated a human relationship to the animals, including, "She loves him!" "He's just like a human!" and "Don't you want to see your little cousin?" Overall, the observations indicated that at least some visitors were using their trip to the zoo to strengthen social bonds and to extend those bonds to at least loosely encompass the zoo animals.

This focus on positive experiences at the zoo should not obscure some of the tensions and ethical concerns associated with zoos. As Milstein (2009) recently described, zoos embody a dialectic between human mastery of nature and harmony with nature; between exploitation and idealism of nature; and, most centrally to this essay, between othering of animals and connection to animals. Thus we should be mindful of the extent to which certain zoo practices may promote disconnected and exploitative attitudes toward animals. However, as Milstein states, "ending zoos may miss an important opportunity to transform zoos and their reflection and production of nature-human relations" (31). Zoo practices have the potential, as I have found in my work, to encourage admiration, respect, and connection between humans and animals.

Such practices involve going beyond merely presenting the animal to situate it in contexts that are ecological, informational, and interactional. Situating it ecologically, as is already done by many zoos, involves providing a physical context of plants, geographical features, and in some cases other organisms. This allows the animal to behave in a way that is more normal, putting the focus on what the animal does as opposed to simply what it is, and may highlight the animal's needs in terms of food, shelter, and experience. Situating it informationally requires providing explicit information about the animal, not only a description of its physical appearance and habits but also information about its status in the wild. Finally, situating it interactionally means allowing zoo visitors to understand the ways in which people affect and are affected by the animal, giving them opportunities to make connections with the animal, and explaining how human activities intersect with, infringe on, or incorporate the species. This could include a discussion of how people are working to protect the animal in the wild or the things zoos do to keep the animals healthy.

From Caring to Conservation: The Need for Research

Appreciation and affection for animals, as pets or in the zoo, is quite common, and by itself it is clearly not enough. Many animal lovers find no dissonance between pampering their pets and eating animals kept in cruel conditions, or between admiring the elephants at the zoo and contributing to the destruction of their habitat. But the appreciation is a first step. We may not protect all the things we care for, but we are unlikely to protect the things we don't care for. What is important is the recognition, first, that people have the potential and even the proclivity to form emotional connections to nature—what E. O. Wilson (1984) famously characterized as biophilia—and, second, that a recognition of similarity enables empathy, which is associated with helping behavior. What we need is more research not only on ways to promote the connection but also on ways to translate the connection into actions that protect animals. Both of these links require attention to social networks and contexts and the ways in which social interactions nurture and sustain our respect for the natural world.

Zoos seem to be increasingly aware of the importance of fostering a sense of shared identity. Exhibits encourage people to recognize the similarities between themselves and other animals by, for example, describing shared physical traits and behavioral patterns. Bexell and colleagues (2009) encouraged participants in their wildlife conservation camp to take the perspective of animals, with promising effects on attitudes and behavior. But other signs may try to attract interest by stressing how bizarre and different animals are, or promote a sense of distance from the animal through the physical design of the exhibit, a strategy that may be associated with reduced interest in conservation (Clayton, Fraser, and Saunders 2009). Research can help provide information about the best way to present and describe animals in order to promote caring and concern rather than contempt or indifference.

Zoos may also be missing the opportunity to help visitors link their concerns to effective action. Although zoo visitors appear to trust and value the zoo as a source of information about environmental and conservation issues (Fraser 2009), results from at least one survey showed that visitors felt confused about what actions they could personally take to address these issues (Milstein 2009).

Conclusion: Conservation Psychology

Conservation psychologists are trying to use the methodological tools, theoretical understandings, and research results from psychology to encourage a healthy relationship between humans and nature, one in which people obtain the benefits that nature can provide for physical and mental well-being and people, in turn, protect ecosystems in which animal species can flourish. By participating in social interactions as well as representing the unknowable Other, animals bridge the gap between humans and nature and provide a powerful route for promoting both human and environmental health. By their charismatic attributes and behavior, animals attract our limited attention and remind us that nature deserves notice.

References

American Pet Products Association (APPA). 2010. "Industry Statistics and Trends." http://www .americanpetproducts.org/press_industrytrends.asp.

American Psychological Association Task Force on the Interface between Psychology and Global Climate Change. 2009. "Psychology and Global Climate Change: Addressing a Multi-Faceted Phenomenon and Set of Challenges." Washington, DC: Author. http://www.apa.org/science/ about/publications/climate-change.aspx.

Batson, C. D. 1987. "Prosocial Motivation: Is It Ever Truly Altruistic?" In *Advances in Experimental Social Psychology*, vol. 20, edited by L. Berkowitz, 65–122. San Diego: Academic Press.

Berenguer, J. 2007. "The Effect of Empathy in Proenvironmental Attitudes and Behaviors." *Environment and Behavior* 39: 269–83.

Bexell, S., O. Jarrett, X. Ping, and R. Feng. 2009. "Fostering Humane Attitudes toward Animals." *Encounter* 22: 25–27.

Bixler, R., and M. Floyd. 1997. "Nature Is Scary, Disgusting, and Uncomfortable." *Environment and Behavior* 5: 202–47.

Carr, L., M. Iacoboni, M-C. Dubeau, J. Mazziotta, and G. L. Lenzi. 2003. "Neural Mechanisms of Empathy in Humans: A Relay from Neural Systems for Imitation to Limbic Areas." *Proceedings of the National Academy of Sciences* 100: 5497–502.

Chawla, L. 1986. "The Ecology of Environmental Memory." *Children's Environments Quarterly* 3 (4): 34–42.

Clayton, S. 2003. "Environmental Identity: A Conceptual and an Operational Definition." In *Identity and the Natural Environment*, edited by S. Clayton and S. Opotow, 45–65. Cambridge, MA: MIT Press.

———. 2008. "Attending to Identity: Ideology, Group Membership, and Perceptions of Justice." In *Advances in Group Processes: Justice*, edited by K. Hegtvedt and J. Clay-Warner, 241–66. Bingley, UK: Emerald.

Clayton, S., J. Fraser, and C. Burgess. 2008. "Defining Wild Animals: Visitor Conversations in Response to Zoo Exhibits." Paper presented at the meeting of the Society for Human Ecology, Bellingham, WA. September.

Clayton, S., J. Fraser, and C. Saunders. 2009. "Zoo Experiences: Conversations, Connections, and Concern for Animals." *Zoo Biology* 28: 377–97.

Clayton, S., and G. Myers. 2009. *Conservation Psychology: Understanding and Promoting Human Care for Nature.* Oxford: Wiley-Blackwell.

Crompton, T., and T. Kasser. 2009. *Meeting Environmental Challenges: The Role of Human Identity.* Surrey, UK: WWF.

Filippi, M., G. Riccitelli, A. Falini, F. Di Salle, P. Vuilleumier, et al. 2010. "The Brain Functional Networks Associated to Human and Animal Suffering Differ among Omnivores, Vegetarians, and Vegans." PLOS ONE 5 (5): e10847. doi:10.1371/journal.pone.0010847.

Fraser, J. 2009. "The Anticipated Utility of Zoos for Developing Moral Concern in Children." *Curator* 52: 349–61.

Fraser, J., S. Clayton, J. Sickler, and A. Taylor. 2009. "Belonging at the Zoo: Retired Volunteers, Conservation Activism, and Collective Identity." *Ageing and Society* 29: 351–68.

Fuller, R., K. Irvine, P. Devine-Wright, P. Warren, and K. Gaston. 2007. "Psychological Benefits of Greenspace Increase with Biodiversity." *Biology Letters* 3 (1): 390–94.

Gifford, R. 2007. "Environmental Psychology and Sustainable Development: Expansion, Maturation, and Challenges." *Journal of Social Issues* 63: 199–212.

Hyman, I., S. M. Boss, B. Wise, K. McKenzie, and J. Caggiano. 2009. "Did You See the Unicycling Clown?" *Applied Cognitive Psychology.* doi:10.1002/acp.1638.

Jones, J. 2010. "In U.S., Many Environmental Issues at 20-Year-Low Concern." March 16. http://www.gallup.com/poll/126716/Environmental-Issues-Year-Low-Concern.aspx.

Kafer, R., D. Lago, P. Wamboldt, and F. Harrington. 1992. "The Pet Relationship Scale: Replication of Psychometric Properties in Random Samples and Association with Attitudes toward Wild Animals." *Anthrozoös* 5 (2): 93–105.

Kahn, P. H., Jr. 1999. *The Human Relationship with Nature.* Cambridge, MA: MIT Press.

Kals, E., D. Schumacher, and L. Montada. 1999. "Emotional Affinity towards Nature as a Motivational Basis to Protect Nature." *Environment and Behavior* 31 (2): 178–202.

Kellert, S. 1996. *The Value of Life: Biological Diversity and Human Society.* Washington, DC: Island Press.

Kidd, A., and R. Kidd. 1996. "Developmental Factors Leading to Positive Attitudes toward Wildlife and Conservation." *Applied Animal Behavior Science* 47: 119–25.

Klenosky, D., and C. Saunders. 2008. "Put Me in the Zoo: A Laddering Study of Zoo Visitor Motives." *Tourism Review International* 11: 317–27.

Kuo, F., and A. Faber Taylor. 2004. "A Potential Natural Treatment for Attention-Deficit/Hyperactivity Disorder: Evidence from a National Study." *American Journal of Public Health* 94 (9): 1580–86.

Manfredo, M. 2008. *Who Cares about Wildlife?* New York: Springer.

Milstein, T. 2009. "Somethin' Tells Me It's All Happening at the Zoo: Discourse, Power, and Conservationism." *Environmental Communication* 3: 25–48.

O'Brien, E., C. Hsing, and S. Konrath. 2010. "Changes in Dispositional Empathy over Time in American College Students: A Meta-Analysis." Poster presented at the American Psychological Society meeting, Boston, MA. May.

Papworth, S. K., J. Rist, L. Coad, and E. J. Milner-Gulland. 2008. "Evidence for Shifting Baseline Syndrome in Conservation." *Conservation Letters* 2 (2): 93–100.

Preylo, B., and H. Arikawa. 2008. "Comparison of Vegetarians and Non-Vegetarians on Pet Attitudes and Empathy." *Anthrozoös* 21: 387–95.

Rosenthal, E. 2010. "Climate Fears Turn to Doubts among Britons." *New York Times,* May 24. http://www.nytimes.com/2010/05/25/science/earth/25climate.html?emc=eta1.

Schultz, P. W. 2000. "Empathizing with Nature: The Effects of Perspective Taking on Concern for Environmental Issues." *Journal of Social Issues* 56: 391–406.

Simons, D. J., and C. F. Chabris. 1999. "Gorillas in Our Midst: Sustained Inattentional Blindness for Dynamic Events." *Perception* 28: 1059–74.

Tunnicliffe, S. 1996. "Conversations within Primary School Parties Visiting Animal Specimens in a Museum and Zoo." *Journal of Biological Education* 30: 130–41.

US Fish and Wildlife Service (USFWS). 2006. "2006 National Survey of Fishing, Hunting, and Wildlife-Associated Recreation." http://library.fws.gov/pubs/nat_survey2006_final.pdf.

Vining, J. 2003. "Connection to Other Animals and Caring for Nature." *Human Ecology Review* 10: 87–99.

Vining, J., and M. Merrick. 2006. "Pet Keeping, Environmental Attitudes and Behavior, and Quality of Life." Paper presented at the 14th International Conference of the Society for Human Ecology, Bar Harbor, ME. October 18–21.

Walsh, F. 2009. "Human-Animal Bonds I: The Relational Significance of Companion Animals." *Family Process* 48 (4): 462–80.

Wells, N. 2000. "At Home with Nature: Effects of 'Greenness' on Children's Cognitive Functioning." *Environment and Behavior* 32: 775–95.

Wells, N., and G. Evans. 2003. "Nearby Nature: A Buffer of Life Stress among Rural Children." *Environment and Behavior* 35: 311–30.

Wells, N., and K. S. Lekies. 2006. "Nature and the Life Course: Pathways from Childhood Nature Experiences to Adult Environmentalism." *Children, Youth, and Environments* 16: 1–24.

Wilson, E. O. 1984. *Biophilia*. Cambridge, MA: Harvard University Press.

17

Conservation Social Work

The Interconnectedness of Biodiversity Health and Human Resilience

Philip Tedeschi, Sarah M. Bexell, and Jolie NeSmith

The one process now going on that will take millions of years to correct is the loss of genetic and species diversity by the destruction of natural habitats. This is the folly our descendants are least likely to forgive us. [Edward O. Wilson 1984, 12]

ALONG WITH MOST fields of study pertaining to the human condition, social work has largely been complicit in ignoring nature. Just as many of us grew up in an era when endangered species were not something most humans were aware of and Earth's resources seemed limitless, social work professionals were not aware that human health would be in such grave danger due to our disregard for proper care of nature. Social work as a discipline has useful models and language referencing the concept of "person in environment" but has not integrated these concepts to include more than human-centric meaning and minimization.

The field of psychology was also slow to understand and include human reliance on nature but has recently embraced this connection and is now grappling with the psychological underpinnings of our adverse treatment of our natural environment, and importantly thinking of interventions that may change our destructive ways (Clayton and Myers 2009). Much of the work involves studies at the individual level, and how this might affect the collective. Social work by nature advocates for health and healthy behavior at multiple levels of society and systems, but especially advocates for community health.

Specifically, social work as a discipline must learn to ask and respond to the question of how our threatened biodiversity relates to human health and well-being. It must prepare itself to recognize the human contribution to the problems of climate change, habitat loss such as deforestation, exploitation and misuse of lands, and severe degradation of oceans and fresh water sources. Social workers are also oriented to encourage action and empower communities to the necessity for attitude and behavior change and human-derived solutions. These are the new social work front lines for advocacy of human health, social justice, and quality of life.

It is our belief that the social work profession is perfectly poised to begin a social movement toward a more sustainable, ecological, and humane human presence on Earth. We hope this essay will serve as a call to the social work profession to include preservation of nature and natural processes in professional social work training, education, and practice.

Why Social Work? Rationale for Conservation Social Work: Ecosystems and Human Health

Humans depend completely on Earth's ecosystems and the services they provide, such as clean air, food, water, disease management, climate regulation, spiritual fulfillment, and aesthetic enjoyment. Over the past fifty years, humans have changed these ecosystems more rapidly and extensively than in any comparable period of time in human history, largely to meet rapidly growing (and it is fair to say irresponsible) demands for food, fresh water, timber, fiber, and fuel (Global Humanitarian Forum 2009; Millennium Ecosystem Assessment 2005b). This transformation of the planet has contributed to substantial gains in human well-being and economic development for a few. But not all regions and groups have benefited; in fact, many have been harmed. The costs associated with these gains have only recently become apparent. We now know that we as a species have so degraded and destroyed parts of our life-support system that our health and well-being are at stake.

It is becoming increasingly clear that environmental problems cannot be resolved through technical solutions alone. This essay considers conscious and unconscious neglect and abuse of the environment as a widespread social disorder and will explore the consequences for failure to understand that human behavior has created and continues to accelerate the decline of global biodiversity that imperils all living ecosystems. What are the manifestations of these avoidant practices on human physical and psychological health? This essay encourages the social work profession and social work academic institutions specifically to confront their own intentional avoidance. Failure to do so will result in the erosion of our highest ideals as advocates for humanity. This essay examines the *person-in-environment* construct in social work and the emergent interdisciplinary field of conservation social work that is being developed at the Institute for Human-Animal Connection of the Graduate School of Social Work at the University of Denver.

The defining feature of social work is the profession's focus on human well-being in a social context, often referred to as *person-in-environment*, and our mission is to enhance the health of individuals, families, and communities through action and advocacy. Social workers often conceptualize their intervention response through the model of a systems perspective where the interconnectedness of systems is endorsed as the most accurate way in which to understand responses of people in context. Surprisingly, these important and defining social work models neglect an integrated understanding for human connection to the natural world.

Tackling the Significant Issues Facing People

Social workers are accomplished at addressing many of society's biggest and most intractable concerns as exemplified historically by confronting the social justice issues of poverty during the settlement house movement or discrimination during the civil rights movement. Today, we have the *most* pressing issue facing the human condition: the anthropogenic worldwide decline in biodiversity and subsequent threats to global stability and human health. We must encourage and empower people to gain an understanding of the interrelationships among themselves, their families, and communities, including the natural environment and its nonhuman inhabitants.

Disconnection-Solastalgia and Psycho-Terric Disease

Many colleagues have expressed their exasperation at the idea of taking on the issues of the conservation movement. They might say, "Don't we have enough problems with just a focus on human problems?" Without a doubt we have plenty to be concerned about as social workers. Most social workers report that mental health is their primary practice area with clients. But are we encountering effects on humans related to disconnection and damage to the natural world? As we spend less time in congruent relationships with our environment, does it matter? In 2003 Glenn Albrecht coined the term *solastalgia* to describe the psychological distress and vicarious trauma related to deleterious impacts on our home environment. In 2007 Albrecht examined the possibility that these environmental "traumas" had clinical significance for the well-being of persons, and a likely symptom would be the predictable defense mechanisms of denial and rationalization. He draws into question whether humans under current dire environmental circumstances are able to accurately assess their own self-destructive tendencies. The manifestations as postulated by Albrecht are that stress reactions, derived from environmental traumas and the emergence of new disorders affecting people with causative factors lying in loss of biodiversity and observing environmental violence, seems likely. These can be defined as psycho-terric disease and generally manifest as severe anxiety, depression, and psychological pain.

Global Practice and Sustainable Communities

At the University of Denver, a new model of social work is being formalized in the graduate school of social work curriculum. Through the combination of conservation psychology, community conservation social work, conservation biology, and knowledge of pressing conservation issues and human needs, we will address wildlife and nature conservation, how this affects human welfare and survival, and how communities can work together for a healthier, sustainable future.

Fundamental to equipping social workers with the knowledge and skills to be effective in all areas of the human condition is the inclusion of a comprehensive understanding of the critical importance of healthy environments and environmental forces that create, contribute to, and address problems in the everyday lives of people. This includes policy and enforcement of relevant laws that focus on land use, water, wildlife conservation, and humane treatment and care of animals and nature.

The Ecology of Social Work

Rapidly growing human demands have brought about a substantial and largely irreversible loss in the diversity of life on Earth (Millennium Ecosystem Assessment 2005a). Earth's continued sustainability (for humans) can no longer be taken for granted. Our life support system is in dangerous decline due to our shortsighted behavior. We must acknowledge our willful dismissal of ecological parameters. Climate change, loss of food and water resources and biological diversity, unchecked human population growth and overconsumption now threaten our survival, with the poorest and most vulnerable on Earth impacted most significantly, making this a most obvious social justice issue facing social work practitioners. Mawle (2010, 272) states it well, without mincing words:

> Unbridled economic development, predicated on unfettered use of resources, ever expanding energy production, and consumption in the nineteenth and twentieth centuries has damaged the world's ecological systems and human health.

Ecosystem services include Earth's provisioning of food, water, timber, and fiber; regulating services that affect climate, floods, disease, wastes, and water quality; cultural services that provide recreational, aesthetic, and spiritual benefits; and supporting services such as soil formation, photosynthesis, and nutrient cycling (Millennium Ecosystem Assessment 2005b). The human species, while buffered against environmental changes by culture and technology, is fundamentally and inextricably dependent on these services.

To think about it slightly differently, let's examine how changes in ecosystem services influence human well-being. Human well-being is assumed to include basic material for a good life, such as secure livelihoods, enough food, shelter, clothing, and access to goods; health, including feeling well and having a healthy physical environment; good social relations, including social cohesion, mutual respect, and the ability to help others and provide for children; security, including stable access to resources, personal safety, and security from disasters; and freedom of choice and action, including the opportunity to achieve what an individual values doing and being (Millennium Ecosystem Assessment 2005b). Freedom of choice and action is influenced by other constituents of well-being (notably education) and is also a precondition for achieving other components of well-being, particularly with respect to equity and fairness—this is where social workers can especially be effective.

Humans are a part of ecosystems. With humans driving, both directly and indirectly, changes in ecosystems and thereby causing changes in human well-being, it thus makes us responsible for understanding our own behavior and its ramifications. We are all connected biologically and ecologically.

Ecosystem services offer a good means for measuring Earth's condition, and we have dangerously damaged Earth's health. In a report, fittingly titled "A Safe Operating Space for Humanity," an international group of scientists revealed that humans have already overstepped three of the nine interlinked planetary boundaries, those being: climate change, biodiversity loss, and the nitrogen cycle (Rockstrom et. al. 2009). As we move further out of our safe operating space, how will we adapt? What further destruction might we be able to avoid with knowledge and, very importantly, willpower? Radical societal and environmental reforms are needed in order to ensure human and planetary health; the fields of social work and social development are needed and paramount. Next we discuss some of the main ecological disturbances impacting human health.

Climate Change and Human Health

Today, millions of people are already suffering because of climate change. The deathly silence of this crisis is a major impediment for international action to end it. [Kofi A. Annan 2009, i]

In developed countries, climate change is often still considered only an environmental problem. It is seen as a distant and intangible threat that might affect our future. A viewpoint reinforced by pictures of disappearing glaciers and polar bears, not humans, but the human images and the suffering are already provocatively and abundantly present. Ninety-nine percent of all climate-change-induced casualties occur in developing countries (Global Humanitarian Forum 2009). This is in stark contrast to the 1 percent of global emissions attributable to some fifty of the least-developed nations. If all countries polluted so little, there would be no climate change.

Here are some of the ways humans are impacted from events directly attributed to climate change.

- Food security: primarily poor people, especially children, suffer from hunger due to reduced agricultural yield, livestock, and fish supply.
- Health: diarrhea, malaria, asthma, and stroke affect more people when temperatures rise. Climate change threatens to slow, halt, or reverse progress toward reducing the spread of diseases, especially in the poorest parts of the world.
- Poverty: livelihoods are destroyed when income from agriculture, livestock, tourism, and fishing is lost due to weather-related disasters and desertification.
- Water: increased water scarcity results from more frequent and severe floods and droughts.
- Displacement: An increase in climate-displaced people has started due to sea level rise, desertification, and floods.
- Security: More people live under continuous threat of potential conflict due to migration, weather-related disaster, and water scarcity.
  (Adapted from the Millennium Ecosystem Assessment 2005b)

For those living on the brink of survival, climate change is very real and dangerous, and for many is a final step of deprivation. Where can a fisherman go when warmer sea temperatures deplete his fish stocks? How can a farmer keep animals or grow crops when the water is gone? How can a family be provided for when fertile soils and fresh water are contaminated with salt from rising seas? The impact of climate change is happening right now and requires urgent attention.

The ramifications of our shortsightedness will be mass starvation, mass migration, and mass sickness. Climate change has been called the greatest ongoing silent crisis in human history (Annan 2009). New social work policy and practice must empower vulnerable communities to cope with these challenges. We must support the wider drive for a dignified existence for all, in harmony with the environment and in safety from it.

Currently over 2.8 billion people (that's one-third of the world's population) live in areas of the world prone to more than one type of the physical manifestations of climate change (Global Humanitarian Forum 2009). Scientists have already determined that climate change causes 300,000 deaths annually (Global Humanitarian Forum 2009). This same study projects that ramifications of unchecked climate change, including severe heat waves, floods, storms, and forest fires, will cause as many as 500,000 human deaths a year by 2030. Additionally, if carbon emissions are not brought under control within twenty-five years, 310 million people will suffer health implications due to temperature rise; 20 million more will fall into poverty; and 75 million more will be displaced (Global Humanitarian Forum 2009). The number severely affected by climate change is more than ten times greater than for those injured in traffic accidents each year, and more than the global annual number of new malaria cases (Global Humanitarian Forum 2009). Those seriously affected are in need of immediate assistance.

Global Environmental Injustice

It is a grave global justice concern that those who suffer most from climate change have done the least to cause it. Developing countries bear over nine-tenths of the climate change burden: 98 percent of the seriously affected and 99 percent of all deaths from weather-related disasters, along with over 90 percent of the total economic losses (Global Humanitarian Forum 2009). Additionally, global pollutants contributing to climate change do not adhere to national or regional boundaries; they harm people regardless of where they were produced or by whom. While the poorest are hardest hit, developed nations are also affected, and increasingly so. The human cost of recent heat waves, floods, storms, and forest fires in rich countries has been alarming (Global Humanitarian Forum 2009).

Climate change threatens all eight millennium development goals (UN Millennium Development Goals 2010), all goals of which social work already plays an integral role for success to be possible. These goals are to (1) eradicate extreme hunger and poverty; (2) achieve universal primary education; (3) promote gender equality; (4) reduce child mortality and improve child health; (5) improve maternal health; (6) combat HIV/AIDS, malaria, and other diseases; (7) ensure environmental stability; and (8) develop a global partnership for development (Global Humanitarian Forum 2009). These goals all need for adaptation efforts to be developed and implemented. Humanity is facing a rare challenge, and there are no sides in the fight for climate justice.

Food Security and Human Health

Major factors impeding food security have already been discussed above but include human population and development pressures, climate change, ecosystem alteration, and the risk of deleterious tipping points within our safe operating space (Rockstrom and Karlberg 2010). The degrees of freedom for sustainable human use of planet Earth have been severely constrained. Malnutrition, starvation, and diarrhea are exceedingly common health problems worldwide and are being exacerbated by climate change, irresponsible water and soil usage, herbicide and pesticide use, and annihilation of our natural pollinators (e.g., bees and other insects, bats, and many more). Also, uneven distribution, international shipment of foodstuffs instead of local consumption, and water shortages further threaten global food security. Social work professionals can help to counter these threats by learning and imparting agricultural skills, responsible consumption, and networking with local agricultural experts.

Biological Diversity and Human Health

The loss of biodiversity is thought to be more of a threat to life on this planet than climate change (Millennium Ecosystem Assessment 2005a). Whereas we still may have a chance to reverse some of the damages we have induced through climate change, we cannot bring back extinct organisms. It is exceedingly important for global citizens to understand and accept is that biodiversity on Earth is what regulates our climate. We cannot reverse the damages to the climate without the dizzying array of organisms on Earth, each with their specific functions to keep ecosystems healthy and viable.

Most of the biodiversity tipping points will be accompanied by large negative regional- or global-scale impacts on ecosystem services and human well-being (Millennium Ecosystem Assessment 2005a). For example, the widespread and irreversible degradation and loss of natural coastal habitats due to pollution, habitat destruction, changes in sedimentation, and sea level rise will be accompanied by increased risk of coastal damage by waves and storm surges and the loss of productivity of coastal fisheries.

The links between species extinctions and reduced capacity to deliver ecosystem services are in some cases elusive, while we know this happens. The loss of biodiversity is also worrisome due to the intrinsic value of biodiversity and its pivotal role in building Earth's resilience to climate change. Examples include: Species diversity assists in strengthening the ability of fishing resources to sustain stress from overfishing. The creation of natural carbon sinks removes carbon dioxide from the atmosphere. Natural coastline barriers like mangrove forests protect coastal inhabitants from storms and soil erosion. And having an assortment of traditional seeds to help identify more drought resistant crop varieties is increasingly critical to survival in drought-prone areas. There is great cause for concern, as the Intergovernmental Panel on Climate Change (IPCC) estimates that 20–30 percent of global species are at risk of extinction this century (Millennium Ecosystem Assessment 2005a).

Human Population Growth and Human Health

Clearly, our numbers and our behaviors that hinder our use of foresight have put us in danger. Human population is predicted to peak at about eleven billion. Research shows that many environmental problems will be easier to address if world population peaks at eight billion (we reached seven billion in 2011) rather than the projected eleven billion (Cincotta, Wisnewski, and Engelman 2000; Engelman 2000; O'Neill et. al. 2010). Good news holds that there is already a global consensus on how to slow population growth, with programs that improve human well-being at very little cost (Population Action International and Population Justice Project 2010). We are at a pivotal moment for human numbers because nearly half the world's population (approximately three billion people) is under the age of twenty-five and entering their childbearing years (Population Action International and Population Justice Project 2010). Their childbearing choices, and the information and services available to them, will determine whether our numbers rise to anywhere from eight to eleven billion by midcentury. The impact of population growth on the environment is provoked or alleviated by factors such as consumption, technology, urbanization, and more. The important point being that slower population growth could reduce pressure on natural systems that are already overtaxed.

Slower population growth would help people adapt to climate change, reduce the scale of human vulnerability to these impacts, and give nations a chance to make essential investments in health care, education, and sustainable economies (Population Action International and Population Justice Project 2010). The field of social work needs programming, with related health care professionals, to provide accurate and unbiased information to families and young women on their reproductive health and well-being—for them and their children.

Animal Welfare and Human Health

The *biophilia* hypothesis, introduced by world-renowned biologist E. O. Wilson, suggests that humans are innately interested in animals due to evolutionary forces that made attention to animals beneficial for survival (Wilson 1984). Biophilia, however, does not mean that humans are instinctively kind to animals, but that they are attuned to animals. Anecdotally, we know the close connections of humans to their companion animals, such that sometimes a pet owner will prioritize the animal over their own basic needs. In 2005 some people refused to evacuate prior to Hurricane Katrina citing inability to take their pets as the reason they risked their lives and remained in New Orleans (McCulley 2007). Several Meals on Wheels programs have partnered with organizations to provide pet food along with meals delivered for low-income, homebound seniors after realizing that many were sacrificing their own nutritional needs and feeding their pets the human food (Ribeiro 2007). These examples illustrate the extent of the human-animal bond, wherein individuals feel so connected with particular animals that they disregard their own health or safety. These behaviors show that humans are certainly capable of empathy with animals and may need these relationships for optimum human health.

Spiritual, religious, and philosophical viewpoints suggest obligations to treat animals kindly, without exception of those in poverty. Faver (2009) suggests that humans recognize the connectedness of all forms of life through animal relationships and that this forms the foundation for compassion. In "A Religious Proclamation for Animal Compassion," authors from twenty-one faiths declared a need for wildlife preservation, action against pet overpopulation, homeless pets, and questionable animal practices in research, sports, entertainment, and as food (Haley 2007). A symbiosis of religion and animal compassion may be forming wherein the theological doctrine of dominion over animals is viewed as care of and for animals, rather than power over them. Besthorn (2002) takes a social work viewpoint and suggests we turn toward our ecological self-identity. He argues that people are realizing they can no longer think of themselves as separate from their environment, but rather, "humanity is part of a complex totality of interconnected relationships, and that these connections among both humans and non-humans are the very essence of existence" (Besthorn 2002, 61). From the affluent to the most impoverished, a connection with animals seems morally, philosophically, and spiritually needed for optimum functioning.

An important applied perspective comes from the field of humane education and research into the practice of cruel or abusive behavior toward animals. Findings suggest animal abuse is associated with many forms of antisocial behaviors (Arluke et al. 1999). The purposes of humane education are "instilling, reinforcing, and enhancing young people's knowledge, attitudes, and behavior toward the kind, compassionate, and responsible treatment of human and animal life" (Ascione 1997, 58). The goal is to increase empathy and compassion, which is a theme of great significance. Poverty can often create situations where cooperation and caring for others is critical and replacing animal cruelty with compassion toward animals is crucial in creating cooperative human relationships and strong communities. Increasingly, evidence would suggest that people are sustained by relationships, even more when personal vulnerability, poverty, and hunger abound. With cruelty to animals so prevalent today in terms of factory farms, the pet trade, pet overpopulation, wildlife consumption, habitat destruction, species extinctions, and more, increasing compassion for all members of our communities, including animals, will only lead to healthier humans, physically and psychologically.

Solutions and Ideas

Theory is being translated into action within the curriculum at the Graduate School of Social Work at the University of Denver. Conservation is integrated into social work courses to study the intricate relationship between healthy environments for people and animals. Students participate in an internship led by Sarah Bexell and her colleagues who run an educational camp in China, where inhumane animal practices as well as loss of wildlife and natural areas are prevalent (Bexell et al. 2009). They have found that teaching children through humane and conservation education creates attitude change and a realization among children of the importance of caring for both animals and people (Bexell et al. 2009).

The World Society for the Protection of Animals (WSPA) outlined one of their projects in "Combating Poverty: The Role of Animal Welfare Education and New Skills" (WSPA 2009). Animals near Tsavo National Park in Kenya were being poached for meat, with animals being killed illegally and painfully. The organization worked to determine the underlying factors, concluding that poverty and hunger were root causes (WSPA 2009). WSPA implemented humane education programs, helped set up beekeeping as a sustainable income source, employed former poachers as antipoaching rangers, and helped provide chili pepper fences to prevent elephants from destroying crops (Redford et al. 2008). Understanding the interconnection of animal welfare and poverty allowed the agency to establish meaningful and effective programs of change. In a similar vein, the University of Denver's Graduate School of Social Work and the African Network for Animal Welfare (ANAW), a nonprofit agency based in Kenya, team with communities to establish effective sustainable alternatives to bushmeat trade, habitat destruction, and poaching (African Network for Animal Welfare 2012). Like WSPA, promoting sustainable alternatives must be provided to the areas where human-animal conflict arises as impoverished people kill animals for income. In addition to removing snares, the program works with communities to promote conservation and animal welfare. New brick-making machines that provide a sustainable livelihood have been introduced. A cultural center was established with the Waatha community to generate income from ecotourism as locals demonstrate and preserve a wide range of cultural practices, native dances, and basket weaving skills to provide alternate sources of income. Young men who might have turned to poaching are now transforming used wire snares into art, which provide a source of income.

Social Work and Impetus for a Paradigm Shift

Despite major accomplishments in environmental justice reforms over the past decade, many millions of Americans and billions of people worldwide continue to be negatively affected by living in unhealthy and unsafe environments. Person-in-environment does not lose any potency in its application to conservation social work practice—just the contrary. At this critical juncture in our collective efforts to advocate for biodiversity protection, we must recognize how change occurs. Empowerment of local communities occurs from the bottom up and must emphasize the practical realities of where people work, grow food, eat and drink, and go to school. Social work and the supporting academic infrastructure can support these efforts by establishing best practice and evidence-based approaches to influence the way scientists, researchers, policy makers, and businesses go about thinking and interacting with the environment.

The most powerful point of intervention in the environmental movement is occurring with people on a local level, a level well suited for social worker participation. The readiness of social work to support well-informed, grassroots community resilience will be needed. Just as we are prepared to recognize the ugly face of discrimination and prejudice and openly label racial discrimination as immoral and illegal, we must prepare the profession to do the same on behalf of the living world that is our only home.

References

African Network for Animal Welfare. 2012. "Biodiversity Conservation and Animal Welfare." http://www.anaw.org/index.php?option=com_content&view=article&id=49&Itemid=66.

Albrecht, G. A. 2003. *Solastalgia: A New Concept in Human Health and Identity*. The Harvey Club (Physicians) of London, Ontario.

Albrecht, G., G.-M. Sartore, L. Connor, N. Higginbotham, S. Freeman, B. Kelly, H. Stain, A. Tonna, and G. Pollard. 2007. "Solastalgia: The Distress Caused by Environmental Change." *Australasian Psychiatry* 15 (1): S95–S98.

Annan, K. A. 2009. Introduction. *Human Impact Report: Climate Change—The Anatomy of a Silent Crisis*. Geneva: Global Humanitarian Forum.

Arluke, A., J. Levin, C. Luke, and F. Ascione. 1999. "The Relationship of Animal Abuse to Violence and Other Forms of Antisocial Behavior." *Journal of Interpersonal Violence* 14 (9): 963–75.

Ascione, F. 1997. "Humane Education Research: Evaluating Efforts to Encourage Children's Kindness and Caring towards Animals." *Genetic, Social, and General Psychology Monographs* 123 (1): 57–77.

Besthorn, F. 2002. "Radical Environmentalism and the Ecological Self: Rethinking the Concept of Self-Identity for Social Work Practice." *Journal of Progressive Human Services* 13 (1): 53–72.

Bexell, S. M., O. S. Jarrett, P. Xu, and X. Feng Rui. 2009. "Fostering Humane Attitudes toward Animals: An Educational Camp Experience in China." *Encounter* 22 (4): 25–27.

Cincotta, R., J. Wisnewski, and R. Engelman. 2000. "Human Population in the Biodiversity Hotspots." *Nature* 404: 990–92.

Clayton, S., and G. Myers. 2009. *Conservation Psychology: Understanding and Promoting Human Care of Nature.* Hoboken, NJ: Wiley-Blackwell.

Engelman, R. 2000. *People in the Balance: Population and Natural Resources at the Turn of the Millennium.* Washington, DC: Population Action International.

Faver, C. A. 2009. "Seeking Our Place in the Web of Life: Animals and Human Spirituality." *Journal of Religion and Spirituality in Social Work* 28 (4): 362–78.

Global Humanitarian Forum. 2009. *Human Impact Report: Climate Change—The Anatomy of a Silent Crisis.* Geneva: Global Humanitarian Forum.

Haley, K. 2007. "A Religious Proclamation for Animal Compassion." http://network.bestfriends .org/religion/news/20029.html.

Mawle, A. 2010. "Climate Change, Human Health, and Unsustainable Development." *Journal of Public Health Policy* 31: 272–77.

McCulley, R. 2007. "Saving Pets from Another Katrina." *Time.* June 6. http://www.time.com/time/ nation/article/0,8599,1629962,00.html.

Millennium Ecosystem Assessment. 2005a. *Ecosystems and Human Well-Being: Biodiversity Synthesis.* Washington, DC: World Resources Institute.

———. 2005b. *Ecosystems and Human Well-Being: Synthesis.* Washington, DC: Island Press.

O'Neill, B. C., M. Dalton, R. Fuchs, L. Jiang, S. Pachauri, and K. Zigova. 2010. "Global Demographic Trends and Future Carbon Emissions." *Proceedings of the National Academy of Sciences* 107 (41): 1–6.

Population Action International and Population Justice Project. 2010. *Population and the Environment: Where We're Headed and What We Can Do.* Washington, DC: Population Action International and Population Justice Project.

Redford, K., M. Levy, E. Sanderson, and A. de Sherbinin. 2008. "What Is the Role of Conservation Organizations in Poverty Alleviation in the World's Wildest Places?" *Oryx* 42 (4): 516–28.

Ribeiro, A. 2007. "Sharing the Eats: The Roanoke Valley's Meals on Wheels Program Has Added 33 Pets of Homebound Seniors to Its Distribution List." *Roanoke Times* (VA). Retrieved from Newspaper Source database.

Rockstrom, J., and L. Karlberg. 2010. "The Quadruple Squeeze: Defining the Safe Operating Space for Freshwater Use to Achieve a Triply Green Revolution in the Anthropocene." *AMBIO* 39: 257–65.

Rockstrom, J, W. Steffen, K. Noone, A. Persson, F. S. Chapin, E. F. Lambin, T. M. Lenton, M. Scheffer, C. Folke, H. J. Schellnhuber, B. Nykvist, C. A. de Wit, T. Hughes, S. van der Leeuw, H. Rodhe, S. Sorlin, P. K. Snyder, R. Costanza, U. Svedin, M. Falkenmark, L. Karlberg, R. W. Corell, V. J. Fabry, J. Hansen, B. Walker, D. Liverman, K. Richardson, P. Crutzen, and J. A. Foley. 2009. "A Safe Operating Space for Humanity." *Nature* 461: 472–75.

UN Millennium Development Goals. 2010. Fact Sheet. http://www.un.org/millenniumgoals/pdf/ MDG_FS_1_EN.pdf.

Wilson, E. O. 1984. *Biophilia: The Human Bond with Other Species.* Cambridge, MA: Harvard University Press.

World Society for the Protection of Animals (WSPA). 2009. "Case Study: Combating Poverty: The Role of Animal Welfare Education and New Skills." http://media.animalsmatter.org/media/ casestudies/en/Combatingpovertywitheducation.pdf.

18

The War on Nature—Turning the Tide?
Lessons from Other Movements and Conservation History

David Johns

The Biological and Political Landscape

AT A RECENT INTERNATIONAL MEETING on wildlands someone mused that if conservationists had led the struggle against apartheid it would likely still exist in some form. The sixth great extinction is not slowing but gaining momentum (Butchart et al. 2010; Brashares 2010). The human footprint is growing, not shrinking (Ewing at al. 2009). Our species is drawing down or destroying natural capital and commandeering a huge proportion of the interest, or net primary product (NPP)—30 percent of marine NPP and 40 percent of terrestrial (Pimm 2001).

Although important species, land, and water protections have been achieved and the ideology of conquest has waned, most human societies behave as conquerors. Leaders pay lip service to biodiversity but humans are consuming the Earth, converting more and more of the living world into commodities and collateral damage. Influential people talk about balancing growth and ecological integrity as if they were equivalents—as if the Earth, its ecosystems, and species were just another factor in production, rather than the basis for all life.

Any opportunity for balancing passed at the beginning of the Neolithic, when our population was about 10 million. And human-caused extinctions predate this period, going back to our migration out of Africa 60,000 years ago (Barnosky et al. 2004; Miller et al. 1999).

Human societies are built on growth. Capitalist, state capitalist, socialist, communist, and premodern economies are and were organized around growing human numbers and growing consumption per capita (Harris 1977, 1979; Cohen 1977; White 1987 [1975]; Wright

2004). Political and business leaders—those conservationists call on to make wise decisions—know that their power and positions depend on maintaining growth. Indeed, when growth falters and the pyramid scheme-nature of the basis for social organization is revealed, they rush to shore it up rather than call for alternatives (Wright 2004). Even conservation organizations depend on the growth of donors' stock portfolios (the most recent Giving USA report [2010] again shows donations tracking economic growth). Many conservation leaders resist acknowledging this dependence and the need to address the structural dynamic of growth, fearing doing so would be unpopular and alienate donors, because it appears unnecessary as near term goals can be accomplished without addressing the issue, and because taking growth on seems a lost cause.

Changing What's Politically Possible

For living things "survival is nothing if not biological . . . [and] perpetuating economic or political institutions at the expense of biological well-being of man, societies, and ecosystems may be considered maladaptive" (Rappaport 1976, 65). Only in the realm of magical thinking can a growing human footprint be reconciled with ending anthropogenic extinctions and ecosystem decline.

Changing societal structures is daunting. As was the case with ending apartheid, there are many powerful interests blocking the road, interests unafraid to use violence (e.g., Boykoff 2007; Davenport, Johnson, and Mueller 2005; Helvarg 2004). Many who are victimized by the status quo are frightened of change. Long before psychologists noted that some people are fearful of leaving destructive personal relationships, the US Declaration of Independence noted that people also have trouble leaving bad political relationships—they are more inclined to accept oppression than rebel because the familiar is preferable to the unknown. Moreover, many people's sense of well-being depends upon identifying with the social order and regarding it as legitimate; to think and feel otherwise calls one's own sense of worth into question (Jost, Burgess, and Mosso 2001). Human hubris also stands in the way: we resist admitting that our species is not smart enough to manage Nature (Ehrenfeld's [1979] critique remains the most cogent).

Conservationists also resist taking on structural change because it seems to require that people make material sacrifices, and this is unrealistic. But is giving up mindlessly complex social hierarchies that infantilize people a sacrifice? Emerson (1994 [1847], 63) observed that "things are in the saddle, and ride mankind." Is giving up such servitude a sacrifice? Is giving up the drudgery of so much labor to pay for toys that distract us from the drudgery a sacrifice? Despite the fantastical promises of material progress for all, there are about twice the number seriously hungry today than were alive at the beginning of the modern period (UNFAO 2010).

Despite the enormous obstacles structural change has been accomplished many times. Abolishing slavery; toppling tyrannies and apartheid; achieving labor rights, women's rights, and many other basic economic and political changes have required the transformation of societal structure. Conservationists can and must do the same if they are serious about sustaining and restoring life on Earth. Lessons from conservation's history and from the successes and failures of other movements suggest seven major attributes of successful movements for major change.

1. Clear, Bold Vision

Movement success is more likely and less costly when the groups in the movement forge a common, overarching vision and invest in maintaining it, commit to good interorganizational communication, and place strategic thinking over self-aggrandizing leaders (Aminzade, Goldstone, and Perry 2001).

Successful movements share broad common themes such as equality in the US civil rights movement or popular elections and freedom of speech in pro-democracy movements (Staggenborg 2011). Without clear and common themes internal divisions cannot be subordinated and effective strategies developed. Shared visions are never seamless—passionate advocates differ and too many leaders have fragile egos and must have their way. Maintaining unity of vision requires much effort and is a moving target. But the effort is cheap compared to the costs of allowing opponents to exploit differences (Polletta and Ho 2006; Benford and Snow 2000). Predemocratic South African security forces cost the anti-apartheid movement much blood and treasure by successfully exploiting Inkatha Freedom Party leaders' divisive ambitions (Wood 2000).

The injuries that give birth to movements—injustice, oppression, and brutality—do not by themselves generate resistance and support for alternatives (Staggenborg 2011; McAdam, Tarrow, and Tilly 2001). Cohorts of people join movements in response to calls that present a compelling and credible vision that appeals to their needs, emotions, and intellect (Johns 2005; Brader 2005; Goodwin, Jasper, and Polletta 2001; Stern 2000). The vision must describe what is wrong, what a better world looks like, and outline the path forward (Ingram and Fraser 2006; Harkin 2004; Wallace 1970; Rambo 1993). The vision must be conveyed by a powerful story in which people can find themselves (Cornog 2004; Cox 2006; Polletta 2006).

Obviously, wild places and animals cannot fashion a vision for conservation-compatible human societies; humans must. But this differs little from other movements in which the vision is fashioned by those who understand the situation of the injured, who are best able to see a more just alternative, *and who understand the dynamics of power.* Visionary leaders such as Jefferson, Mandela, Stanton, and many others, if not always the children of privilege, were not among the most victimized and have a much wider range of experience and sense of possibilities than those they spoke for and served (Morris and Staggenborg 2004). Nonetheless, conservation presents a special challenge of speaking for the needs of countless species with whom humans understand and empathize imperfectly. A sound vision will emerge from those who care most about wild places and creatures and less about the approval of other humans.

Bold vision is also a hallmark of successful movements. Movements emerge in response to the failure of existing societal structures. Structurally rooted failures can only be addressed by bold solutions, not band-aids. Visions of the end of racism, economic and political domination, or the lethal exploitation of the natural world are visions of fundamental change. Boldness is also a tactical imperative—one can bargain down, but not up.

Conservation has a history of bold visionaries. Some have understood the challenges of 12,000 years of institutional inertia. But too often conservationists' vision has not recognized that politics is based not on reasoned analysis of information, testing of biological or social hypotheses, or the careful evaluation of the likelihood of success of alternative courses of action. Instead, politics is about the pursuit of power, influencing the powerful, the threat or exercise of violence, and often appeals to delusional or magical thinking. Many conservationists are from educated, middle-class, professional backgrounds that do not provide a good experiential education in contentious politics. Nor are they prepared to engage opponents who are willing to do almost anything to keep their advantages (e.g., mass murder to protect access to strategic resources such as oil [Klare 2001; Paskal 2010]), nor the degree to which many victimized people will go to protect their ideological certainty (e.g., persecute or kill those who disagree with them [Edelman 1988]).

A bold and shared vision does not guarantee success, but it is a condition for intelligent and purposeful action. In 1963 Martin Luther King regalvanized a movement and much of a nation by giving voice to a bold vision of equality and an end to racial hatred. He also understood there was more to a bold vision than a dream; it must encompass a sound understanding of power, strategy, and organization.

2. Combine Insider and Outsider Strategies (Good Cops and Bad)

Insider strategies understand politics is the art of the possible. Outsider strategies are about changing what is possible. Combining both strategies is necessary to achieve structural change.

Throughout much of the world conservation politics relies overwhelmingly on insider strategies such as lobbying within established and normalized channels, on personal connections with elites, and on the largesse or personal inclinations of some leaders. Insider strategies have led to the creation of national parks and other protected areas, and limiting the trade in rare, threatened, and endangered species. But insider strategies are inherently limited, as demonstrated by the lax enforcement of domestic and international conservation laws and the primacy of economic and trade law. Elites—the wealthy and powerful—seldom abandon their material interests; their support is always conditional on truncated conservation goals (e.g., Kraft and Kamieniecki 2007; Kamieniecki 2006; Gonzalez 2001). In the United States the fingerprint of elites is evident in inadequate Endangered Species Act funding under both Democratic and Republican administrations, the deference to resource extraction and the ranching industries on public lands, and biologically irrational borders for protected areas. Conservationists, though capable and skilled, typically lack the resources to use insider strategies effectively with decision makers when they face powerful opponents. Opponents are often part of the elite (they control wealth, votes, media or are themselves the target decision makers [Domhoff 2009; Dye 2002]), are better connected to decision makers and administrators (including police and private repressive entities [e.g., Drohan 2003; Helvarg 2004; Donner 1990]), and have greater access to expert political resources (to frame issues, attack conservationists' credibility [Libby 1999; Helvarg 2004; Ewen 1996]).

Insider strategies do not yield fundamental change. Fundamental change invariably requires breaking the rules imposed by the elites for their benefit, and creating new rules. No major societal change has been achieved without outsider strategies that include mass mobilization. Mass mobilization provides the credible threat of disrupting business as usual until demands are met (Meyer and Tarrow 1998; McCarthy and McPhail 1998). Slavery was not abolished, Jim Crow laws were not dismantled, nor did women and labor obtain rights by playing by the rules. Structural change is neither the product of the timid nor amateurs, but of efforts led by professionals (Meyer and Tarrow 1998). Outsider strategies are high risk and require people willing to take on the risks of repression (McAdam 1986).

To be effective, groups pursuing outsider strategies must forge alliances and coalitions with those pursuing insider strategies and other groups whose support prevents opponents inside or outside the state from isolating outsider, that is, "radical" groups (Davenport 2005; Meyer and Tarrow 1998). It took decades of patient organizing for the US civil rights and labor movements to achieve broad support via allies and the media for their goals if not always their disruptive tactics. In extremely repressive countries gaining international attention and support is critical (Clifford 2005). Successful outsider strategies also depend on correctly anticipating the mix of concessions and repression that disruptive action will trigger from elites (McAdam 1997 [1983]). Gauging elite responses can be difficult because decision makers are often divided, uncertain, and irrationally fearful (Goldstone and Tilly 2001).

When existing structures or foes are strong, nonviolent protest may be the only way to avoid crushing repression. Nonviolent but disruptive protest was successfully used by groups in the US civil rights and anti–Vietnam War movements, but their success was owed in part to other groups in the movement espousing revolution. Although unrealistic, calls for revolution shifted the political center, making nonviolent disruption more acceptable. On the one hand, the US American Indian Movement, though not revolutionary, engaged in armed defense of Indian communities against corrupt tribal governments backed by the US government, and it was crushed by repression (Matthiessen 1983) as were many prodemocracy movements around the world (e.g., China). On the other hand, it was primarily the effective threat of civil war in South Africa—a civil war the elites knew they could not win—that ultimately brought authorities to the bargaining table (Wood 2000). Democratic struggles played out in a similar way in El Salvador and Guatemala.

Outsider strategies encounter repression not only because they are disruptive but also because they aim at creating new centers of political power thereby diminishing existing decision-makers' options. Permanent changes in power relations are a considerable threat to elites, and many regimes go to great lengths to forestall the creation of any autonomous centers of power.

Some repression is inevitable. Successful movements prepare for it and minimize it by exploiting elite divisions and finding sympathizers within the elite who may limit its use, by demonstrating to those using it that it won't work or will backfire, and by gaining broad recognition that repression is unjustified and indicates elite malevolence (Davenport 2005; Koopmans 2005; Zwerman and Steinhoff 2005; Giugni 2004; Goldstone and Tilly 2001).

Combining outsider and insider strategies allows movements to simultaneously create new centers of power that work to force structural changes and openly challenge elites, and use insider connections to negotiate with elites. Outsider and insider approaches must usually be pursued by different organizations within a movement to enable decision makers to save face when they make concessions—they concede to representatives of the insider strategy even though they are responding to disruptive pressure.

Obstacles to coordination among insiders and outsiders include insiders' protectiveness of their access to the powerful, their desire to be players above all else, and their tendency to temper their demands to avoid making powerful friends uncomfortable—though these "friends" may give them very little in return (e.g., Michels 1962 [1915]; Piven and Cloward 1977; Dryzek et al.

2003; cf. Rootes 2004). Change is not made by movements afraid of upsetting powerful people. The primary outsider obstacle to coordination is their sense of purity—they won't work with compromisers.

3. Create a Strong Community

Movements consisting mostly of organizations whose members are check writers supporting professional staff lack the passion and energy to create fundamental change (Meyer and Tarrow 1998; McCarthy and McPhail 1998). Such organizations often cannot even mobilize enough member support for insider strategies because check writers do not follow organizational leaders (Shaiko 1999). In many countries organizations are also precluded by law from participating in processes that choose leaders—which is where lobbying begins—by their dependence on tax-deductible contributions.

In contrast, successful movements are embedded in a strong, mass community or network of communities. Community is critical for a number of reasons. The bonds of community extend beyond politics, to friendship, family, ritual, marriage, sex, love, play, music, and other cultural relationships. Such bonds create feelings of belonging and forestall attrition resulting from the uncertainty of outcomes, the often multigenerational path to realizing significant change, the oppressive asymmetry of power relationships, the potential for demobilization following major interim successes, and the vilification of movement members by defenders of the status quo (Aminzade and Perry 2001; Goodwin, Jasper, and Polletta 2001; Lofland 1997 [1978]; Staggenborg 2011). Personal relationships afforded by community buffer against isolation, fatigue, and fear; the more developed they are the more resilient the movement organizations (Taylor 1989).

Fostering systemic change is high risk. The bonds of community, not just bonding with a cause or with leaders, sustain political action in the face of repression (Taylor 1989). Trust and loyalty are built upon strong interpersonal ties, a commitment to a common cause, and ritual. Ritual reinforces belonging, emotional connections, identity, and purpose, that is, community (Goodwin and Pfaff 2001; Barker 2001; Kertzer 1988; Moore and Myerhoff 1977). Virtual social networks are effective at recruitment for mass events, but inadequate to support the organization building necessary to sustain over time active involvement of large numbers of people (Cassen 2004; Tilly 2004).

Movements create a core community of activists, but to be effective this small community must be embedded in a broader one—usually preexisting communities (McAdam, Tarrow, and Tilly 2001). The US civil rights movement relied on black churches, universities, and fraternal orders (Skocpol, Liazos, and Ganz 2006); the anti-apartheid movement on the townships, labor organizations, and simultaneously on ethnic ties and transcendence of ethnicity (Wood 2000); the US labor movement on fraternal orders, mutual aid organizations, the Grangers, and neighborhood ties (Foner 1975); and the US antiwar movement on liberal churches, universities, feminist women's networks, and many old left and labor networks (Wittner 1984; DeBenedetti 1990). The US conservation movement has generally not extended its community as these other movements have. Earth First! was extraordinarily creative in the 1980s, generating an ecocentric culture, but it lacked the capacity to reach broader audiences (Lee 1995). The bulk of conservation supporters are check writers who lack conservation-centered community ties (Shaiko 1999), leaving conservation a sideshow or an afterthought, not a society-changing movement.

4. Uncompromising on Goals but Flexible Means

Movements are energized and sustained by organization, real and perceived progress toward goals, threats, leaders, relationships among movement participants, and the inspiration imparted by vision. Compromise on core goals— those essential to achieving an organization's or movement's vision—drains the energy and determination that purpose generates. Opponents and decision makers do not take seriously those who compromise on important goals.

Because movements consist of diverse organizations having different goals, the same position may mean compromise for one but not for others. Nevertheless, successful movements share key elements of a vision, such as equality in the US civil rights movements, autonomous political organizations and free speech in prodemocracy movements, an end to the loss of biodiversity in the conservation movement. If these are compromised then the movement's capacity to achieve its core goals is diminished (Staggenborg 2011).

One of the greatest enticements to compromise on basic goals is partial success. When an organization attains a seat at the table with decision makers, it comes under significant internal and external pressure to compromise (Michels 1962 [1915]). Leaders like being players and will too often "go along to get along." Decision makers exert strong pressure on organizations to limit demands. Some observers think these tendencies are not a problem for the conservation movement because it is decentralized and new, uncompromising organizations are constantly emerging that keep pressure on more staid organizations (Piven and Cloward 1977; Dryzek et al. 2003; cf. Rootes 2004). Nonetheless, the dominating presence of larger, more conservative organizations makes it difficult for smaller, activist groups to gain attention and attract resources.

Although differences among the goals of organizations within a movement create tension, they are key to success. Organizational variety is necessary if movements are to attract those with different levels of commitment, different views about what needs to be done, and different risk tolerance. Variety also provides a pathway for people to move among organizations as commitment and political sophistication grows (Shaiko 1999). Different organizational approaches also coincide with different policy options, for example, influencing legislators or agencies and working with private landowners.

In any movement a few organizations pursue high-risk outsider approaches and many pursue low-risk approaches. Diversity becomes a weakness and movements falter when low-risk groups allow opponents or decision makers to divide them from the high-risk groups that play such a pivotal role in defining the political landscape. It is those pushing the envelope that define the political center.

If unwillingness to compromise on goals is critical to achieving them, so is flexibility in the means employed. Many paths may lead to a goal and being open to taking the most advantageous one can make all the difference (see the discussion of crises below). Sometimes goals and paths are confused. Some US civil rights organizations were criticized for compromising although they had not abandoned their demand for full equality, but merely pursued a strategy that first sought equality in public education, and then in other areas one by one.

5. Persevere

Achieving significant change depends on opportunities and on advocates with uncompromising vision, adequate resources, capable leaders, constant tactical innovation, and perseverance as well as other factors (McAdam, Tarrow, and Tilly 2001; Kreisi and Wisler 1999). When defenders of the status quo think those who demand change will tire and fade the former will try to wait out the latter or encourage their waning. When a movement's organizational strength and commitment of participants makes clear that it will pursue its goals indefinitely, decision makers are more likely to bargain—or resort to repression.

Sustained action, although not sufficient, has always been necessary to achieve systemic change (Giugni 2004). Perseverance depends on harnessing people's emotions (e.g., anger, outrage, affection for other participants), needs (e.g., for others, for a sense of efficacy and purpose), and deepest beliefs (what constitutes justice and the highest good) (Wood 2001; Clayton 2000).

Many movements' goals take decades and longer to achieve. Conservation has few milestones that are as easily defined as the eight-hour workday, women's suffrage, or the end of legal segregation (Rucht 1999). Results, such as whether protecting an area will secure its species and ecological function, may only be clear in the distant future. The immediate costs of achieving protection may be high. In the face of such uncertainties the role of ritual—which presupposes a community—is critical to sustaining mobilization. Rituals define, declare, and celebrate achievements that are otherwise obscure. Just as most cultures recognize people as adults at a particular age by declaring them to be adults even though adulthood is achieved gradually, so the milestones of societal change must be marked. Ritual is also important in reaffirming commitment to action in the face of achievements, which can otherwise lead to notions that the struggle is over (Taylor 1989; Bevington 2010).

If perseverance depends on feelings of effectiveness, effectiveness depends on making progress toward goals. Progress depends on many factors, but one of the most important is constant tactical innovation that keeps the movement ahead of opponents and authorities (Sellars 2004). When groups do not innovate, their actions become easy to counter.

Leadership is critical to perseverance. Successful movements and organizations are led by a group of people that collectively are able to inspire, organize, and implement effective strategies (Aminzade, Goldstone, and Perry 2001). No one person possesses all the attributes of a good leader.

Ideology expresses vision in greater detail and contributes to perseverance in a number of ways. It explains the nature of the struggle and its importance, fulfills supporters' need to make sense of things, and sustains people by sanctifying purpose, not just by providing it. Religion has often played an ideological role in movements, but there are secular variants as well: notions of inevitable historical development, or that the universe unfolds in ways that favor justice or progress (e.g., Aminzade and Perry 2001; Skocpol, Liazos, and Ganz 2006).

Mobile telephones and the Internet have made it possible, without direct physical interaction, to organize mass events on short notice. In a few cases such events have led to important political consequences. But observers also note that most participants show little inclination for organization building or discipline, a prerequisite to creating a base of political power (Cassen 2004). Without such a base, movements cannot persevere long enough to bring systemic change. Participants in events and virtual social networks do not develop the trusting personal relationships and organizational bonds that underlie community and encourage ongoing involvement (Tilly 2004). The failure of 2008 Obama supporters to turn out in 2010 is partly attributable to the weakness of virtual ties.

6. Exploit Divisions within Elites and Crises

Divisions within decision-making elites and among a movement's opponents, and crises that weaken opponents and delegitimate dominant ideologies and institutions are important opportunities for movements if they are recognized and acted on decisively.

When national or global elites are united they usually get their way (McAdam, Tarrow, and Tilly 2001; Giugni 2004). When they are divided there is greater potential for alternative definitions of problems and solutions, and more room for action by nonelite actors. When elites are divided their factions depend on nonelite allies to bolster their power, which gives nonelites greater leverage to advance their goals. Movements that exploit these divisions, sometimes by first exacerbating them, can win concessions.

It is no coincidence that some of the strongest US conservation laws were passed by the governing elite faction that was divided from factions over continuing to wage an aggressive war and for resisting dismantling racist institutions and other failings (Repetto 2006). These divisions, combined with pressure from energized mass environmental and conservation movements, caused those holding the government's reins to support laws including the Endangered Species Act. Conservationists won greater protection for dolphins by using and encouraging divisions among big US tuna companies, tuna fishermen, and at times congressional leaders (Layzer 2005). Division between economic and political elites was a major proximate cause for negotiations between rebels and the governments of Guatemala and South Africa (Wood 2000). Elite divisions do not generate desired results on their own—they must be exploited aggressively. *Nature* (2010)—not noted for politically feisty editorials—advised climate scientists that they were in a street fight, not an academic debate over greenhouse gas emissions.

Although elites share an interest and desire to maintain the social order that rewards them so generously, they are also divided into factions by differing interests. In constantly seeking advantage they often are at odds with each other. Understanding potential and actual divisions makes opportunities easier to see and exploit. For example, elites tend to be more divided in states that have been at the top of the international pyramid (e.g., United States, United Kingdom, Germany, France) than elites in rising states (e.g., Brazil, China) because of the greater need for unity in the face of stronger powers (Chase-Dunn and Hall 1997).

Crises present greater or lesser opportunities depending on how deeply rooted they are. Random crises, generated by the idiosyncratic behavior of political, business, or cultural leaders, may provide opportunities to replace an unfriendly decision maker or make them more receptive to demands. Scandals and similar may also weaken sympathetic leaders. This type of crisis usually presents limited opportunities for significant change.

Incremental change is the norm in most policy areas, alternating with rare periods of significant policy change resulting from the concatenation of factors, including media and "public" attention cycles, temporary shifts in the relative power of opposing groups, new knowledge that contributes to new definitions of issues and problems, a catastrophe, and the unexpected consequences of legislation or court decisions (Repetto 2006; Baumgartner 2006).

Structural crises provide opportunities for much greater change, and by identifying them as such and giving them greater attention, much can be achieved. By analogy, once a train is headed down the track, options are limited—changing the engineer or speed is of minimal account when the train is headed in the wrong direction. But when the train is in the switching yard it can be set in a new direction, overhauled, and refitted, and this is the option structural crises provide.

Although structural crises and deep elite division may seem to appear suddenly, they are invariably a long time in the making (McAdam and Sewell 2001). A good understanding of societal dynamics can provide early warning and suggest points of maximal political leverage. With early warning conservationists are also better able to prepare to counter elite efforts to spin a crisis for their benefit by manipulating perception away from real causes and solutions (e.g., Brock 2006; Ewen 1996).

Structural crises present differing opportunities. Modern economies undergo major changes in technology and infrastructure about every fifty-five years (e.g., Berry 1991; Berry et al. 1998). When existing technologies cannot generate enough profits for investors, growth and prices drop, and new, more profitable technologies are sought; once selected (mostly by venture capitalists or their equivalent, for example, enterprise and party bureaucrats in China) and put in place they are difficult to replace for the next fifty to sixty years. Many societies are in such a technological and infrastructure transition now, the potential ecological consequences of the choices are enormous. Although some conservationists are working on the margins—for example, in support of so-called green energy—they mostly lack the capacity and understanding to exert much influence. In the past revolutionary and reform movements in developing countries have intervened at technology/infrastructure replacement junctures through their influence and leapfrogged ahead of more developed countries in many technological areas (e.g., Chase-Dunn and Hall 1997). There are lessons for conservationists and a very big question—can efforts to influence such transformations results in technologies and infrastructure that shrink rather than increase the human footprint?

Once more profitable new technologies and infrastructure are adopted growth takes off and a new cycle begins. New technologies usually require new types of resources, including new energy resources, and as growth accelerates resource wars often result (Goldstein 1988; Klare 2001; Paskal 2010). The opportunities war presents for conservation are complex. Wars can evoke patriotism, pushing other issues such as conservation off the political agenda. Because wars rarely go as planned they can also weaken leaders, generate antiwar movements that are critical of the causes of war—which often are also the causes of ecological degradation—and enhance support for changes in societal direction.

These cycles also are associated with changes in political leadership, offering some insight into the most useful alliances and strategies (Berry et al. 1998). When growth is slow and prices are high (stagflation)—around the middle of the fifty-five-year cycle—liberal elite factions give way to more conservative ones. When growth and prices are both stagnant (recession or depression), there is usually a shift from conservative to more liberal elite factions.

There are longer-term oscillations and developments that offer potential opportunities to movements. As dominant powers wax and wane and ultimately weaken in comparison to countries whose power is increasing, the former's reach shrinks in their regions of influence and globally, inevitably resulting in challenges (Goldstein 1988; Thompson 2000, 2009; Vasquez 2009). Global wars can result when the challenger to the dominant power seeks a different world order, rather than just to replace the dominant power (Kugler and Tammen 2009). On the one hand, the US civil rights movement and national liberation movements around the world took advantage of the need of the Allies for their cooperation in World War II and gained important concessions from the war's victor (e.g., Rosenberg 2006). On the other hand, ebbing US dominance may present a challenge for conservation.

The periods of modern history dominated by the unbridled pursuit of growth by Britain and the United States have resulted in the destruction of biodiversity and whole ecosystems on a heretofore unprecedented scale. Both countries, however, have a countervailing tradition of valuing nature that has produced species and wildlands protection. As power shifts—with or without major war—from North America and perhaps the European Union to China and India, conservationists must grapple with a world order where influence is wielded by growth even less constrained by conservation values. Moreover, China does not tolerate autonomous nongovernmental organizations that might act to check the destruction of species.

On a much longer time horizon conservationists confront 12,000 years of societal inertia beginning with the "Neolithic Revolution" that gave rise to the growth-focused hierarchical institutions so familiar today (Johnson and Earle 2001). The drive of societies for more and more resources and the biological consequences presents conservationists with an issue as thorny as disarmament. No country wants to weaken its competitive position by foregoing the control of resources, especially energy resources that allow for the enhanced capture of all resources (White 1987 [1975]; Johns 2002).

The conservation movement as a whole operates with an understanding and a capacity that ignores societal inertia and the growing human footprint. A forest is saved here, but logging increases elsewhere. World leaders such as Lula da Silva of Brazil have proclaimed a new path forward that stresses conservation goals, but have not delivered (Hochstetler and Keck 2007). If conservationists want to realize their vision they must commence building a much stronger movement—one strong enough to effect societal change.

There are some clues about where to begin such a monumental task. For example, societies situated between the poorest and richest and that are gaining in wealth and power are often the source of important societal and technological innovations that drive the evolution of the world system (Chase-Dunn and Hall 1997). Historically, both types of innovation were aimed at intensifying exploitation of the natural world to boost efforts to break free from long-standing domination by other countries. But their role as global rudder may harbor the potential for forms of societal organization compatible with conservation goals.

7. Understand Power

Former US Secretary of the Interior Bruce Babbitt (pers. comm.) said, "Don't expect me to do the right thing, make me do it." Prevailing in the choice of policy or leaders has little to do with reasoned arguments and facts, though they may provide public justification for actions. Instead, political outcomes depend on the ability of contesting parties to effectively mobilize more money, votes, media, and other resources than their opponents. Decision makers must care about an issue before information about it matters. Although some issues are near to their heart, many care most about continuing to be in a position to make decisions (Johns 2009). Even sympathetic decision makers need to feel systematic pressure—it allows them to resist counterpressure.

Successful movements and organizations understand who holds power, how decisions are made, who can directly influence the outcomes, and how to mobilize those groups or individuals (Johns 2009, 2010). They understand that the process of influencing decision makers starts with influencing decisions about who makes decisions. They understand that reaching goals depends on the willingness to use to the fullest their capacity to reward, punish, and otherwise influence, despite its crudeness and imprecision. Timidity is ineffective.

Movements sometimes recognize that achieving their goals requires more than influencing existing decision makers or replacing them. Achieving their goals requires a new societal and political structure. Some movements become revolutionary when other avenues of obtaining basic change are blocked.

No one more elegantly expressed a grasp of power than abolitionist and former slave Frederick Douglass (1985 [1857], 204):

> Power concedes nothing without a demand. It never did and it never will. Find out just what any people will quietly submit to and you have found out the exact measure of injustice and wrong that will be imposed . . . and these will continue till they are resisted with words or blows or with both. . . .
>
> If there is no struggle there is no progress. Those who profess to favor freedom but depreciate agitation are men who . . . want rain without thunder and lightning. They want the ocean without the awful roar of its many waters.

Those who want above all else a quiet ocean are part of the problem, not part of the solution.

Acknowledgments

The author is indebted to Erica Fleischman for a careful critical reading of the manuscript. Dominique Bessee and Patty Burk caught many errors and contained much verbosity.

References

Aminzade, Ronald R., Jack A. Goldstone, and Elizabeth Perry. 2001. "Leadership Dynamics and Dynamics of Contention." In *Silence and Voice in the Study of Contentious Politics*, edited by Ronald R. Aminzade, Jack A. Goldstone, Doug McAdam, Elizabeth J. Perry, William H. Sewell, Sidney Tarrow, and Charles Tilly, 126–54. Cambridge: Cambridge University Press.

Aminzade, Ronald R., and Elizabeth J. Perry. 2001. "The Sacred, Religious, and Secular in Contentious Politics." In *Silence and Voice in the Study of Contentious Politics*, edited by Ronald R. Aminzade, Jack A. Goldstone, Doug McAdam, Elizabeth J. Perry, William H. Sewell, Sidney Tarrow, and Charles Tilly, 155–78. Cambridge: Cambridge University Press.

Barker, Colin. 2001. "Fear, Laughter, and Collective Power: The Making of Solidarity at the Lenin Shipyard in Gdansk, Poland, August 1980." In *Passionate Politics*, edited by Jeff Goodwin, James M. Jasper, and Francesca Polletta, 175–94. Chicago: University of Chicago Press.

Barnosky, Anthony D., Paul L. Koch, Robert S. Feranac, Scott L. Wing, and Alan B. Shabel. 2004. "Assessing the Causes of Late Pleistocene Extinctions on the Continents." *Science* 306 (October 1): 70–75.

Baumgartner, Frank R. 2006. "Punctuated Equilibrium Theory and Environmental Policy." In *Punctuated Equilibrium and the Dynamics of U.S. Environmental Policy*, edited by Robert Repetto, 24–46. New Haven, CT: Yale University Press.

Benford, Robert D., and David A. Snow. 2000. "Framing Processes and Social Movements." *Annual Review of Sociology* 26: 611–39.

Berry, B. J. L. 1991. *Long Wave Rhythms in Economic Development and Political Behavior*. Baltimore: Johns Hopkins University Press.

Berry, B. J. L., E. Elliot, E. J. Harpham, and H. Kim. 1998. *The Rhythms of American Politics*. Lanham, MD: University Press of America.

Bevington, Doug. 2010. *The Rebirth of Environmentalism*. Washington, DC: Island Press.

Boykoff, Jules. 2007. *Beyond Bullets*. Oakland, CA: AK Press.

Brader, Ted. 2005. *Campaigning for Hearts and Minds*. Chicago: University of Chicago Press.

Brashares, Justin S. 2010. "Filtering Wildlife." *Science* 329 (July 23): 402–3.

Brock, William A. 2006. "Tipping Points, Abrupt Opinion Changes, and Punctuated Policy Change." In *Punctuated Equilibrium and the Dynamics of U.S. Environmental Policy*, edited by Robert Repetto, 47–77. New Haven, CT: Yale University Press.

Butchart, Stuart H. M., et al. 2010. "Global Biodiversity: Indicators of Recent Declines." *Science* 328 (May 28): 1164–68.

Cassen, Bernard. 2004. "Inventing ATTAC." In *A Movement of Movements*, edited by Tom Mertes, 152–74. London: Verso.

Chase-Dunn, Christopher, and Thomas D. Hall. 1997. *Rise and Demise*. Boulder, CO: Westview Press.

Clayton, Susan. 2000. "Models of Justice in the Environmental Debate." *Journal of Social Issues* 56 (3): 459–74.

Clifford, Bob. 2005. *Marketing of Rebellion*. Cambridge: Cambridge University Press.

Cohen, Mark Nathan. 1977. *The Food Crisis in Prehistory*. New Haven, CT: Yale University Press.

Cornog, Evan. 2004. *The Power and the Story*. New York: Penguin Press.

Cox, Robert. 2006. *Environmental Communication and the Public Sphere*. Thousand Oaks, CA: Sage.

Davenport, Christian. 2005. "Introduction: Repression and Mobilization: Insights from Political Science and Sociology." In *Repression and Mobilization* edited by Christian Davenport, Hank Johnson, and Carol Mueller, vii–xli. Minneapolis: University of Minnesota Press.

Davenport, Christian, Hank Johnson, and Carol Mueller, eds. 2005. *Repression and Mobilization*. Minneapolis: University of Minnesota Press.

DeBenedetti, Charles. 1990. *An American Ordeal*. Syracuse, NY: Syracuse University Press.

Domhoff, G. William. 2009. *Who Rules America?* 6th ed. Boston: McGraw-Hill.

Donner, Frank. 1990. *Protectors of Privilege*. Berkeley: University of California Press.

Douglass, Frederick. [1857] 1985. "The Significance of Emancipation in the West Indies." Speech, Canandaigua, New York, August 3, 1857; collected in pamphlet by author. In *The Frederick Douglass Papers. Series One: Speeches, Debates, and Interviews*. Vol. 3: 1855–63, edited by John W. Blassingame, 183–208. New Haven, CT: Yale University Press.

Drohan, Madelaine. 2003. *Making a Killing*. Toronto: Random House Canada.

Dryzek, John S., David Downes, Christian Hunold, and David Schlosberg. 2003. *Green States and Social Movements*. Oxford: Oxford University Press.

Dye, Thomas R. 2002. *Who's Running America?* 7th ed. Upper Saddle River, NJ: Prentice-Hall.

Edelman, Murray. 1988. *Constructing the Political Spectacle*. Chicago: University of Chicago Press.

Ehrenfeld, David. 1979. *Arrogance of Humanism*. New York: Oxford University Press.

Emerson, Ralph Waldo. [1847] 1994. "Ode, Inscribed to W. H. Channing." In *Emerson: Collected Poems and Translations*, 61–64. New York: Library of America.

Ewen, Stuart. 1996. *PR! A Social History of Spin*. New York: Basic Books.

Ewing, B., S. Goldfinger, A. Oursler, A. Reed, D. Moore, and M. Wackernagel. 2009. *The Ecological Footprint Atlas 2009*. Oakland, CA: Global Footprint Network.

Foner, Philip S. 1975. *History of the Labor Movement in the United States*. Vol. 1. New York: International Press.

Giugni, Marco. 2004. *Social Protest and Policy Change*. Lanham, MD: Rowman and Littlefield.

Giving USA/Center on Philanthropy. 2010. *Giving USA 2010: Annual Report on Philanthropy for 2009*. Glenview, IL: Giving USA.

Goldstein, J. S. 1988. *Long Cycles*. New Haven, CT: Yale University Press.

Goldstone, Jack A., and Charles Tilly. 2001. "Threat (and Opportunity): Popular Action and State Response in the Dynamics of Contentious Action." In *Silence and Voice in the Study of Contentious Politics*, edited by Ronald R Aminzade, Jack A. Goldstone, Doug McAdam, Elizabeth J. Perry, William H. Sewell, Sidney Tarrow, and Charles Tilly, 179–94. Cambridge: Cambridge University Press.

Gonzalez, George A. 2001. *Corporate Power and the Environment*. Lanham, MD: Rowman and Littlefield.

Goodwin, Jeff, James M. Jasper, and Francesca Polletta. 2001. Introduction. In *Passionate Politics*, edited by Jeff Goodwin, James M. Jasper, and Francesca Polletta, 1–24. Chicago: University of Chicago Press.

Goodwin, Jeff, and Steven Pfaff. 2001. "Emotion Work in High-Risk Social Movements: Managing Fear in the US and East German Civil Rights Movements." In *Passionate Politics*, edited by Jeff Goodwin, James M. Jasper, and Francesca Polletta, 282–302. Chicago: University of Chicago Press.

Harkin, Michael E., ed. 2004. *Reassessing Revitalization Movements*. Lincoln: University of Nebraska Press.

Harris, Marvin. 1977. *Cannibals and Kings*. New York: Random House.

———. 1979. *Cultural Materialism*. New York: Random House.

Helvarg, David. 2004. *The War against the Greens*. Rev. ed. Boulder, CO: Johnson Books.

Hochstetler, Kathryn, and Margaret E. Keck. 2007. *Greening Brazil*. Durham, NC: Duke University Press.

Ingram, Helen, and Leah Fraser. 2006. "Path Dependency and Adroit Innovation: The Case of California Water." In *Punctuated Equilibrium and the Dynamics of U.S. Environmental Policy*, edited by Robert Repetto, 78–109. New Haven, CT: Yale University Press.

Johns, David. 2000. "Biological Science in Conservation." In *Proceedings: Wilderness Science in a Time of Change*, edited by David N. Cole and Stephen F McCool, 223–29. Proc. RMRS-P-000. Ogden, UT: U.S. Department of Agriculture, Forest Service, Rocky Mountain Research Station.

———. 2002. "Wilderness and Energy." *Wild Earth* 14 (3): 12.

———. 2005. "The Other Connectivity: Reaching Beyond the Choir." *Conservation Biology* 19 (6): 1681–82.

———. 2009. *A New Conservation Politics*. Oxford: Wiley-Blackwell.

———. 2010. "Adapting Human Societies to Conservation." *Conservation Biology* 24 (3): 641–43.

Johnson, Allen W., and Tim Earle. 2001. *The Evolution of Human Societies*. 2nd ed. Stanford, CA: Stanford University Press.

Jost, John T., Diana Burgess, and Cristina O. Mosso. 2001. "Conflicts of Legitimacy among Self, Group, and System." In *The Psychology of Legitimacy*, edited by John T. Jost and Brenda Major, 363–88. Cambridge: Cambridge University Press.

Kamieniecki, Sheldon. 2006. *Corporate America and Environmental Policy*. Stanford, CA: Stanford University Press.

Kertzer, David I. 1988. *Ritual, Politics, and Power*. New Haven, CT: Yale University Press.

Klare, Michael T. 2001. *Resource Wars*. New York: Metropolitan Books. Rev. ed. 2002, New York: Owl Books.

Koopmans, Ruud. 2005. "Repression and the Public Sphere." In *Repression and Mobilization*, edited by Christian Davenport, Hank Johnson, and Carol Mueller, 159–88. Minneapolis: University of Minnesota Press.

Kraft, Michael E., and Sheldon Kamieniecki, eds. 2007. *Business and Environmental Policy*. Cambridge, MA: MIT Press.

Kriesi, Hanspeter, and Dominique Wisler. 1999. "The Impact of Social Movements on Political Institutions." In *How Social Movements Matter*, edited by Marco Giugni, Doug McAdam, and Charles Tilly, 42–65. Minneapolis: University of Minnesota Press.

Kugler, Jacek, and Ronald L. Tammen. 2009. "Implications of Asia's Rise to Global Status." In *Systemic Transitions*, edited by William R Thompson, 161–86. New York: Palgrave Macmillan.

Layzer, Judith. 2005. *The Environmental Case*. 2nd ed. Washington, DC: CQ Press.

Lee, Martha F. 1995. *Earth First! Environmental Apocalypse*. Syracuse, NY: Syracuse University Press.

Libby, Ronald T. 1999 *Eco-Wars: Political Campaigns and Social Movements*. New York: Columbia University Press.

Lofland, John. [1978] 1997. "Becoming a World Saver Revisited." In *Social Movements: Readings on Their Emergence, Mobilization, and Dynamics*, edited by Doug McAdam and David A. Snow, 284–89. Los Angeles: Roxbury Publishing.

Matthiessen, Peter. 1983. *In the Spirit of Crazy Horse*. New York: Viking.

McAdam, Doug. 1986. "Recruitment to High Risk Activism." *American Journal of Sociology* 92 (1): 64–90.

———. [1983] 1997. "Tactical Innovation and the Pace of Insurgency." In *Social Movements: Readings on Their Emergence, Mobilization, and Dynamics*, edited by Doug McAdam and David A. Snow, 340–56. Los Angeles: Roxbury Publishing.

McAdam, Doug A., and William H. Sewell. 2001. "It's about Time: Temporality in the Study of Social Movements and Revolutions." In *Silence and Voice in the Study of Contentious Politics*, edited by Ronald R Aminzade, Jack A. Goldstone, Doug McAdam, Elizabeth J. Perry, William H. Sewell, Sidney Tarrow, and Charles Tilly, 89–125. Cambridge: Cambridge University Press.

McAdam, Doug, Sidney Tarrow, and Charles Tilly. 2001. *Dynamics of Contention*. Cambridge: Cambridge University Press.

McCarthy, John D., and Clark McPhail. 1998. "The Institutionalization of Protest in the United States." In *The Social Movement Society*, edited by David S. Meyer and Sidney Tarrow, 83–110. Lanham, MD: Rowman and Littlefield.

Meyer, David S., and Sidney Tarrow. 1998. "A Movement Society: Contentious Politics for a New Society." In *The Social Movement Society*, edited by David S. Meyer and Sidney Tarrow, 1–28. Lanham, MD: Rowman and Littlefield.

Michels, Robert. [1915] 1962. *Political Parties*. Glencoe. IL: Free Press.

Miller, Gifford H., John W. Magee, Beverly J. Johnson, Marilyn L. Fogel, Nigel A. Spooner, Malcolm T. McCulloch, and Linda K. Ayliffe. 1999. "Pleistocene Extinction of *Genyornis newtoni*: Human Impact on Australian Megafauna." *Science* 283 (January 8): 205–8.

Moore, Sally Falk, and Barbara G. Myerhoff, eds. 1977. *Secular Ritual*. Assen, Netherlands: Van Gorcum.

Morris, Aldon D., and Suzanne Staggenborg. 2004. "Leadership in Social Movements." In *The Blackwell Companion to Social Movements*, edited by David A. Snow, Sarah A. Soule, and Hanspeter Kriesi, 171–96. Oxford: Blackwell Publishing.

Nature editors (unsigned). 2010. "Closing the Climategate." *Nature* 468 (345).

Paskal, Cleo. 2010. *Global Warring*. New York: Palgrave Macmillan.

Pimm, Stuart. 2001. *The World According to Pimm*. New York: McGraw-Hill.

Piven, Francis Fox, and Richard Cloward. 1977. *Poor People's Movements*. New York: Pantheon.

Polletta, Francesca. 2006. *It Was Like a Fever*. Chicago: University of Chicago Press.

Polletta, Francesca, and M. Kai Ho. 2006. "Frames and Their Consequences." In *Oxford Handbook of Contextual Political Analysis*, edited by Robert E. Goodin and Charles Tilly, 187–209. New York: Oxford University Press.

Rambo, Lewis R. 1993. *Understanding Religious Conversion*. New Haven, CT: Yale University Press.

Rappaport, Roy A. 1976. "Adaptations and Maladaptations in Social Systems." In *The Ethical Basis of Economic Freedom*, edited by I. Hill, 39–79. Chapel Hill, NC: American Viewpoint.

Repetto, Robert. 2006. Introduction. In *Punctuated Equilibrium and the Dynamics of U.S. Environmental Policy*, edited by Robert Repetto, 1–23. New Haven, CT: Yale University Press.

Rootes, Christopher. 2004. "Environmental Movements." In *The Blackwell Companion to Social Movements*, edited by David Snow, Sarah Soule, and Hanspeter Kriesi, 608–40. Oxford: Blackwell.

Rosenberg, Jonathan. 2006. *How Far the Promised Land?* Princeton, NJ: Princeton University Press.

Rucht, Dieter. 1999. "The Impact of Environmental Movements in Western Societies." In *How Social Movements Matter*, edited by Marco Giugni, Doug McAdam, and Charles Tilly, 204–24. Minneapolis: University of Minnesota Press.

Sellars, John. 2004. "Raising a Ruckus." In *A Movement of Movements*, edited by Tom Mertes, 174–91. London: Verso.

Shaiko, Ronald G. 1999. *Voices and Echoes for the Environment*. New York: Columbia University Press.

Skocpol, Theda, Arianne Liazos, and Marshall Ganz, eds. 2006. *What a Mighty Power We Can Be*. Princeton, NJ: Princeton University Press.

Staggenborg, Suzanne. 2011. *Social Movements*. New York: Oxford University Press.

Stern, Paul. 2000. "Toward a Coherent Theory of Environmentally Significant Behavior." *Journal of Social Issues* 56 (3): 407–24.

Taylor, Verta. 1989. "Social Movement Continuity: The Women's Movement in Abeyance." *American Sociological Review* 54 (5): 761–75.

Thompson, William R. 2000. "K-Waves, Leadership Cycles, and Global War." In *A World Systems Reader*, edited by T. D. Hall, 83–104. Lanham, MD: Rowman and Littlefield.

———. 2009. "Structural Preludes to Systemic Transitions Since 1494." In *Systemic Transitions*, edited by William R. Thompson, 55–73. New York: Palgrave Macmillan.

Tilly, Charles. 2004. *Social Movements, 1768–2004*. Boulder, CO: Paradigm Publishers.

UNFAO (United Nations Food and Agricultural Organization). 2010. Press Release of September 14. Rome.

Vasquez, John A. 2009. "When and How Global Leadership Transitions Will Result in War." In *Systemic Transitions*, edited by William R. Thompson, 131–60. New York: Palgrave Macmillan.

Wallace, A. F. C. 1970. *Culture and Personality*. 2nd ed. New York: Random House.

White, Leslie. [1975] 1987. "The Energy Theory of Cultural Development." In *Ethnological Essays*. Albuquerque: University of New Mexico Press.

Wittner, Lawrence. 1984. *Rebels against War*. Philadelphia, PA: Temple University Press.

Wood, Elisabeth Jean. 2000. *Forging Democracy from Below*. Cambridge: Cambridge University Press.

———. 2001. "The Emotional Benefits of Insurgency in El Salvador." In *Passionate Politics*, edited by Jeff Goodwin, James M. Jasper, and Francesca Polletta, 267–81. Chicago: University of Chicago Press.

Wright, Ronald. 2004. *A Short History of Progress*. New York: Carroll and Graf.

Zwerman, Gilda, and Patricia Steinhoff. 2005. "When Activists Ask for Trouble." In *Repression and Mobilization*, edited by Christian Davenport, Hank Johnson, and Carol Mueller, 85–107. Minneapolis: University of Minnesota Press.

19

Consuming Nature

The Cultural Politics of Animals and the Environment in the Mass Media

Carrie Packwood Freeman and Jason Leigh Jarvis

AT THE END of his nonfiction book *Collapse*, after outlining how the destructive practices of most human societies are steering them toward ecological collapse and causing extinction of species, Jared Diamond (2005, 522) pins his hope for change on the global awareness-raising potential of the media: "Our television documentaries and books show us in graphic detail why the Easter Islanders, Classic Maya, and other past societies collapsed. Thus we have the opportunity to learn from the past mistakes . . . an opportunity that no past society enjoyed to such a degree."

Diamond indicates the two most important factors to prevent collapse are "long-term planning, and willingness to reconsider core values" (522), both of which, we contend, can be instigated by the media, as the agenda setter of public policy, as the cultivator of national identity and values, and as the primary cultural storyteller. The stories media choose to tell matter. Scientists can discover all kinds of problems and solutions to species issues, but if the media fail to convey and frame these discoveries productively, and if people's media-cultivated value systems don't allow them to care, then all the information in the world won't matter.

The commercially driven mass media package human identity and all our surrounding environment for daily consumption in the public sphere. It is of critical importance whether they choose to ignore humanity's responsibility toward the natural world and simply have us consume it as a product, or whether they actively cultivate ecological responsibility and newfound respect toward animals as fellow sentient beings. This chapter explores the necessity, potential, and challenges of relying on the media (journalism, television, advertising, film, radio, Internet, and such) to inspire the social change needed to reverse the destructive behaviors and beliefs that are contributing to our global ecological calamity. We address this both in specific terms related to how media raise awareness about habitat and wildlife protection and also in broader terms of how media could change humanist worldviews and consumptive lifestyles to promote self-awareness of humanity's position as a fellow species in an ecological web in crisis. To begin, we review scholarly literature on the social function of mass media and the way they represent nonhuman animals (NHAs). We then suggest methods for addressing environmental challenges through the news and entertainment media, including ideas for media practitioners as well as concerned citizens.

Media, Culture, and Nature

The role of media in modern industrial society should not be understated. The media are a vital link in systems of information that transmit and create meaning through representation. These representations serve to bind social networks together in a way that is both imaginary and real. Communication is a process of meaning-making that constantly creates, modifies, and maintains a shared culture and reality (Carey 1989). This reflects a view of language as a social construction that attempts to fix meaning and a "truth" to signs so the signification appears natural rather than arbitrary or contrived. According to critical scholar Stuart Hall (1997a, b), media are now the dominant means of social signification that both reflect and manufacture discursive "truths," and as such are the site of much ideological struggle to define meaning within modern capitalist society.

Television, for example, is not just an entertainment medium; it is a cultural artifact with a mainstreaming effect (Gerbner et al. 1978; Earp 2010). Because it is a commercial institution, television's version of social reality conforms to the interests of its owners and sponsors, which ultimately cultivates widespread support for the status quo and resistance to change. This bolsters political economy scholars' critiques of a commercially structured American media system that presents itself as a democratic free marketplace of ideas while ultimately serving the vested interests of owners and advertisers (McChesney 1999).

Newspapers and books combine narratives about events in the world into a daily consumer product that serves to bind social groups together and maintain community (Anderson 2006). When an individual privately reads about tangible things in the world, he or she is also confident in the knowledge that other citizens publicly undertake the same action in the same way. Therefore, consumption patterns related to the maintenance of society are normalized into routine—a process critical to the creation of national consciousness.

Western media representations of NHAs provide a fascinating look inside the cultural practices and attitudes of society toward the fellow members of biotic communities. As Maxwell Boykoff (2009) explains:

> Media representations are convergences of competing knowledges, framing environmental issues for policy, politics, and the public and drawing attention to how to make sense of, as well as value, the changing world. Emanating out from these processes, public perceptions, attitudes, intentions, and behaviors, in turn, often link back through mass media into ongoing formulations of environmental governance. (434)

The mass media are the tools institutions and members of society use to create intellectual frames that are critical to the dissemination of environmental values and attitudes. The media transmit and reflect these beliefs, functioning as a link between different discourses of knowledge. Boykoff (2009) notes the media face many important issues in regard to the environment, particularly questions of "fairness, accuracy, and precision" (440). As the link between the environment, science, and government, the media participate in a "cultural politics of the environment" that perpetuates a system of domination and control.

Predators, Prey, or Friends?

A substantial amount of scholarship has focused on media coverage of types of species, such as those portrayed as threats or pests. Judy Cohen and John Richardson (2002) trace the development of public fear of pit bulldogs, something that they claim was exacerbated by the news media coverage of pit bull attacks on humans. They claim this coverage is part of a traditional story of man versus beast: "The deviant incident of beast attacking man has fascinated and horrified people for millennia" (295). Indeed, the theme of evil animals is detailed by the work of Rod Giblett (2006) on alligators and Jan-Christopher Horak (2006) on sharks. Colin Jerolmack (2008) also identifies the *New York Times'* vilification of pigeons as pests. Then there are animals designated as prey or "game." For example, hunting video games encourage resistance to environmental protection because they embody a conception of nature as an object for pleasure and personal use (Sawers and Demetrious 2010). Dolphins present an interesting case because they fall into a category that is different from "attack" or "prey" animals. A study found popular literature portrays dolphins in four primary ways: (1) as friends to humans, (2) as a symbol of freedom and peace and a romanticized view of nature, (3) as innocent and in need of our protection, and (4) as superior to humans (Fraser et al. 2006).

Advertising

Television advertisements use six primary frames to portray NHAs: (1) as an object of affection and love, (2) as a symbol of something else, (3) as a tool or object that can be used in a practical way, (4) as wildlife in their natural habitat, (5) as an allegorical creature, and (6) as a pest (Lerner and Kalof 1999, 574). The researchers found the most common frame was an animal identified as a loved one who provides assistance or participates in family life. The second most common frame was animals as symbols, where animals become logos, transferring their traits to the product advertised.

Companies use animal imagery to symbolically convey their green nature (Spears and Germain 2007). As attitudes toward nature have shifted to value NHAs more inherently, ads less often depict NHAs with humans and more often portray them in a natural setting. The role of green advertising is an increasingly important issue, as many companies try to brand themselves to appear environmentally friendly, which can be called "greenwashing" if it is misleading. Others, like fast food chains, often resist ecological framing to emphasize masculine hedonistic consumption (in this case, men eating lots of animals) unrestrained by "feminine" concerns about health, ethics, or ecological consequences (Freeman and Merskin 2008).

News

The need for profit impacts the structure of news programming and framing of environmental issues. Like other forms of programming, the news is a product, and consumers must be made to watch it for profits to be maintained. Subsequently, the needs of the market outweigh the needs of the environment, even when it is the subject of programs. According to Boykoff (2009), while news media sometimes give voice to the environment, sensationalism or support for exploitation are just as likely as support for protection because media "articulations may take on varied roles over time, from watchdog to lapdog to guard dog" (435).

Initial news reports on environmental controversies are frequently negative and narrowly focused, as shrinking environmental news budgets and the demands of modern journalism create a need to start with a "news hook" that is often linked to the portrayal of humans in a dramatic struggle with nonhuman nature (Boykoff 2009, 445–46). The resulting stories frequently conflate multiple issues into simplistic explanations that fail to reflect the complexity of problems, thereby skewing public debate about scientific issues (433). An example of this problem can be seen in news coverage of black bears in New York State. The news relied more on episodic rather than thematic framing, resulting in predominantly negative coverage that characterizes the environmental issue as a personal conflict between bears and humans (Zavestoski et al. 2004; Siemer, Decker, and Shanahan 2007).

There is an inherent danger in the way the news media provide information, according to Jacquelin Burgess (1990, 155). She suggests that environmental news coverage runs the risk of alienating the public because it can create feelings of helplessness in the face of staggering problems. She argues for new types of research into the intertextual nature of mass media and its production of environmental knowledge in mainstream society in order to better address environmental challenges.

More productive environmental news coverage is vital, as the news plays a leading role in influencing the public's political consciousness and priorities, setting the agenda for public policy discussions in terms of what issues the public deems most important (McCombs 2005). While the news doesn't tell people exactly what to think, it tells them what to think *about*; and because of the power of framing, it often tells people *how* to think about it. Robert Entman (1993) acknowledges the power of news framing to identify problems and solutions: "to frame is to select some aspects of a perceived reality and make them more salient in a communicating text, in such a way as to promote a particular problem definition, causal interpretation, moral evaluation, and/or treatment recommendation" (52).

The agenda-setting role of the news was illustrated by a study of news coverage in the 1980s that demonstrates the convergence of coverage of several environmental issues: ozone depletion, the greenhouse effect, species extinction, and rain forest destruction (Mazur and Lee 1993). As coverage of these issues grew, a general agenda-setting effect was witnessed as problems facing the environment were placed firmly on the national political landscape.

Documentaries

While beneficial in many respects, nature documentaries can increasingly be pragmatically conceived as scientific tools to record animals and ecosystems before they become extinct (Horak 2006, 459–60). Cinema and film thus serve as crypts for animals that are "perpetually vanishing" but never die (Lippit 2000, 1). In the words of Akira Lippit (2000), "The cinema developed, indeed embodied, animal traits as a gesture of mourning for the disappearing wildlife. The figure for nature in language, animal, was transformed in cinema to the name for movement in technology, animation" (197). Technology (media and machinery) itself is inspired by animals, just as it comes to replace them—transferring their bodily energy into a virtual life animated in metal and imagery.

Nonetheless, nature documentaries are a contested area for environmentalists. Many feel they productively generate sympathy for environmental causes, while others feel they are more about profit and entertainment (Burgess 1990, 153; Barbas, Paraskevopoulos, and Stamou 2009). In his study of Swedish television, Hillevi Ganetz (2004) argues that nature documentaries contain images and storylines that reinforce traditional gender norms and often focus on incredibly violent images of animals in conflict or hunting and catching prey, as opposed to the idyllic pastoral scenes common in Disney films, reflecting changes in cultural attitudes more accepting toward violence. These problems relate back to the fundamental fact that documentaries are framed, mediated interpretations of nature that reflect narratives dominant within the culture that created them.

Similarly, in analyzing the Discovery Channel, David Pierson (2005) outlines four major themes running through its nature programming: (1) nature and gender; (2) anthropomorphism; (3) nature and social structure or hierarchy; and (4) social conceptions of nature, of which there are three—nature as object of scientific control, nature as threatened, and nature as sacred. Pierson concludes that the Discovery Channel creates some understanding of nature but ultimately does so through a lens that anthropomorphizes nature, such that it is understood through human characteristics and traits. Nature is seen as having complex, hierarchical, social organizations that mirror human families and reinforce existing human social and gender hierarchies.

Also criticizing documentaries is Jan-Christopher Horak (2006), who notes that the increase in animal programming on television is taking place during a time of dramatic species extinction. Yet rather than producing a desire to protect wilderness, animal shows primarily serve to create a desire to consume more images of nature, as viewers identify with anthropomorphized animal depictions. Governmental and political solutions to environmental problems are notably absent from the majority of programs in favor of personal solutions, such as a kindhearted individual rescuing a stranded animal. For example, the Animal Planet network has altered its programming to focus on human-centered interventions into the environment (Umstead 2009). Its highest-rated program, *Whale Wars*, follows the Sea Shepherd activists on their attempts to stop whale hunting. It has been argued that a whale-centered show would be preferred to the anthropocentric focus this program uses in primarily chronicling the exploits of human activists against hunters (Besel and Besel 2010).

Another thought-provoking critique of the animal documentary is provided by Brett Mills (2010) who argues quite persuasively that the documentary it-self is an inherently anthropocentric instrument that denies NHAs the same right to privacy that many people, particularly in the West, cherish. Nature documentaries pride themselves on enabling audiences to experience nature without harming it, yet nobody questions if it is harmful to film NHAs engaging in their most intimate actions. "To look at an animal—and to decide that humans have a right to look at animals because animals don't have a right to privacy—is an act of empowerment, reinforcing the moral hierarchy which legitimizes the act in the first place" (199). Mills's argument creates a troubling double bind for environmentalists and documentarians because even if nature documentaries did result in a desire to protect the environment, they do so from an anthropocentric perspective that legitimizes human superiority and justifies management practices that are rooted in surveillance and an assumption of inequality.

An Ecologically Sensitive Media

Despite the power of mass media, audiences maintain the agency to interpret media messages in a variety of ways: as dominant readings reinforcing the status quo, as negotiated readings partially transformed by personal meaning-making, or as oppositional readings. Oppositional readings reject the dominant perspective, and for Stuart Hall (1980), this represents a moment in which political alternatives are formed. It is through challenging dominant discursive formations and the creation of oppositional meanings that changes can be made in media systems and representation of the nonhuman world. With hope for addressing the crises facing our planet, in this section we suggest potential solutions for producers of entertainment programming, news, and activist campaigns.

One issue that affects both entertainment and news programming is the increasing portrayal of human-caused violence. Human-to-human violence on television can produce what Jeremy Earp calls a "mean world syndrome" where heavy viewers think society is a more violent and mean place than it is, which can cause fear, mistrust, and a desire for authoritative security measures (Earp 2010). Human violence toward NHAs is also prevalent across all media platforms, especially in the form of ubiquitous messages supporting meat eating and hunting/fishing (consider dining and outdoor sections of the news, advertisements for meat and hunting equipment, and a plethora of hunting and cooking shows). This normalizes the unnecessary killing and death of NHAs, cheapening their lives and status in human society. While killing certain NHAs is legal and therefore understandably reflected in the media, responsible media practitioners may see it as their duty to counteract the speciesist bias by actively questioning the killing and use of NHAs for food or sport and providing contradictory, nonviolent representations (Freeman 2009).

Film and Television

While many fictional movies have illustrated an apocalyptic future due to war or environmental devastation, it may be more useful for storytellers to use their imagination to help viewers envision a less speciesist future world governed by ecological principles that enforce sustainability, including a smaller human population. Movies can show us what that world would look like, how it would be structured, and explain the path to get there.

Fictional films are also noted for their ability to destroy human/nature dualisms through the portrayal of romantic relationships between humans and nonhumans. Gothic romance, science fiction, and horror are powerful genres because they allow the articulation of a "zoocentric perspective" (Swan 1999; Creed 2006). Werewolf and vampire films and epics such as *King Kong* all demonstrate the themes of human evolution and destroy the lines that have been drawn between humans and nonhumans.

Whether in film or television programs, viewers learn from what they witness in media and often identify with favorite characters (Bandura 1994), therefore, we believe producers should incorporate environmentally responsible themes and have main protagonists model sustainable behaviors and attitudes daily. As a matter of course, characters could be less materialistic, less consumption oriented, less wasteful, more civic minded, and respectful and nonviolent toward fellow animals, such as eating a plant-based diet. Parents could be shown adopting children or giving birth to only one or two.

When NHAs are the main "characters" in film and television documentaries, viewers' desires to relate to characters creates a disproportionate emphasis on social animals who are charismatic, cute, exciting, or beautiful (according to human cultural standards). Therefore, NHA programming tends to privilege mammals and birds at the expense of fish, amphibians, reptiles, and invertebrates, even though the latter categories comprise the vast majority of the species on Earth—such as worms, insects, and plankton who are ecologically valuable in their own right. For a positive example of a documentary representing the diversity of life, the BBC's *Planet Earth* series' episode on jungles features leaf frogs and bullet ants in addition to elephants, monkeys, and birds of paradise.

Yet *Planet Earth*'s behind-the-scenes peek at the videographer's struggle to be the first person to capture the bird of paradise's mating ritual demonstrates Mills's (2010) conundrum about cameras invading NHA privacy and humans' sense of entitlement in witnessing the intimate details of NHA life. Here utilitarian pragmatism and idealism collide in terms of short- and long-term solutions to save species, as on a practical level, humans believe they come to know and respect NHAs by learning about them. Yet philosophically, documentaries may be inadvertently perpetuating the human/animal dualism by treating other animals as unwitting actors and objects of curiosity.

A major question remains: can consuming animals in documentaries keep people from consuming their actual habitats, thus saving their lives? This addresses Horak's (2006) and Lippit's (2000) concerns in wondering how to keep documentaries from being mere historical records or virtual habitats for animals going extinct. If there is hope for change, it lies in a reorientation of documentaries toward an emphasis on the role of the individual acting within a community: both human and nonhuman. The key is an emphasis on the way that personal action functions within larger institutional and organizational approaches as well as the biotic community where the individual lives.

In addition to promoting animal rescue, documentaries could focus more holistically on promoting social change, human self-critique, and personal responsibility in terms of living sustainably. Shows on green living in both urban and rural settings are warranted in terms of showcasing human cultures that are embracing ecological principles. Related nonfiction programming could connect production and consumption, showing how products and services are made, going all the way back to the source—the "natural resources" impacted or used (the deforestation, pollution, killing, or displacing of NHAs, and such).

Media narratives need to place humans in an interconnected web to avoid a dichotomous "us and them" perspective. If people begin to appreciate their own animality, this should foster further respect for fellow animals as persons/individuals. It only perpetuates the nature/culture and human/animal dualisms if documentaries characterize "the wild" as dangerous, harsh, unethical, and inhuman while implicitly privileging so-called developed human society as humane, civilized culture. This denies humans' place in nature, the wild justice of social animal cultures, and the rational sustainability of natural systems (Bekoff and Pierce 2009; Freeman 2010a). When nature programs frequently emphasize dramatic predator-prey chase and attack scenes for purposes of heightening the action for viewers, it can perpetuate the stereotype of nature as primarily "red in tooth and claw"—a brutal aggression implying an innately violent animalistic nature that civilized humans seek to repress.

It would be beneficial to originate a TV network that represents a nonspeciesist, biocentric perspective in its documentaries, news, public affairs, drama, comedy, and lifestyle programming—linked to the web for free viewing. Channels such as Discovery, Planet Green, Animal Planet, and National Geographic

Wild are often too conservative and anthropocentric. Additionally, to supplement fine series such as PBS's *Nature*, public broadcasting should integrate a biocentric perspective across the spectrum of its kids and adult programming (as it has a mandate to be educational and socially responsible), including starting a weekly prime-time show dedicated to critical environmental issues such as mass extinction and climate change.

News

As part of their propensity for episodic frames, American news media do not focus much on nature and environmentalism unless there is a catastrophic episode like an oil spill. Non-event-oriented catastrophes that are ongoing and chronic (such as species extinction, climate change, or factory farm pollution) are harder to fit into narrative news story formats that value drama, timeliness, and visual spectacle. Unfortunately, episodic frames create incomplete understandings (Siemer, Decker, and Shanahan 2007). News stories that are thematic rather than episodic in nature are important at the beginning of a controversy, as they tend to emphasize solutions while providing a broader context for understanding issues in a way that is public rather than personal.

We contend that if the environment were made a regular news beat like politics, sports, or business, then issues such as mass extinction could be covered thematically as a daily crisis. This dedicated environmental beat should view ecological issues as more than just scientific, as they are also sociopsychological, ethical, and political. A new less human-biased perspective is called for that is more biocentric and values the interests of the other living beings regardless of their usefulness or charm to humans.

One way this can be achieved is for journalists to view NHAs as a legitimate news source whose perspective and interests deserve a voice in stories that affect their lives (Freeman, Bekoff, and Bexell 2011). When voices are absent, those beings appear as if they do not matter, reinforcing a speciesist privileging of human interests. As part of journalism's commitment to truth and justice, they should represent other animals accurately and fairly, discussing them as individuals on their own terms, and avoiding discussing them primarily in terms of human-centered utilitarian calculations. This would mean journalists should avoid stereotyping certain species as "natural" pests, threats, game, or tools for humans (for food, research, skins, entertainment, etc.) (Freeman 2009).

As human spokespeople for NHA interests, environmental and animal activists deserve to have their perspectives respectfully incorporated into news instead of being marginalized as radicals in favor of more "reasonable" government or industry sources. One way for this to develop is at the university level, where incorporation of environmental classes into journalism curriculum would provide a foundation for lasting changes in the attitudes and practices of professionals in the industry.

Conclusion

Activists understandably rely on media exposure as their main tool for raising awareness and leveraging power against governments and exploitative industries. As an example, one of the last scenes in the Academy Award–winning documentary *The Cove* shows activist Ric O'Barry standing in the crowded, neon-lit commercial heart of downtown Tokyo wearing a DVD screen on his chest to showcase undercover footage of the dolphin slaughter in Taiji, Japan. While emphasizing the media's importance, this scene also reveals the marginalization of proanimal discourse competing for attention among a clutter of images in a commercially dominated public sphere (Freeman and Tulloch in press). So while the media have the *potential* to be a major force in preventing ecological collapse (as Jared Diamond asserts), we acknowledge the need to demand a paradigm shift in mainstream media values that currently put profit, consumerism, and amusement before the long-term planning, problem solving, and reassessment of core values that will be required to save life on Earth.

We call upon citizens to monitor and reward ecologically responsible media by supporting media watchdog groups such as Dawnwatch (for animals) and media awards presented by the Environmental Media Association and the Humane Society of the United States (the Genesis Awards). Citizens should write to media producers to express praise or criticisms and should financially support media that stand up for animals and nature. Because commercially funded programming has less financial incentive to produce ecologically responsible messages, citizens must also use and support public, noncommercial, and nonprofit media, including emerging nonprofit journalism organizations that may require donations to produce investigative reports. Local citizens should take advantage of public access channels, community radio, websites, blogs, and social media as ways to start producing their own public affairs or advocacy programming or airing documentaries that commercial media tend not to show. In crafting messages, ecologically minded citizens, scientists, and activists should present the hard facts while also openly speaking to their ideals and moral vision (without watering it down). The challenge to speaking candidly is to do so in strategic ways that still resonate culturally with target audiences who may be speciesist or environmentally unsavvy (Freeman 2010b; Lakoff 2004).

This chapter demonstrates that society is bombarded with images and ideas about nature in both fiction and nonfiction media platforms. These networks must recognize the shared collective burden we have as humans inextricably linked to the ecosystem where we live. It is the connection we have as individuals both to our human community as well as to our ecological bioregion that is often lost in modern media. To avoid simply consuming nature, we must view ourselves as more than media consumers (or as consumers in general): as media reformers, media producers, and engaged ecological citizens. This chapter highlights both the strengths and weaknesses in modern media formats and conventions with the belief that change is possible to prevent ecological collapse but must begin now.

References

Anderson, Benedict. 2006. *Imagined Communities: Reflections on the Origin and Spread of Nationalism*. New York: Verso.

Bandura, Albert. 1994. "Social Cognitive Theory and Mass Communication." In *Media Effects: Advanced in Theory and Research*, edited by J. Bryant and D. Zillman. Hillsdale, NJ: Erlbaum.

Barbas, Tasos A., Stefanos Paraskevopoulos, and Anastasia G. Stamou. 2009. "The Effect of Nature Documentaries on Students' Environmental Sensitivity: A Case Study." *Learning, Media, and Technology* 34 (March): 61–69.

Bekoff, Marc, and Jessica Pierce. 2009. *Wild Justice: The Moral Lives of Animals*. Chicago: University of Chicago Press.

Besel, Richard, and Renee Besel. 2010. "Whale Wars and the Public Screen: Mediating Animal Ethics in Violent Times." In *Arguments about Animal Ethics*, edited by G. Goodale and J. E. Black, 163–77. Lanham, MD: Lexington Books.

Boykoff, Maxwell T. 2009. "We Speak for the Trees: Media Reporting on the Environment." *Annual Review of Environment and Resources* 34 (November): 431–57.

Burgess, Jacquelin. 1990. "The Production and Consumption of Environmental Meanings in the Mass Media: A Research Agenda for the 1990s." *Transactions of the Institute of British Geographers* 15 (2) new series: 139–61.

Burgess, Jacquelin, Carolyn Harrison, and Paul Maiteny. 1991. "Contested Meanings: The Consumption of News about Nature Conservation." *Media, Culture, and Society* 13: 499–519.

Carey, James W. 1989. "A Cultural Approach to Communication." In *Communication as Culture: Essays on Media and Society*, edited by James W. Carey, 13–35. London: Unwin-Hyman.

Carmack, Betty J. 1997. "Realistic Representations of Companion Animals in Comic Art in the USA." *Anthrozoös* 10 (2): 108–20.

Cohen, Judy, and John Richardson. 2002. "Pit Bull Panic." *Journal of Popular Culture* 36 (Fall): 285–317.

Creed, Barbara. 2006. "A Darwinian Love Story: Max Mon Amour and the Zoocentric Perspective in Film." *Continuum: Journal of Media and Cultural Studies* 20 (March): 45–60.

Crystal, Richard. 2003. *Animals in the News*. New York: Warner Books.

Diamond, Jared. 2005. *Collapse: How Societies Choose to Fail or Succeed*. New York: Penguin.

Earp, Jeremy. 2010. *The Mean World Syndrome: Media Violence and the Cultivation of Fear*. DVD. Documentary. Media Education Foundation. http://www.mediaed.org/cgi-bin/commerce .cgi?preadd=action&key=143.

Entman, Robert. 1993. "Framing: Toward Clarification of a Fractured Paradigm." *Journal of Communication* 43 (4): 51–58.

Fraser, John, Diana Reiss, Paul Boyle, Katherine Lemcke, Jessica Sickler, Elizabeth Elliott, Barbara Newman, and Sarah Gruber. 2006. "Dolphins in Popular Literature and Media." *Society and Animals: Journal of Human-Animal Studies* 14 (4): 321–49.

Freeman, Carrie P. 2009. "This Little Piggy Went to Press: The American News Media's Construction of Animals in Agriculture." *Communication Review* 12 (1): 78–103.

———. 2010a. "Embracing Humanimality: Deconstructing the Human/Animal Dichotomy." In

Arguments about Animal Ethics, edited by G. Goodale and J. E. Black, 11–30. Lanham, MD: Lexington Books.

———. 2010b. "Framing Animal Rights in the 'Go Veg' Campaigns of U.S. Animal Rights Organizations." *Society and Animals* 18 (2): 163–82.

Freeman, Carrie P., Marc Bekoff, and Sarah Bexell. 2011. "Giving Voice to the 'Voiceless': Incorporating Nonhuman Animal Perspectives as Journalistic Sources." *Journalism Studies* 12: 1018.

Freeman, Carrie P., and Debra Merskin. 2008. "Having It His Way: The Construction of Masculinity in Fast Food TV Advertising." In *Food for Thought: Essays on Eating and Culture*, edited by L Rubin, 277–93. Jefferson, NC: McFarland.

Freeman, Carrie P., and Scott Tulloch. In press. "Was Blind but Now I See: Animal Liberation Documentaries' Deconstruction of Barriers to Witnessing Injustice." In *Screening Nature: Cinema beyond the Human*, edited by A. Pick and G. Narraway. Oxford and New York: Berghahn Books.

Ganetz, Hillevi. 2004. "Familiar Beasts: Nature, Culture, and Gender in Wildlife Films on Television." *NORDICOM Review* 25 (September): 197–213.

Gerbner, George, L. Gross, M. Jackson-Beeck, S. Jeffries-Fox, and N. Signorielli. 1978. "Cultural Indicators: Violence Profile No. 9." *Journal of Communication* 28: 176–206.

Giblett, Rod. 2006. "Alligators, Crocodiles, and the Monstrous Uncanny." *Continuum: Journal of Media and Cultural Studies* 20 (3): 299–312.

Hall, Stuart. 1980. "Encoding/Decoding." In *Culture, Media, Language: Working Papers in Cultural Studies, 1972–1979*, 128–38. London: Hutchinson.

———. 1997a. "Representation and the Media." Media Education Foundation, Northampton, MA. http://www.mediaed.org/assets/products/409/transcript_409.pdf.

———. 1997b. *Representation: Cultural Representations and Signifying Practices*. London: Sage Publications.

Harrison, Carolyn M., and Jacquelin Burgess. 1994. "Social Constructions of Nature: A Case Study of Conflicts over the Development of Rainham Marshes." *Transactions of the Institute of British Geographers* 19 (3) new series: 291–310.

Herzog, Harold A., and Shelley L. Galvin. 1992. "Animals, Archetypes, and Popular Culture: Tales from the Tabloid Press." *Anthrozoös* 5 (2): 77–92.

Hirschman, Elizabeth, and Clinton Sanders. 1997. "Motion Pictures as Metaphoric Construction: How Animal Narratives Teach Us to Be Human." *Semiotica* 115 (1): 53.

Horak, Jan-Christopher. 2006. "Wildlife Documentaries: From Classical Forms to Reality TV." *Film History* 18 (4): 459–75.

Jerolmack, Colin. 2008. "How Pigeons Became Rats: The Cultural-Spatial Logic of Problem Animals." *Social Problems* 55 (1): 72–94.

Lakoff, George. 2004. *Don't Think of an Elephant! Know Your Values and Frame the Debate*. River Junction, VT: Chelsea Greene Publishing.

Lerner, Jennifer E., and Linda Kalof. 1999. "The Animal Text: Message and Meaning in Television Advertisements." *Sociological Quarterly* 40 (Autumn): 565–86.

Lippit, Akira Mizuta. 2000. *Electric Animal: Toward a Rhetoric of Wildlife*. Minneapolis: University of Minnesota Press.

Mazur, Allan, and Jinling Lee. 1993. "Sounding the Global Alarm: Environmental Issues in the US National News." *Social Studies of Science* 23 (November): 681–720.

McChesney, Robert. 1999. *Rich Media, Poor Democracy: Communication Politics in Dubious Times*. Urbana: University of Illinois Press.

McCombs, Maxwell. 2005. "A Look at Agenda Setting: Past, Present, and Future." *Journalism Studies* 6 (4): 543–57.

Mills, Brett. 2010. "Television Wildlife Documentaries and Animals' Right to Privacy." *Continuum: Journal of Media and Cultural Studies* 24 (April): 193–202.

Pierson, David P. 2005. "'Hey, They're Just Like Us!' Representations of the Animal World in the Discovery Channel's Nature Programming." *Journal of Popular Culture* 38 (May): 698–712.

Roll-Hansen, Nils. 1994. "Science, Politics, and the Mass Media: On Biased Communication of Environmental Issues." *Science, Technology and Human Values* 19 (Summer): 324–41.

Sawers, Naarah Catherine, and Kristin Demetrious. 2010. "All the Animals Are Gone? The Politics of Contemporary Hunter Arcade Games." *Continuum: Journal of Media and Cultural Studies* 24 (April): 241–50.

Siemer, William F., Daniel J. Decker, and James Shanahan. 2007. "Media Frames for Black Bear Management Stories during Issue Emergence in New York." *Human Dimensions of Wildlife* 12 (March): 89–100.

Spears, Nancy, and Richard Germain. 2007. "A Note on Green Sentiments and the Human-Animal Relationship in Print Advertising during the 20th Century." *Journal of Current Issues and Research in Advertising* 29 (Fall): 53–62.

Swan, Davina, and John C. McCarthy. 2003. "Contesting Animal Rights on the Internet: Discourse Analysis of the Social Construction of Argument." *Journal of Language and Social Psychology* 22 (3): 297–320.

Swan, Susan Z. 1999. "Gothic Drama in Disney's Beauty and the Beast: Subverting Traditional Romance by Transcending the Animal-Human Paradox." *Critical Studies in Mass Communication* 16 (3): 350.

Umstead, R. Thomas. 2009. "Animal Turns to Reality." *Multichannel News* 30 (March 16): 2–18.

Wenner, Kathryn S., and Jill Rosen. 2002. "Dog Gone, Not Forgotten." *American Journalism Review* 24 (May): 14.

Zavestoski, Stephen, Kate Agnello, Frank Mignano, and Francine Darroch. 2004. "Issue Framing and Citizen Apathy toward Local Environmental Contamination." *Sociological Forum* 19 (June): 255–83.

20

Children, Animals, and Social Neuroscience

Empathy, Conservation Education, and Activism

Olin E. "Gene" Myers Jr.

A Paradox of Ignored Duty and Animal Ubiquity

WHAT IS the interest—the stake—of children in animals, in conservation? I start with this question because we cannot know what to ask about psychology and society unless we first are ethically clear on our direction. Children—a "future" generation that is already present— have a huge stake in animals, or more appropriately, in biodiversity conservation. Philosophers back this up regarding future generations generally. On the most obvious level, different animals and plants are humanity's inheritance, in John Locke's terms the common patrimony, naturally held by all people in all generations. According to Weiss (1989), intergenerational equity requires observance of three principles regarding this patrimony: conservation of options, conservation of quality, and conservation of access. Each generation must leave the next as much, as good, and as accessible resources, including living things. Alternately, applying Rawls's (1971) logic of equal basic liberties and fair equality of opportunity to intergenerational justice, we should arrange institutions so that no generation may claim the right to reduce the richness of the panoply of biota for later-born generations. These claims can be read as recognitions of the inherent value of every species to every person, no matter the historical period into which they are born. They can just as strongly be interpreted to refer to nonsubstitutable ecosystem functions and thereby to the foundations of human welfare that biodiversity makes possible (Naeem et al. 2009). Thus children's future is as tenuous as other species': they are united by each having everything to lose. In short, children's interest in biological conservation is complete, fundamental, and prerequisite for the satisfaction of their other interests.

Do adults in general, including those concerned directly with children's interests, appear to recognize the implied duties? On the one hand, survey data show recognition of obligations to the future, as does the potency of concern with one's legacy. But on the other hand, the duties are glaringly ignored in practice. Parents and society both invest greatly in children—in their education, health care, and material, spiritual, and moral well-being. We try to create "small worlds" (usually indoors) that serve as models and incubators for the ideals we hope children to carry with them. But we seem blind to the fact that however much we pour into our offspring, their individual futures are perilous unless we also ensure the wide living world will be adequate.

Ironically, members of modern developed societies flood their children with wild animals—beautiful, wonderful animals, baby animals, mysterious animals, humanized animals, fantasized animals, dangerous animals, heroic animals, smarter and holier-than-people animals. Animal media and discourses are varied, ubiquitous, and extravagantly costly too: pet animals, pictured and fictionalized animals, stuffed animals, robotic animals, filmed animals, manually or digitally animated animals, spectacularly housed captive land and water animals. For the lucky few, witnessed wild animals, rescued animals, animals one can really connect with. Nonetheless, these varied and pervasive animal experiences are fleeting, filling only a fraction of children's time but feeding some great hunger. Do we know, gnawingly, that this facsimile or faux animal-full-world deception hides our failure at our legitimate duties?

In a remarkable public address, the interests of children in real conservation action was articulated by twelve-year-old Severn Suzuki, David Suzuki's daughter, at the 1992 Rio Earth Summit conference:

> In my life, I have dreamt of seeing the great herds of wild animals, jungles and rain forests full of birds and butterflies, but now I wonder if they will even exist for my children to see. Did you have to worry about these little things when you were my age? All this is happening before our eyes and yet we act as if we have all the time we want and all the solutions. I'm only a child and I don't have all the solutions, but I want you to realize, neither do you! You don't know how to fix the holes in our ozone layer. You don't know how to bring salmon back up a dead stream. You don't know how to bring back an animal now extinct. And you can't bring back forests that once grew where there is now desert. If you don't know how to fix it, please stop breaking it! (Severn Suzuki 1992)

The Developmental Psychology of Children's Connections to Animals

I would like to think that all children naturally see and collaborate with animals (and adults) as allies in conservation. Some do, such as the young Ms. Suzuki.

But what is critical is to understand developmental routes to effective care and action as well as the obstacles on such paths. Unfortunately, the topic has not been central in psychology. A search for "extinction" in PsycINFO brings up many articles on learning theory; "conservation" is a Piagetian logical feat; and "animals" seldom refers to real wild creatures. But a growing and diverse body of evidence shows how children perceive, experience, express, connect with, and care about animals. I am sure that a connection to animals is a developmentally primed way by which some become passionate conservationists. Describing that path and some of its vulnerabilities is what I want to do here.

Children recognize and respond to animals in reliably patterned ways. Infants distinguish animals very early. In adults, specific brain injuries eliminate the ability to identify or name animals but not artifacts, and vice versa (Warrington and Shallice 1984; Kurbat 1997). PET scans of normal brains show areas specialized for categorizing animals versus artifacts (Martin et al. 1996). And single-neuron activation studies show a high specialization in response to pictures of animals in the right amygdala (Mormann et al. 2011). Such centers are likely the basis for what psychologist Howard Gardner (1999) calls the "naturalist intelligence"—the only addition he has made to his original seven forms of intelligence. Parts of the brain that process animal stimuli are undoubtedly foundations for cultural folk biologies and children's "naive biology"—beliefs that emerge early in childhood and robustly persist into adulthood. Both sorts of biological belief system show two broad conceptual regularities: ranked taxonomy (categorizing living things into branching hierarchies), and essentialism (the belief that species are defined by enduring internal essences inherited from parents). How much the brain specifies biological beliefs versus provides an architecture that generates these beliefs is intensely debated (Astuti, Solomon, and Carey 2004; New, Cosmides, and Tooby 2007; Atran and Medin 2008), but it is fairly clear that the ways children think about animals and about people are not fully separate.

My own research suggests that children experience live, present animals as very immediately potent individual beings that are the authors of their own actions. Thus an animal's "agency" may have consequences for the self, good or bad, and "reading" it is important. Further, animals display coherence: coordinated movement of a stable bodily form, a characteristic of selfhood. And they evince affective states—moods, patterns of arousal, or rest. Children also gauge an animal's attention and intention. And if they have the chance to interact with an animal repeatedly, there is a basic sense of continuity or relationship. These are in fact the same basic categories of self and other that underlie human social interaction from early infancy (Stern 1985, 2007). What I found is that children use their relational social intelligence not to simply anthropomorphize but to understand the animal as a subjective other while also registering its difference in degrees and type in agency, affect, coherence, and relationship. Many consequences flow from this. Primary among them: the self—always built from the dynamics of self and other in relationship—is formed in the available mixed species community. And we come to understand what it is to be human in part by this process of relating with nonhuman creatures (Myers 2007).

The insight can be expressed abstractly: children relate to animals because they recognize them as another animate being—a self-organizing living thing. At the very edge of awareness an animal resonates with children because it feels implicitly the same aliveness and a similar subjective interiority as the self possesses. Animals, even common wild ones such as dragonflies encountered at summer camp, are a point of entry for children into the "more-than-human" world of nature (Watson 2006). Dogs and horses have proven effective in therapy with at-risk youth, PTSD patients, and individuals with psychiatric and developmental challenges in part because their immediacy is palpable, and their responses emotionally astute. A living animal confirms a child's selfhood simply by responding; it clarifies and enriches identity by implicit comparison; it symbolizes living qualities and vitality; and it can carry forward the child's own living process into new experience by providing a sense of connection and companionship—real or imagined—across difference.

This connection reaches into the child's thoughts, feelings, and actions. I found that young children are acutely aware of the animate qualities of animals they know. They show this in interactions, and in imitation and pretend play, when they imaginatively translate their bodily form into an animal's. They see animals first as individuals, often named, not as representatives of types or species. And they may be delicately sensitive to harm to animals, responding with moral emotions such as concern, worry, sadness, sympathy, and wanting to take action or feeling outrage at the cause. Other researchers have found children likely to also include animals in their moral sphere. For example, Peter Kahn (1999) reported that in several cultures, harm to animals is one of the main reasons children give for why they judge it wrong to pollute a waterway.

New ideas and findings provide a biopsychological way to explain this human or child resonance with the animal. The social brain hypothesis (Dunbar 1998, 2009) proposes that in a close social group, the key selective forces on an individual's success are the interactions with the other group members. The intensity of social bonds within hominin and human groups (such as would be necessary for rearing very dependent young) explain the growth in size and complexity of the human brain. Most specifically, in an interdependent, dynamic, power-pervaded group, one needs a way to understand and predict other members' actions. Meeting this need to understand others is arguably the function of the exceptionally developed "mirror neuron" system in the human brain that allows us to use the self as a rough model of the other. In this system, the same areas of the brain that we use to do an action or have a feeling are activated when observing another do the action or experience the feeling. Other areas that may help distinguish self and other are also activated. This spontaneous "natural" empathy lets us internally "match" another's actions, feelings, and intentions (Sommerville and Decety 2006). But empathy can be further elaborated by perspective-taking, understanding cognitively more about the other, thus potentially correcting ways the other is different than the self (Decety 2005), and calibrating a moral response, including to animals (Schultz 2000).

In recent studies of brain-area activation across children to adults, subjects watched simulations of people harming themselves and of someone else harming the targets (intentionally stepping on their foot). In children (age seven) empathic responses to both scenarios were significantly dominated by emotional centers of the brain (Decety and Michalska 2010), whereas adolescents and adults showed comparatively more activation in cognitive processing areas. Mirroring areas showed up in all ages, but the amygdala, an emotional center that is also concerned with threat response, was relatively more active in the younger children. The child's ability to self-regulate emotions increases as the brain's frontal lobes mature (a process characterizing adolescence), allowing more cognitive processing of the events. But studies have shown that children's dispositions vary in how well they regulate their emotions. In vicarious empathy-inducing situations, some children are able to avoid overaroused distress and thus respond with more other-directed sympathy. Less well self-regulated children experience more empathic distress, where self-concern swamps empathy, and thus show a lesser tendency to respond prosocially (Eisenberg et al. 1994). Nonetheless, the core of empathy is emotion-matching.

Despite the explosion of social cognitive neuroscience that has revealed the workings of the mirroring system in recent years, researchers have yet to present animal emotion and feeling to subjects to see how the brain processes them. Probably the mirroring system is involved, partly because it is so widespread among other animals (Preston and de Waal 2002) and likely involved in behavioral ecology. More to the point, I have observed children playfully imitating dogs, birds, frogs, and even snakes with their bodies. This phenomenon could reflect spontaneous selective recruitment of parts of the mirror system to approximate motor/emotional state-matching, or it could be cognitively mediated, or both. In either case, from the above we could speculate that some children (low self-regulators) may be overwhelmed by exposure to vivid portrayals of harm to animals, whereas others may be more moved to act.

Mirroring is key in making us a moral creature (Hastings, Zahn-Waxler, and McShane 2006). With development children may move from natural empathy to perspective-taking and sympathy (concern for the other's welfare) and then to higher levels of moral reasoning about animals in their more complex situations. Facing these challenges, they need an emotionally supportive environment for many years. This account gives us a primary understanding of why children and animals can be natural allies in conservation.

Obstacles to the Expression of Natural Care

Why do we then observe so much ignoring of animals' plights? For one, the path outlined above, while developmentally prepotent, is, like many other developmental achievements, a result of person-environment interactions. Some environments do not support a thoroughgoing compassion for animals, just like some do not support seeing some groups of humans as worthy of moral consideration. The literature indicates a number of patterns that can block caring.

Traits of the animal make a difference. Charismatic traits pull toward caring. Attractiveness, aesthetic beauty, similarity to humans, cute features, ascriptions of intelligence, and positive symbolic values all increase the likelihood of caring about an animal. Humbler animals fare less well. Natural history and ecological literacy are important in building caring awareness of how animals meet their needs in nature, also weaving the easily ignored small creatures that drive ecosystem functions into a more complete valuing of habitat, biome, and biodiversity. Ethological information on a species, especially if framed in narratives that create a sympathetic frame of reference, also helps because the functional meanings of behaviors are made apparent.

When an animal is perceived to be a threat, caring is likely to be reduced; fear is more likely. If people perceive conflicts between their interests and a species, not surprisingly they value it less, as seen in studies in many cultures. We noted above how the children in Kahn's (1999) studies worried about harm to animals—but such biocentric reasoning is weak when there is a countervailing human benefit. However, Severson and Kahn (2010) devised a hypothetical scenario that removed human-animal conflicts of interest and found much more biocentric reasoning.

Children age ten to thirteen years often express a strong moral urgency about wrongs they learn of, insisting that something be done. This urgency attenuates, however, in later adolescence (Bardige 1988). Part of the reason may be greater cognitive appreciation of complexity—being able to see different sides of the story. This may explain, for example, the decreased commitment to act on behalf of sea turtles following an education program that included stakeholder role-modeling, for eleven- to thirteen-year-olds on the Greek island of Zakynthos (Dimopoulos, Paraskevopoulos, and Pantis 2008). Indeed, it is easy to be dumbfounded over biodiversity value conflicts: the rights of Makah Indians to traditional subsistence, versus the rights of the gray whales hunted; the urge to control inconvenient deer populations where humans have exterminated predators; the hazing or killing of sea lions and terns that devour salmon deterred by the Columbia River dams. Our conclusions in such cases need not amount to paralysis, but rather the resolve to creatively integrate the human-centered and animal-centered values in more adequate ways.

Where people exploit animals we tend to find psychological defenses that reduce uncomfortable conflicting feelings and beliefs. The animals' suffering may be kept out of sight; the animals may be out-grouped, denigrated, and demonized; the harm rationalized against human benefits. Sometimes the balance is kept by cultural representations that affirm the animals' subjective status—usually through ritual and belief in a human-animal spiritual continuity powerful enough to bring bad luck if people do not propitiate the spirits and follow strict rules. Such elaborate rationalizations can become vicious, proposing that the animal wishes to be sacrificed. Or they may produce a respectful and ecological coexistence. But few have looked at the developmental aspects. Melson (2001) interviewed children and families who participate in 4-H animal projects, where the children raise baby farm animals for several months, show them in competitions, and then relinquish the animals to be led to slaughter. Younger participants (ages eight to ten) find this very difficult but may hide their feelings in public. Later they come "to understand the whole cycle, how they *have* to go to market. . . . Life has to go on, after all. But still, it's always hard" (68, twelve-year-old girl, emphasis in original quote). Older children cease naming their animals, focus on the money earned, or the competitions won. The children's elders emphasize the benefits: learning, responsibility, camaraderie. Melson notes, however, that this avoids the children's struggle of "how to reconcile emotional attachment to an animal cared for almost since birth with the exigencies of turning that animal into meat products" (69). Farmers appear to balance caring for their animals with their economic fate, a route that involves some psychic trade-offs.

Such psychological defenses may become "naturalized" as shared cultural norms of belief and emotion management. In one intervention, even providing youths with a chance to reevaluate negative beliefs about prairie dogs (aided by potent tools like a field-based experiential approach, interaction with experts and other positively leaning stakeholders, and learning relevant ethology) failed to overcome highly negative community feelings toward the species (Fox-Parrish and Jurin 2008). Perhaps the curriculum could have gone further to provide empathy-inducing experiences, such as Sarah Bexell (2006) designed in China. Chinese cultural norms, rooted in a long agrarian tradition and utilitarian attitude toward animals, are not known for a humane ethic. Bexell's conservation education curriculum provided young middle-class Chinese participants and adult helpers with opportunities to carefully observe animal behaviors and relate them to animals' needs. They were allowed to personalize animals as subjective individuals, to validate perceptions of animal feeling and mentality, to touch, pet, and hold pet animals, and to learn about appropriate pets and their care and about wild animal conservation needs. After these experiences, children, and most notably, adults, said they would never look at animals in the old way again. Such examples should give us pause, because it may be too easy to accept another culture's indifference to animals relativistically, as a matter of inevitable and unchangeable cultural patterns. Certainly we cannot cavalierly criticize others' cultures. But it is possible that the attitudes that justify ignoring animals are less deeply rooted than we have thought, that liberating empathy is just under the surface.

Stress and Coping

Another challenge is how children cope with stresses related to biodiversity loss and other dimensions of environmental damage. Here I will address not so much the physical survival stresses that already confront two-fifths of humanity, but the psychological ones that attend learning about ecological trends affecting children's own futures. The main conclusion is that if adults provide a culture of support, children will be equal leaders in movements to remember animals.

One theory has been termed "ecophobia" by Sobel (1996). In his sense it combines two notions: it suggests that lacking direct experience in nature, children will be afraid of it and never become attached. It also suggests that exposing children to information about faraway environmental disasters—such as habitat and species loss in the developing world—will make them fear environmental losses and feel disempowered. Although these are really separate issues, his dictum—no environmental tragedies before age ten—has been widely accepted. He recommends that nearby direct nature experiences and involvement in local community efforts should be offered to build children's affinity with nature, and their self-efficacy. These prescriptions are supported by developmental and environmental education research. Direct experience in nature develops lasting affective ties, as shown in studies of early influences in the lives of environmental educators and activists (Chawla 1998, 1999). The declines of children's access to nature, and unstructured time spent in it (Louv 2008), and of natural history knowledge and role models, may undermine attempts to motivate caring about animals and their homes. And, true also, people of any age can be defeated by overwhelming odds. Instead people gain competence, confidence, and make real progress by pursuing a strategy of "small wins" (Weick 1984) and collective efficacy.

But despite such endorsement of positive nature experience, what is interesting is that the negative effects predicted from exposure to biological losses have been accepted on slim grounds. Sobel (1996) offers little evidence that studying far-off disasters is damaging to children. Ironically, more anecdotes to this effect were pro-offered by the discredited environmental education backlash book *Facts Not Fear* by Sanera and Shaw (1996). But virtually no serious research has isolated the affective consequences of such instruction. The one exception points away from Sobel's (and Sanera and Shaw's) theory. A study of four classes of fourth graders by Matteson (2008) found no increase in fear of or dissociation from nature following a weeklong rain forest curriculum highlighting deforestation (compared to the four control group classes), and no effect on empowerment.

The broader issue here is how children deal with threats, and more particularly how we adults help them deal with them. For informative comparisons we can look to times children have been thrown into a problem much larger than themselves. In *Children of Crisis*, psychiatrist Robert Coles (1964) documented how black children in the American South took roles, even leading ones, in nonviolent activities surrounding desegregation. They persisted despite retaliatory attacks, but showed no "discernible psychiatric harm or collapse" (319). His celebrated description of six-year-old Ruby Bridges, the first child to integrate William Frantz Public School in New Orleans, is one case of many. An acute observer of children in many exceptionally psychologically and socially challenging situations, Coles comments that "many children the world over have revealed a kind of toughness and plasticity under far from favorable conditions that make the determined efforts of some parents to spare their children the slightest pain seem quite ironic" (323–24).

The question of whether to expose children to the fight against loss of animals and involve them in it differs from desegregation. For while as we argued earlier children do have a fundamental interest in biodiversity, that interest is not as direct and personal as one's civil rights: in the latter, one is the direct beneficiary, and can speak from that position. In the animal case, the animal is the direct beneficiary (or victim) and many can claim to speak for its interests—or for their own interests, which counter the animal's. Nonetheless, one can imagine looming social confrontations over biocollapse as intense as those over desegregation. The case of desegregation is helpful, because it points to factors that psychology can help elucidate, factors that can help children stand up for animals, nature, and themselves. These factors include coping resources, multigenerational participation, and resilience. Marshaling these will, however, require some redirection in adults' strengths, thinking, and activities.

Supporting Practical and Emotional Coping

When people are confronted with a threat, protection-motivation (PM) theory suggests they go through several steps in arriving at a response (Rippetoe and Rogers 1987; Gardner and Stern 2002). The first steps include determining whether something one values is threatened, with what severity and probability, and generating an emotion such as fear. Also, one's ability to do something effective and the cost of doing it lead to an assessment of one's ability to cope. The balance of these two appraisals—of threat and coping—is important. It determines how the person is likely to respond. Considering just the cases of high perceived threat, if one feels able to respond, an adaptive and problem-focused response is likely. If, however, one feels helpless, then emotion-focused coping is more likely. Children generally have less coping skills—of both practical and psychological sorts—than adults. They are likely to use emotion-focused coping if confronted with large challenges: they will be overwhelmed, ignore, deny, feel defeated, or turn to happier thoughts.

It would be a mistake, however, to stop here. First, we need to look more deeply at emotion coping, which may be underrated in PM theory. Several measures of emotion-focused coping are confounded with maladjustment. In contrast, contemporary theories of affect regard emotion as adaptive. While intense emotion may reduce thoughtful responding, attending to and expressing emotion is an important part of dealing with stress of any sort, as in many therapeutic practices. Studies using a new construct, "emotional approach coping" (EAC) (Stanton, Sullivan, and Austenfeld 2009) have shown that emotion processing and expression benefit people suffering from depression, chronic pain, anxiety, consequences of sexual assault, infertility, and cancer. EAC works best when significant others are receptive, and when there are appropriate outlets for emotions. It may have its positive effects through clarifying goals, rethinking the cause of the stress, affirming positive personal qualities, and helping the person draw maximally on their social environment (Stanton, Sullivan, and Austenfeld 2009).

Combining children's emotional responding with the EAC findings brings to mind the everyday ways parents help children cope with challenges as well as develop self-regulation and concern for others. For example, suppose ordinary events lead to a "meltdown" where emotions are rawly expressed. The parent provides a safe environment for all these feelings, helps the child think back over events and process feelings, and supports and affirms the inner character development occurring. This may not resolve or remove the cause of the threat;

temporary "shutdown" may continue, but the coping assistance is directly help-ful and sets the child up for successful future coping. This is familiar ground. Adults can, and indeed must, help children (and one another) this way.

Second, children are capable of practical action, of problem-focused coping. Sources Sobel overlooked reveal children already do know about large environ-mental problems. And they act on what they learn. Wals (1994) and King (1995) both provide qualitative portraits of various patterns in how children construct environmental problems. Expectably incomplete, children's views reflect the major discourses in society. These include American culture's emphasis on individual actions such as littering and recycling, technocratic solutions, and one-dimensional conceptions of "human nature." But they also include, for some seventh graders in Wals's study, a politicized understanding of the prob-lems. These children see current problems in a historical context, including institutional actors and collective actions as both causes and solutions. I am reminded by King's (1995) work that while few children have Severn Suzuki's access to influential forums, many children have communicated their concerns to officials. They respond sincerely and urgently to their awareness of what is at stake for them. Among King's subjects, several stand out for their overt ac-tion. One is Benjamin, a nine-year-old who sees what's at stake, understands a bit of the pathways of power, and tries to organize a few friends to take action. King points out that his mode is not to nag adults as in the caricature child activist. Rather, he is in harmony with his parents, who are politically active; they scheme and work together. It is not at all surprising a child might be good at and interested in the life pursuits of his parents.

Adults and Children Learning to Participate In and Lead Change

But here is a telling difference between the civil rights movement and society's biodiversity-crisis responses. Both cases involve social conflict, but for civil rights there was an overt, principled, well-organized movement that used the personal and institutional powers of many to determinedly pursue its goals. Like Ruby Bridges, Benjamin has dependable and seasoned social-emotional support for confronting the much-larger-than-himself problems his future holds. But most children do not. What, however, if they did?

Adults tend to reproduce the ways they themselves were raised in how they raise and regard children. Since few adults were raised or schooled in an envi-ronment that systematically supported authentic democratic participation, it is little surprise that there are few Benjamins out there. There is no need for that to continue, however. David Johns (2009) provides many useful lessons for generating effective political movements for conservation. Adults could then truly support children's participation via a renewed collective activism. Case studies, principles, and tools for nonmanipulative children's participation are available, for example in Roger Hart's book *Children's Participation* (1997), and in journals such as *Children, Youth, and Environments*. Some of the best examples include child-adult initiated projects, decision making, action, and evaluation. The next generation needs to be raised in an entirely different and pervasively empowering social environment.

What would an adults-and-children movement to acknowledge and stop extinction and biocollapse look like? Beyond providing exposure to animals and nature, and education for ecological literacy, it would affirm perceptions of animals' experienced welfare. It would also embrace *all* a child's emotional reactions to our dilemmas with animals, and work through the obstacles to moral commitment listed above, including defense mechanisms, ecophobia, and disempowerment. For comparison, in programs to educate about the Holocaust, Bardige (1988) reported, high schoolers encountered confusing human complexities, and the basic facts prompted some disbelief and denial. Nonetheless, the educators didn't allow students to stop there, avoiding the essential lessons. This powerful emotional work was aided by success—the Nazi regime was defeated—and driven by the potential of fascism to rise again. In the case of biocollapse, we have some assurance from past successes, and we have a clear warning for the future. Lacking strong assurance, we need the conviction that solutions are possible.

Psychological Resilience

The concept of resilience arose in the field of developmental psychology in the 1960s and 1970s. As anticipated in the passage by Coles (1964) quoted above, children have multiple pathways toward normal development, utilizing biological, psychological, and social resources to navigate sometimes daunting terrain. There are many lessons from what has become a large research area. Children increase resilience by developing: self-control of attention, arousal, and impulses; positive self-perceptions; self-efficacy; a positive outlook on life; and faith or a sense of meaning in life. Adults can employ multiple strategies to promote their own resilience: increasing community assets, organizing activity centers, building graduated successes, fostering secure relationships, encouraging friendships with prosocial peers (Masten et al. 2009). Resilience is a key part of the realignment from a focus on human weaknesses to what is now called "positive psychology." While it may seem off-key to be positive in the context of ignoring animals, the approaches listed here hint at some rethinking. We might think in terms of realigning emotional lives, of the ways we develop self-regulation, of creating settings that boost confidence. We might revision our faiths and our sense of meaning—our sense of being part of something much larger than ourselves—around a new collective project: saving our home and its many diverse dwellers.

Of course, none of this can happen if parents and educators pretend that there is really no threat, if adults collude in denying and ignoring animals and children's stake in them. But could it be a legitimate extension of adult-child paternalism to guard children from a threat so important to their future but so large? Yes, but only if it is motivated out of sincere desire to help children realize their interests and long-run independence, and effective cooperation in helping this happen. In this connection, Douglas notes, "Even when children believe that their parents are completely wrong about what is in their best interest, they rarely rebel and will go along with parents' desires . . . as long as they perceive that the parents love them" (1983, 174). The danger is that children will suppress their own misgivings and go along with their socialization into ignoring animals. There are grave risks to overprotection. Exposure to stresses, and learning that one can overcome them, builds resilience. Whether we "protect" our children or not, we must help them develop a better response capability than the current adult culture demonstrates. This requires us to change too, to face realities whose implications we would rather ignore, to model—to live—effective responses, to be partners with children in defending their interest in animals.

Philosopher Sissela Bok (1982) examines a pervasive human phenomenon: secrecy. There is no prima facie case against secrets, she points out, although we seldom question our justifications for whether we keep or reveal a secret, or probe into or leave one alone. Secrecy is sometimes needed, but also it is important to see how secrecy highlights our ignorance and vulnerability. On matters of such import as the ones this book treats, plus matters of such personal concern as our children's development, we should question what is gained by furthering children's ignorance. We should question what we fear if we told them the truth. This will lead to us grasping what we must do to help our youth act now—perhaps after we discover that the only ones we have kept in the dark are ourselves, that our ignoring of animals has been in plain sight to the young all along.

References

Astuti, R., G. Solomon, and S. Carey. 2004. "Constraints on Conceptual Development: A Case Study of the Acquisition of Folkbiological and Folksociological Knowledge in Madagascar." *Monographs of the Society for Research in Child Development, Series no. 277* 69 (3).

Atran, S., and D. Medin. 2008. *The Native Mind and the Cultural Construction of Nature*. Cambridge, MA: MIT Press.

Bardige, B. 1988. "Things So Finely Human: Moral Sensibilities at Risk in Adolescence." In *Mapping the Moral Domain*, edited by C. Gilligan, J. V. Ward, and J. M. Taylor, 87–110. Cambridge, MA: Harvard Graduate School of Education.

Bexell, S. 2006. "Effect of a Wildlife Conservation Camp Experience in China on Student Knowledge of Animals, Care, Propensity for Environmental Stewardship, and Compassionate Behavior toward Animals." Unpublished PhD diss., Georgia State University.

Bok, S. 1982. *Secrets: On the Ethics of Concealment and Revelation.* New York: Pantheon.

Chawla, L. 1998. "Significant Life Experiences Revisited: A Review of Research on Sources of Environmental Sensitivity." *Journal of Environmental Education* 29 (3): 11–21.

———. 1999. "Life Paths into Effective Environmental Action." *Journal of Environmental Education* 31 (1): 5–26.

Coles, R. 1964. *Children of Crisis: A Study of Courage and Fear.* Boston: Little, Brown.

Decety, J. 2005. Perspective Taking as the Royal Avenue to Empathy. In *Other Minds: How We Bridge the Divide between Self and Others,* edited by B. F. Malle and S. D. Hodges, 143–57. New York: Guilford.

Decety, J., and K. J. Michalska. 2010. "Neurodevelopmental Changes in the Circuits Underlying Empathy and Sympathy from Childhood to Adulthood." *Developmental Science* 13 (6): 886–99.

Dimopoulos, D., S. Paraskevopoulos, and J. D. Pantis. 2008. "The Cognitive and Attitudinal Effects of a Conservation Educational Module on Elementary School Students." *Journal of Environmental Education* 39 (3): 47–61.

Douglas, J. D. 1983. "Cooperative Paternalism versus Conflictual Paternalism." In *Paternalism,* edited by R. Sartorious, 171–200. Minneapolis: University of Minnesota Press.

Dunbar, R. I. M. 1998. "The Social Brain Hypothesis." *Evolutionary Anthropology* 6 (5): 178–90.

———. 2009. "The Social Brain Hypothesis and Its Implications for Social Evolution." *Annals of Human Biology* 36 (5): 562–72.

Eisenberg, N., R. A. Fabes, B. Murphy, M. Karbon, P. Maszk, M. Smith, C. O'Boyle, and K. Suh. 1994. "The Relations of Emotionality and Regulation to Dispositional and Situational Empathy-Related Responding." *Journal of Personality and Social Psychology* 66: 776–97.

Fox-Parrish, L., and R. R. Jurin. 2008. "Students' Perceptions of a Highly Controversial yet Keystone Species, the Black-Tailed Prairie Dog: A Case Study." *Journal of Environmental Education* 39 (4): 3–14.

Gardner, G., and P. Stern. 2002. *Environmental Problems and Human Behavior.* Boston: Pearson Custom Publishing.

Gardner, H. 1999. *Intelligence Reframed: Multiple Intelligences for the 21st Century.* New York: Basic Books.

Hart, Roger. 1997. *Children's Participation.* London/New York: Earthscan/UNESCO.

Hastings, P. D., C. Zahn-Waxler, and K. McShane. 2006. "We Are, By Nature, Moral Creatures: Biological Bases of Concern for Others." In *Handbook of Moral Development,* edited by M. Killen and J. G. Smetana, 483–516. Mahwah, NJ: Lawrence Erlbaum.

Johns, D. 2009. *A New Conservation Politics.* West Sussex, UK: Wiley-Blackwell.

Kahn, P. H., Jr. 1999. *The Human Relationship with Nature: Development and Culture.* Cambridge, MA: MIT Press.

King, D. L. 1995. *Doing Their Share to Save the Planet: Children and Environmental Crisis.* New Brunswick, NJ: Rutgers University Press.

Kurbat, A. 1997. "Can the Recognition of Living Things Really Be Selectively Impaired?" *Neuropsychologia* 35 (6): 813–27.

Louv, R. 2008. *Last Child in the Woods.* Chapel Hill, NC: Algonquin Books.

Martin, A., C. L. Wiggs, L. G. Ungerleider, and J. V. Haxby. 1996. "Neural Correlates of Category Specific Knowledge." *Nature* 379: 649–52.

Masten, A. S., J. J. Cutuli, J. E. Herbers, and M. J. Reed. 2009. "Resilience in Development." In *Oxford Handbook of Positive Psychology,* edited by C. R. Snyder and S. J. Lopez, 117–31. Oxford: Oxford University Press.

Matteson, K. A. 2008. "Too Young for Rain Forests? The Impact of a Rain Forest Curriculum on Elementary Students' Affective Responses to Nature." Dissertation abstract, Fielding Graduate University. Dissertation Abstracts International Section A: Humanities and Social Sciences, vol. 69 (4-A), 1267.

Melson, G. 2001. *Why the Wild Things Are: Animals in the Lives of Children*. Cambridge, MA: Harvard University Press.

Mormann, F., J. Dubois, S. Kornblith, M. Milosavljevic, et al. 2011. "A Category-Specific Response to Animals in the Right Human Amygdala." *Nature Neuroscience* 14: 1247–49.

Myers, O. E., Jr. 2007. *The Significance of Children and Animals: Social Development and Our Connections to Other Species*. 2nd rev. ed. West Lafayette, IN: Purdue University Press. Original work published 1998.

Naeem, S., D. E. Bunker, A. Hector, M. Loreau, and C. Perrings, eds. 2009. *Biodiversity, Ecosystem Functioning, and Human Wellbeing: An Ecological and Economic Perspective*. New York: Oxford University Press.

New, J., L. Cosmides, and J. Tooby. 2007. "Category-Specific Attention for Animals Reflects Ancestral Priorities, Not Expertise." *PNAS* 104 (42): 16598–603.

Preston, S. D., and F. B. M. de Waal. 2002. "Empathy: Its Ultimate and Proximate Bases." *Behavioral and Brain Sciences* 25: 1–20.

Rawls, J. 1971. *A Theory of Justice*. Cambridge, MA: Belknap Press of Harvard University Press.

Rippetoe, P., and R. Rogers. 1987. "Effects of Components of Protection-Motivation Theory on Adaptive and Maladaptive Coping with a Health Threat." *Journal of Personality and Social Psychology* 52: 596–604.

Sanera, M., and J. Shaw. 1996. *Facts Not Fear*. Washington, DC: Regnery Publishing.

Schultz, W. P. 2000. "Empathizing with Nature: The Effects of Perspective Taking on Concern for Environmental Issues." *Journal of Social Issues* 56 (3): 391–406.

Severson, R. L., and P. H. Kahn Jr. 2010. "In the Orchard: Farm Worker Children's Moral and Environmental Reasoning." *Journal of Applied Developmental Psychology* 31: 249–56.

Sobel, D. 1996. *Beyond Ecophobia*. Great Barrington, VT: Orion Society.

Sommerville, J. A., and J. Decety. 2006. "Weaving the Fabric of Social Interaction: Articulating Developmental Psychology and Cognitive Neuroscience in the Domain of Motor Cognition." *Psychonomic Bulletin and Review* 13 (2): 179–200.

Stanton, A. L., S. J. Sullivan, and J. L. Austenfeld. 2009. "Coping through Emotional Approach: Emerging Evidence for the Utility of Processing and Expressing Emotions in Responding to Stressors." In *Oxford Handbook of Positive Psychology*, edited by C. R. Snyder and S. J. Lopez, 225–35. Oxford: Oxford University Press.

Stern, D. 1985. *The Interpersonal World of the Infant*. New York: Basic Books.

———. 2007. "Applying Developmental Neuroscience Findings on Other-Centered Participation to the Process of Change in Psychotherapy." In *On Being Moved: From Mirror Neurons to Empathy*, edited by S. Bråten, 35–47. Philadelphia: John Benjamins.

Wals, A. E. J. 1994. *Pollution Stinks*. De Lier, the Netherlands: Academic Book Center.

Warrington, E. K., and T. Shallice. 1984. "Category Specific Semantic Impairments." *Brain* 107: 829–54.

Watson, G. 2006. "Wild Becomings: How the Everyday Experience of Common Wild Animals at Summer Camp Acts as an Entrance to the More-Than-Human World." *Canadian Journal of Environmental Education* 11 (1): 127–42.

Weick, K. E. 1984. "Small Wins: Redefining the Scale of Social Problems." *American Psychologist* 39: 40–49.

Weiss, E. B. 1989. *In Fairness to Future Generations*. Dobbs Ferry, NY: Transnational Publishers.

PART FIVE

CULTURE, RELIGION, AND SPIRITUALITY

USING EMPATHY AND COMPASSION TO DEVELOP

A UNIFIED GLOBAL MOVEMENT

TO PROTECT ANIMALS AND THEIR HOMES

HUMANS ARE an extremely heterogeneous lot. There are large differences about attitudes toward animals and the environment within cultural and religious groups and more significant variations among different groups. It's easy to understand why people who are struggling to have even their most basic needs met don't give priority to the well-being of animals or to the integrity of shared environments. Indeed, the vast number of people around the world falls into this group. Also, as Dale Peterson pointed out in his essay on the bushmeat crisis (part 2), local poor people do not really benefit from this industry or attempts to fix it. Nor will most of the people in need likely benefit as animal and environmental issues are tackled and hopefully solved in many poor countries. The situations faced in the four countries represented here, Australia, China, India, and Kenya, highlight most of the problems other countries face in trying to deal with animal/wildlife and habitat protection and conservation, and the well-being of human residents. Fortunately, natives of these countries who are "working on the ground" were able to write for this collection.

The first essay in this section by Daniel Ramp and his colleagues reviews current science and policy guiding conservation and wildlife management in Australasia, especially in relation to lethal actions in interspecies relationships. They highlight the trend in Australasia (as embodied by the Australasian Wildlife Management Society) to treat symptoms (invasive or presumed overabundant species) rather than examining causes (habitat change, loss of connectivity), restricting avenues of decision making that lead to nonlethal solutions. Lack of consideration for individual well-being is rife from pest management to environmental restoration projects. To emphasize the need for a paradigm shift, they present a range of examples of wildlife management of macropodid marsupials (predator control to recover threatened rock-wallabies, commercial killing under the guise of Conservation through Sustainable Use, issues about connectivity, fertility control, translocation, and killing, and the recognition of indigenous hunting rights and potential clashes in animal welfare standards) to tease out the implications of adopting individual well-being in conservation and management. They go on to present guidelines for policy/paradigm shifts necessary to bring about compassionate conservation in Australasia and set out future objectives.

There is no better representation of acts that ignore nature than the runaway exploitation and abuse of wildlife animals on the Chinese mainland. Many wildlife species are composed of intelligent and sentient beings who should be left alone in their natural habitats. The same hands-off principle applies to nonsentient beings. As Peter Li writes, commercial exploitation of "wildlife resources" has reached an unprecedented level in China. Each one of the wildlife-related business practices in China is nothing but a wanton violation of the nature of the wildlife species concerned. Tigers are majestic carnivores who do not belong in small barren cages. Yet China's controversial tiger farming sends some five thousand tigers to a life of solitary confinement. Caging more than ten thousand live Asiatic black bears for bile extraction from an open wound cut in their abdomen is the most brutal mode of production driven by a crude greed for profit. Every one of the farming practices on the bear farms, such as denying the bears their favorite foods, caging them for life, and inserting metal catheters or creating an artificial bile dripping duct inside their bodies, is singularly inhumane. China also is the world's largest consumer of wildlife. Each year tens of millions of sharks die a slow and agonizing death after their fins are cut off. Tens of millions of other animals, dead or alive, are trafficked to Mainland China as food. Outdated enclosure design and poor management show a total lack of respect on the part of the Chinese zoo industry. Worse still, they impart to visitors the outdated idea that humans are superior to other animals.

Does China have a legacy of respect for nature? Or is the Chinese nation culturally programmed to act in defiance of nature? By going back to China's ancient religious and philosophical ideas, Li argues forcefully that Chinese cultural tradition does not sanction assault on wildlife animals and on nature in general. At least, it does not approve the level of wildlife exploitation we have seen in China's contemporary era. Daoism, indigenous to China, and Buddhism, an imported religion, both call on the society to hold nature in respect and have mercy for other nonhuman lives. Confucianism, the most powerful and long-lasting ideology that has influenced East Asia for more than two thousand years, rejects excessive, unreasonable, and unplanned use of natural resources, including wildlife animals, despite its fundamentally anthropocentric outlook. The unbridled assault on wildlife in contemporary China, like the shortsighted exploitation of many of its other natural resources, is a by-product of the national drive for economic modernization. Li writes: "China has a more complex legacy that includes ideas and practices for compassion and protection of nonhuman animals. Blaming China's past for the contemporary flaws is misleading and misses the real target. . . . It is therefore not Chinese culture but the current 'development first' mind-set that is behind the nation's wildlife crisis."

India also faces major problems when it comes to conserving its natural heritage. Renowned conservationist Vivek Menon envisions a triangular playing fielding in which social, economic, and ethical factors figure into the equation of conserving nature. India is mired in poverty, and Menon notes, as does Ben Minteer in his discussion of the bushmeat crisis (part 2), that it's essential to pay attention to what can actually be done given what local people face. There has to be a pragmatic side to conservation efforts that take into account local lifestyles and history. In India there is a strong historical component for the preservation of nature based on the principles of ahimsa and the spiritual, ethical, and moral code of dharma. Thus in the midst of rampant poverty, where preservation in the name of the inherent value of nature is simply not tenable, protected areas for wildlife had increased by the year 2000 to 5 percent of the country's surface area, a tribute to the strong yet visionary policies and laws of the land.

Josphat Ngonyo and Mariam Wanjala note that there are similar challenges in Kenya, as there are in other poor nations. In Kenya there are a number of projects that center on the protection of endangered species and maintaining biodiversity. Those taking precedence include protecting rhinoceroses and elephants, preserving wetlands and forests, and educating youth. The particular challenges faced include economic and social-cultural ones, the destruction of wildlife habitats, security, inadequate incentives, and climate change. Political corruption is also a major problem. A recent study showed that political corruption and bad governance, rather than human population pressures and poverty, might present the greatest threat to wildlife in developing countries. Researchers found that high levels of corruption in African countries strongly correlate with declining elephant and black rhinoceros populations. Ngonyo and Wanjala stress that community involvement in resource management is now regarded as an essential component of conservation projects, and local people, the ultimate owners and guardians of natural ecosystems, must be the direct beneficiaries of the income that accrues from the use of ecosystems. Ecotourism should also be promoted in reserves and parks because it has all-around benefits to both the people and the environment. They also suggest that "Kenyan scientists and teachers should evolve a new approach to the study of natural sciences that integrates indigenous principles of natural resource conservation and management. A holistic approach to the study of natural organisms and systems should be adopted."

The last two "big picture" essays of this eclectic collection bring to the fore ideas about the preservation of nature and the roles of religion and spirituality. Bron Taylor asks if green religion is an oxymoron and notes, "Religions often lead us to ignore nature, distracting us from what we *can* more easily see and know—namely, our absolute dependence on the biosphere and . . . the rest of the living community." Taking a historical perspective, Taylor writes, "The sensory and sensual spiritualities of these environmental thinkers reflect an approach that is the opposite of ignoring nature, for they all depended on the close observation of it." Weaving in the social sciences, Taylor notes, "when we use the lenses of the social sciences, without worrying about where the boundaries of religion lie, we can see a dramatic increase of religion-resembling beliefs and practices in which people consider nature to be sacred in some way, and time focused in the close observation of nature to be a critical pathway to spiritual truth. Through such an approach we can see how close observation of nature can become a spiritual epistemology revealing the interconnectedness, mutual dependence, and kinship of life, leading to a reverence for life and a desire to protect biotic diversity." There is no doubt that as Taylor puts it, "the long-term trend *must* be a closer understanding of our place in nature, for when we stop ignoring nature will we be able to develop lifeways, livelihoods, spiritualities, and cultures that will be adaptive and resilient."

Taylor's essay leads nicely into Anthony and Gabriela Rose's piece in which they introduce the notion of biosynergy to the discourse. Biosynergy emphasizes the psychosocial and emotional benefits of seeing all beings as kindred spirits in a cohesive universe. Using the award-winning movie *Avatar* as a model, this father and daughter team explores antidotes to the modern human folly of ignoring nature by building on the film's messages of planetary harmony, spiritual and species communion, global conservation, and compassion. *Avatar*, set in the mid-twenty-second century, is the story of a corporation that's trying to drive the indigenous humanoid Na'vi on the planet Pandora away from their ancestral home because it happens to sit on a mineral called unobtanium the company wants to mine for profit.

Rose and Rose, like Taylor and others, focus on the importance of actually *seeing* the world from another's eyes, envisioning another reality, and feeling with compassion nature's magnificence and the unity of all life and landscapes. They use this brief exchange from *Avatar* to make their point (Cameron 2009):

JAKE SULLY: Well, if I'm like a child, then maybe you should teach me.
NEYTIRI: Sky People cannot learn, you do not see.
JAKE SULLY: Then teach me how to see.
NEYTIRI: No one can teach you to see.

Neytiri, a Na'vi, is saying that seeing means feeling, and the sky people, the narrow-minded earthlings who have come to ravage her land—ignoring its beauty and meaning to the indigenous peoples—don't know how to see and feel. They're simply motivated by economic incentives and have no regard for sacred webs of nature in which there is continual give and take among those who are enmeshed in these mysterious and wondrous networks about which we know so little.

Concerning the notions of community, connection, and unity, Jake Sully (played by Sam Worthington) translates for botanist Grace Augustine (Sigourney Weaver) what Neytiri had said to him: "She says that all energy is only borrowed; at some point you have to return it" (Cameron 2009).

Ignoring nature leads to imbalances that have far-reaching consequences not only for the animals and landscapes that are directly affected but also for the humans who have to live in the wake of the incessant and rampant destruction. There are many lessons to be learned from *Avatar*. First and foremost, *Avatar* points out just how invasive and inconsiderate we can be, blind to those who suffer as we relentlessly redecorate nature, be they nonhuman animals, less fortunate humans, or humans we assume to be lesser or lower members of our species. We constantly need to be reminded, "It's not all about us." We're not the only species on the planet and we need to stop behaving as if we are.

Reference

Cameron, J. 2009. Director, *Avatar*. Santa Monica, CA: Lightstorm Entertainment.

21

Compassionate Conservation

A Paradigm Shift for Wildlife Management in Australasia

Daniel Ramp, Dror Ben-Ami, Keely Boom, and David B. Croft

Introduction

HUMANS DIRECTLY AND INDIRECTLY impact the lives of wild animals, primarily by altering landscapes through the removal of habitat for human dwelling or resource production (e.g., agriculture, mining, forestry) (Mathews 2010), but also through changes to the quality of remaining landscapes (e.g., roads, chemical and noise pollution, disease, stress) (Fraser and MacRae 2011). The lives of wild animals are further impacted in the management of remaining natural habitat and human-occupied land (e.g., production landscapes, urban remnants) where wild animals still reside. Wildlife management stems from the need to control species that impinge on human lives and/or livelihood (i.e., where species are defined as pests) or where some form of ecological dysfunction results in what is perceived as an imbalance that requires intervention (i.e., for some higher conservation objective). The subjective and anthropocentric nature of wildlife management, particularly where it relates to the reduction of pest or "overabundant" species, was recognized by Graeme Caughley (1981). As the human population expands and demands more land and resources, the separation of clear conservation goals from the need to protect human livelihoods is likely to prove increasingly difficult. Although the welfare concerns of wild animals have been treated as an unimportant consideration in the development of environmental law and policy, there is considerable benefit in joining animal welfare science and animal conservation science to assist in wildlife management policy. Indeed, under the banner of compassionate conservation, a new paradigm for wildlife management beckons, one where nature has a voice in environmental policy and is no longer ignored.

Management in the form of control is usually directed toward wildlife species that for one reason or another are considered pests (Littin 2010), whether native or alien. Eradication is the common end goal for alien pests (Reynolds 2004), but the feasibility of achieving complete eradication and the resultant suffering this causes are not often explicitly examined. In recognition of the constraints on complete eradication, management objectives are frequently aimed at reducing densities of pest species, the ongoing need for which is implicit in management policy (although this aspect is mostly unspoken). There may be conservation goals that justify the need for eradication (e.g., harm for the greater good of biodiversity); however, this does not preclude welfare objectives within the policy framework. For native species, management objectives are frequently blurry as decision making is confounded by scientific dogma and hyperbole. By considering animal welfare alongside animal conservation it becomes possible to establish wildlife management frameworks that are explicitly oriented toward the lives of individuals and their social groups and not just the species or population as a whole (Fraser 2010).

Here we present a series of case studies of wildlife management in Australasia. In doing so we aim to explore where animal welfare science and animal conservation science have similar and dissimilar goals with respect to wild and free-living animals. Australasia has a distinct history of wildlife management that rightly or wrongly has an international reputation for lethal management. Our objective is to initiate discourse among scientists, environmental managers, welfare organizations, and the general public so that the wealth of knowledge from both welfare and conservation fields of science can be used to define a new paradigm for wildlife management in Australasia and elsewhere for the twenty-first century, one where individuals are treated humanely and with respect in environmental policy and law and resultant management frameworks.

Case Studies of Wildlife Management

Australasia is comprised of a series of large islands (one, Australia, of continental scale) in the western Pacific that represent fragments of the supercontinent, Gondwana (Australia, New Zealand), the result of tectonic upheaval (Papua New Guinea [PNG]), or the remnants of volcanism (e.g., Vanuatu). Given their isolation in an oceanic void and their long subtraction from Gondwana (New Zealand eighty million years, Australia forty-five million years), the floras and faunas of the region have high levels of endemism. Even so, migration from other biomes has occurred such as Asian rodents and snakes into Australia, and Australian birds and bats to New Zealand. The pace of such introductions did not accelerate much with the first peoples in Australia (hunter-gatherers from around 70,000 years ago), however, the entry of the dingo to Australia (about 4,500 years ago), the Maori to New Zealand, and the establishment of agricultural practices in Melanesian societies (PNG) all had widespread consequences to landscapes, including mass faunal extinctions. In recent times, European colonization has wrought massive environmental changes, including a raft of exotic species introduced to produce food and fiber or for aesthetic reasons. In many instances these species "ran wild" (Rolls 1984; Low 1999), and we have seen yet more introductions as biological controls to the former. Indeed, calls for further introductions continue (Bowman 2012).

Australia and New Zealand share similar colonial histories, although the first peoples are quite different in their longevity in the country and culture. European colonists had a less overwhelming effect on the West Pacific islands, where technology transfer and the introduction of market economies have been more influential. From a conservation perspective, the issues that have emerged are often intergenerational and variously include:

1. Introductions of predators to control other species (usually also introduced for their utility to people) causing unforeseen effects cascading through native communities;
2. Abandonment of utilitarian species for one generation (e.g. one-humped camels) to the detriment of ecosystem function for future generations;
3. Ineffectual containment of species introduced for sport (e.g. red fox, European rabbit, deer) to the detriment of native ecosystems;
4. Ecosystem transformation to support food and fiber industries based on species used in colonial cultures;
5. Government-sponsored programs to control or eradicate native species (e.g., marsupial herbivores) in one generation resulting in government-sponsored programs to recover the same species for conservation purposes in future generations; and
6. Technological innovations translating subsistence use to commercial slaughter.

1. A Pest to Control a Pest—The Ultimate Conservation Quandary

Despite the island status of countries within Australasia and strong biosecurity, the globalization of the twentieth and twenty-first centuries has witnessed a tide of exotic species, including pathogens, sweep over Australasia. Those that get in usually serve human interests in which conservation and compassion are not paramount.

The most notorious examples are the introduction of the cane toad (*Bufo marinus*) into Australia to (unsuccessfully) control the gray-back cane beetle, a pest of sugar cane (itself an introduced plant) (Lever 2001). The cane toad is a conservation threat across tropical and subtropical Australia, causing local extinctions of marsupial (quolls: *Dasyurus* spp.) and reptilian carnivores (goannas: *Varanus* spp.) and other consequences cascading through amphibian and mammal communities. The response has largely been ineffectual mass killing (e.g., crushing the skull, inducing hypoxia, depravation from water) of the toad (Florance et al. 2011), and a thus far unsuccessful search for a biological control agent. In New Zealand the rabbit (*Oryctolagus cuniculus*) was introduced for food and sport hunting, and then the weasel (*Mustela nivalis*) and stoat (*Mustela erminea*) were introduced to (unsuccessfully) control the rabbit (King and Powell 2007). The stoat, in particular, has caused localized extinction of iconic birds like the kiwi and kakapo. The stoat is controlled by kill-trapping, baiting with poisons, and hunting with dogs.

The first response to such issues is mass killing, often with inhumane methods, as the once savior and now pest species is demonized. As Caughley and Sinclair (1994) note in their textbook on wildlife management, "the notion of humane treatment is often the first casualty of turning a species into a pest." However, by the time the conservation issue is recognized the option for eradication is lost except for some success on small offshore New Zealand islands (Simberloff 2001). Despite sophisticated eradication programs, successful removal often proves short-lived as capture rates decline with density and behavioral plasticity can mobilize recovery (Sweetapple and Nugent 2009). Additionally, methods of targeting species are not always species specific, having consequences for other pest and natives species, sometimes for generations (Gillies and Pierce 1999). Extensive but ineffective attempts to contain the species fail; for example, cane toads have used the road network to disperse (Brown et al. 2006). With the causes of eradication and containment lost, certain compassion emerges with the recognition that the pest is now an ever-present part of the biota. The species at most risk of extinction from the pest are relocated to offshore islands where the pest is absent (e.g., northern quoll in Australia [Woinarski et al. 2007], threatened bird species in New Zealand [Taylor, Jamieson, and Armstrong 2005]) or protected on a small scale by exclusion barriers to the pest. An alternative or complementary strategy is to build the capacity of the threatened species to resist the pest (e.g., taste aversion training of northern quolls toward cane toads ahead of the invasion front in Australia [O'Donnell, Webb, and Shine 2010], predator recognition training for birds in New Zealand [McLean, Hoelzer, and Studholme 1999]). Capacity to resist the invasive species can also be built within the ecosystem from native predators (e.g., predatory ants and cane toads [Ward-Fear, Brown, and Shine 2010]) or competitors (e.g., native frogs and cane toads [Cabrera-Guzman,

Crossland, and Shine 201)]). Modest anthropogenic interventions to nurture the natives rather than kill the exotics may enhance peoples' engagement with conservation effort.

2. Useful One Day, Conservation Threat the Next

Both Australia and New Zealand have wild herds of horses through abandonment and improper containment (Dobbie, Berman, and Braysher 1993). Pigs are ubiquitous in Australasia and form an important cultural resource in PNG. Feral populations occur in Australia and New Zealand (Choquenot, McIlroy, and Korn 1996). Captain James Cook presented pigs to the Maori and released others in New Zealand on his voyages of exploration in 1773–77. However, there is some speculation that pigs were translocated into Australia across the Torres Strait (Heinsohn 2003) prior to European colonization, and they merely served to augment an already feral population. Australia has an additional burden of one-humped camels, donkeys, goats, water buffalo, and banteng cattle (*Bos javanicus*).

All these species provided utility at the time of their introduction; for instance, camels provided transport through the arid lands before being displaced by vehicles (Edwards et al. 2010), and goats provided milk and meat in remote mining sites before abandonment after the minerals were exhausted (Parkes, Henzell, and Pickles 1996). With the exception of banteng, these species are invasive, some sustain high populations (millions in Australia), and all cause ecological degradation (landscape dysfunction) through actions such as grazing/browsing pressure, soil turnover and wallowing damaging to wetlands (buffalo and pigs), competition with and displacement of native species (goats and rock-wallabies), and predation (pigs). All are subject to shooting, poison baiting, and trapping campaigns to reduce populations, but none are likely to be eradicated because their utility remains and they are broadly distributed.

The conservation threat of these species is well researched and their management through lethal actions may evoke compassion (especially horses) (Nimmo and Miller 2007). There is also strong resistance to "killing for nothing" if an end product (meat or hide) is not consumed. This is particularly true of indigenous Australians who have a strong cultural ethic against killing for waste (Vaarzon-Morel and Edwards 2012), and this ethic and other attachments formed with these animals shape cross-cultural management of buffalo and camels in northern Australia. In contrast, pigs are often maltreated. They are trapped and sometimes left to die, or shot and their carcasses left to attract other pigs. There is a large contingent of pig hunters in Australia with several magazines supporting their activities. They have (illegally) exacerbated the conservation issues with pigs by translocation to increase stocks. They frequently hunt with specially bred dogs that corner and hold the pig occasioning injury and pain to the pig (and sometimes the dog) before the pig is killed by knife incision. The dogs are often inadvertently abandoned and contribute to feral dog attacks on livestock and genetic introgression into dingo populations. Furthermore, the capacity of these dogs to cause harm has led to fatalities in attacks on children. Pig hunters also cause environmental degradation and property loss through setting fires to flush out their game (especially in northern Australia). The population of pigs in Australia is very large (possibly more than twenty million) and typically exceeds the populations of individual kangaroo species under commercial use. Thus this species represents a particular challenge in ensuring its humane treatment in both feral and farm populations.

3. Oh Deer, Out-Foxed and Cat-Napped—Major Drivers of Conservation Management

A large variety of ungulate "game species" were introduced into New Zealand soon after European settlement. These include deer—fallow (*Cervus [Dama] dama*), red or wapiti (*Cervus elaphus*), Timor or rusa (*Cervus timorensis*), sambar (*Cervus unicolor*), white-tailed (*Odocoileus virginianus*), and sika (*Cervus nippon*)—and goats—chamois (*Rupicapra rupicapra*) and Himalayan tahr (*Hemitragus jemlahicus*). Their grazing and browsing has degrading impacts on native vegetation and forestry plantations (Husheer, Coomes, and Robertson 2003), and they are a reservoir of bovine tuberculosis. Their original purpose as game animals for hunting predominates in their management, or they have been domesticated for meat and hide production (Nugent and Fraser 1993). However, given the absence of native grazing and browsing mammals in New Zealand, this choice and subsequent impacts have cascaded through native plant communities. Fallow, red, sambar, and rusa deer were introduced into Australia along with hog deer (*Cervus [Axis] porcinus*) and chital, axis, or spotted deer (*Cervus [Axis] axis*) (Moriarty 2004). Unlike New Zealand, Australian plant communities evolved with grazing and browsing mammals over millions of years, including now-extinct megaherbivores (Flannery 1994). Nonetheless, deer are recognized as threats to native vegetation communities, and such threats may be escalating with translocation of deer to new locations. The response to deer management has been similar to New Zealand—namely, the deer are established in numerous populations, and they have utility in recreational hunting (Finch and Baxter 2007). These factors override their lack of utility in ecosystem function. The resultant paradigm is that ecosystems should be allowed to progress to some new state to accommodate these exotic species and not be restored to their preintroduction states (Hall and Gill 2005).

The management of introduced red foxes (*Vulpes vulpes*) and cats (*Felis catus*) in the conservation of Australia's native fauna has proved particularly vexing. The fox was introduced for hunting with horses and dogs, but this activity is now minimal given the very large geographic range the fox now occupies (Saunders et al. 1995). Hunting for fur has waxed and waned along with demand, and this is no longer a common activity. The fox has been implicated in the local extinction of many marsupials up to 3.5 kilograms in weight (Dickman et al. 1993). There is some argument as to whether the fox is the cause or the symptom resulting from ecosystem transformation following the expansion of European pastoral and agricultural practices and the establishment of rabbit populations (Fisher, Blomberg, and Owens 2003). Regardless, the ongoing presence of the fox inhibits recruitment in remnant fauna populations and the success of reintroductions through captive breeding and translocation from fox-free habitat (e.g., offshore islands) (Kinnear, Onus, and Sumner 1998; Short and Turner 2000). The fox has been subject to shooting and large-scale poison baiting with general landholder support since the fox also preys on valued livestock (e.g., lambs, chickens).

While the fox is limited to southern Australia, the cat is ubiquitous, including arid lands and all parts of Australasia. The cat was assisted in its geographic expansion in the vain hope that it would control mice, rats, and rabbits. Unlike the fox, there is a widespread pool of domestic cats that are nurtured as a companion animal and can recruit into feral populations. Thus eradication has only been possible on remote islands. Like foxes, cats are a symptom of ecological transformation and become a threat in the extinction of native fauna when habitat is simplified by land clearing and/or frequent fire and remnant populations are isolated and exposed. In spite of such experience from southern Australia, these processes accelerate among relatively intact tropical savanna habitat in northern Australia, and the cat has emerged as a major threat to a rapidly declining mammal fauna (McKenzie et al. 2007). Trapping, shooting, and poisoning can reduce cat populations (Short et al. 1997; Short, Turner, and Risbey 2002).

There is concern that suppression of fox populations leads to a competitive release of cat populations that may defeat management for conservation purposes. Construction of exclusion fences has thus far proved the most effective tool to restore smaller marsupials to former habitat (Moseby, Hill, and Read 2009). Such projects have garnered private sector (e.g., http://www .australianwildlife.org) and community support for conservation efforts (e.g., http://www.savethebilbyfund.com) but represent small arks in a very large continent.

Current debate is about whether the ecosystem can strike back by restoration of the apex predator, the dingo (*Canis lupus dingo*) (Glen et al. 2007). There is some merit in restoring the ecosystem to a more equitable state rather than continuing the mass killing of foxes and cats in poisoning campaigns. Evidence has mounted that the absence of a top predator, like the dingo, releases populations of mesopredators, like foxes and cats, which then multiply to the known detriment of the smaller and often threatened native vertebrate fauna. For example, conservation of threatened species like yellow-footed rock-wallabies (*Petrogale xanthopus*) and malleefowl (*Leipoa ocellata*) is favored where dingoes are present, including areas within the continent-spanning Dingo Barrier Fence (Wallach, Murray, and O'Neill 2009). However, if both dingoes and fauna favored by the presence of this top predator are to be conserved, especially within the Dingo Barrier Fence, then the dingo needs protection from genetic introgression from domestic dogs (Claridge and Hunt 2008), and livestock enterprises need protection from dingo predation. Thus in the sheep rangelands that cover much of southern Australia, the suppression of wild dogs (including dingoes) by poisoning, shooting, and trapping is a core practice (Fleming et al. 2001). Reintroduction of dingoes is unlikely to be favored across most landholdings, particularly as attitudes toward dingoes harden with recent confirmation of predation on children (Gorman and Kenneally 2012). Furthermore, evidence from northern Australia where dingoes persist does not fully support the apex predator hypothesis and implicates the dingo itself in faunal decline beyond the rapid mainland extinction some thousands of years ago of the thylacine (*Thylacinus cynocephalus*) and Tasmanian devil (*Sarcophilus harrisii*) (Allen 2011). The capacity for the dingo to improve biodiversity outcomes is contingent on specific circumstances that may not be universally achievable (Fleming, Allen, and Ballard 2012).

4. A Little Slice of England Down Under

Ecological transformation through land clearing for agriculture, conversion to grasslands for grazing, and impoundment and redistribution of water for irrigated horticulture have been key threats to native fauna and flora (Lindenmayer et al. 2008). In spite of recognition of the various processes that cause landscape dysfunction and loss of ecosystem services in relatively intact ecosystems like the rangelands (Ludwig et al. 1997), the threats continue unabated and historical lessons are ignored. In the relatively intact tropical savannas of northern Australia, agronomists introduced 463 exotic grasses and legumes (Whitehead and Wilson 2000) to support the pastoral industry, yet only twenty-one are useful cattle fodder. These tall African grasses (*Andropogon gayanus* and *Pennisetum* spp.) predominate and have colonized a range of habitats, including wetland margins, riparian corridors, open woodlands, and closed forests (Kean and Price 2003). These grasses withstand the long dry season, produce a high biomass, and out-compete native plants for essential resources (Rossiter-Rachor et al. 2008). They create fuel loads three- to five-fold of native grasses (Howard 2002) and thus support late, intensely hot dry season fires that kill the overstory and decimate flora and fauna (Dyer et al. 2002). These altered fire regimes are a dominant threat to biodiversity with the burning of the tall introduced grasses causing ecological transformation of woodlands to grasslands of little diversity.

One suggested solution to the inadvertent creation of African-style grass-lands is to introduce African fauna, like elephants, to control their biomass and spread (Bowman 2012). This is hardly a compassionate option given the raft of lethal actions taken against current introduced fauna (see above) and the well-known conflicts between people and elephants in their countries of origin (O'Connell-Rodwell et al. 2000). Elephants would go forth and multiply, conflicts would emerge with pastoral and horticultural enterprises, and then the elephants would be subject to lethal control, sequestered into some red-meat industry that would likely fail to gain traction (cf. camels in Australia), or be used for "safari" hunting. A more compassionate option is to look to the ecosystem services already provided by native herbivores in reduction of fire fuel loads (Kirkpatrick, Marsden-Smedley, and Leonard 2011), and indigenous knowledge and practices of fire management (Russell-Smith, Whitehead, and Cooke 2009), since indigenous Australians are significant landholders across northern Australia.

5. If It Moves Shoot It, If It Doesn't Then Chop It Down, and Then Try to Put It Back Together Again

In the late eighteenth, the nineteenth, and early twentieth centuries, Australian governments required landholders to clear native vegetation and convert woodlands and forests to grassy pastures for livestock or the growing of crops. They paid extensive bounties to kill hundreds of thousands of the marsupial fauna, especially herbivores (kangaroos, wallabies, and rat-kangaroos) of all sizes (Jarman and Johnson 1977; Croft 2005). This was an era for which Hornadge (1972) used the quip, "If it moves, shoot it; if it doesn't, cut it down." Now Australian governments create "Threatened Species Action Plans" (usually too expensive to implement) and fund landholders to restore native vegetation through various granting programs (e.g., Landcare, Caring for Our Country, and Biodiversity Fund).

The brush-tailed rock-wallaby (*Petrogale penicillata*) is a good exemplar of changing fortunes in the Australian conservation landscape. This species had the misfortune of inhabiting steep rocky habitat in ranges bounding the closely settled eastern seaboard of Australia. Its habitat in "rough country" should have afforded it some protection, but agricultural, pastoral, and forestry enterprises cleared and penetrated surrounding habitat. When the rock-wallaby ventured out of its shelter sites it was declared a pest and bounties were paid for the carcasses of 640,000 in New South Wales (NSW) between 1884 and 1914 (Short and Milkovits 1990). It also had a thick pelage attractive to the fur trade, and so hunters penetrated its refuges and a further 144,000 skins were traded in this period in NSW (Lunney, Law, and Rummery 1997). Incursions by foxes and cats into its habitat negated recruitment into populations, and the species has become extinct over much of its former range with few remaining strongholds in northern NSW and southern Queensland. Recovery plans for brush-tailed rock-wallabies has seen one country's pest become another's potential savior. The species was introduced to Kawau and Motutapu Islands in New Zealand in the 1870s and subsequently colonized Rangitoto Island (Eldridge, Browning, and Close 2001). They have been trapped and exported back to captive breeding colonies in Australia from Kawau Island but eradicated from the other two islands by shooting and poisoning. Their provenance from NSW has been established by genetic studies, and they now provide a potential pool for reintroductions, although their small founding population in New Zealand has created a genetic bottleneck. The recovery actions have also seen one species (yellow-footed rock-wallaby) serve the recovery of another (brush-tailed rock-wallaby) through cross-fostering to augment captive breeding for reintroduction (Taggart et al. 2005). The reintroduction program has had some success (Molyneux et al. 2011). Considerable research effort is now placed on filling knowledge gaps to improve the conservation status of the sixteen species of rock-wallabies (Eldridge 2011).

The long-term outcome likely depends on a formerly hostile landscape returning to one supportive of the recovered species. There is an impatience for this to occur as conservation funding cycles are short and there may be unsupportive surrounding communities when protected areas are excised from what they consider to be productive landscapes. This is exemplified in the rangelands when large pastoral properties have been acquired to form national parks. One cannot expect a landscape impacted by a century of livestock grazing to be restored within a few years of de-stocking. However, with patience and commitment to managing threats to conservation values of such landscapes, the prognosis may be good (e.g., the return of species considered extinct [Croft, Montague-Drake, and Dowle 2007]). However, the new state will hardly be pristine, and compassion in its conservation may be often tested.

6. Conservation through Sustainable Use—But What Use?

Historically, some cases of hunting and gathering by indigenous populations have led to the conservation of target species through exclusion of other land uses. Such cultural affiliations have promoted a willingness to engage in further effort (Muhic, Abbott, and Ward 2012). Conservation benefits can accrue if there is incentive for people to only take sustainably (i.e., below a threshold of negative effect on population persistence) and therefore allows the continued persistence of species in natural ecosystems. Although traditional hunting methods may result in poor welfare outcomes, the welfare of animals killed for meat may be improved by providing access to technology. However, this situation does not always occur: in PNG the bushmeat trade (cf. Africa [Bennett et al. 2007]) has threatened fauna as new technology has better favored the hunter, a market has developed, and old taboos have broken down (see Flannery [1998] and Martin [2005] for tree-kangaroos). For marine mammals like the dugong (*Dugong dugon*), traditional hunting is culturally important across northern Australasia, particularly in the Torres Strait and PNG (Kwan, Marsh, and Delean 2006), but overexploitation has led to population decline in some areas (Marsh et al. 2002), many dugong killed experience poor welfare outcomes, and the species is now listed as vulnerable on the IUCN Red List (IUCN 2011).

In recent times, conservation through sustainable use has been co-opted by wildlife management advocates arguing for the perpetuation or expansion of the use products from wildlife to sustain local economies. Ostensibly, these arguments are used where conservation goals are employed to justify sustaining commercial killing industries, particularly in Australia in relation to kangaroos (Cooney 2008). A perception of grazing competition with livestock, excessive grazing pressure in threatened ecosystems, and damage to crops has led landholders and environmental policy makers to support the need to reduce kangaroo populations, the dominant marsupial herbivores prior to the introduction of placental mammals (ungulates). From the kangaroo bounty of the early twentieth century to the commercial killing industry that began in the 1960s (Lunney 2010), kangaroos have been, and continue to be, killed for commercial and noncommercial purposes, legally and illegally.

Although the commercial kill aims to produce meat (for pet food and human consumption) and skins, the initial impetus was to reduce numbers of kangaroos. Four of the fifty-three mainland kangaroo species in the family Macropodidae are killed commercially (red kangaroo [*Macropus rufus*], eastern grey kangaroo [*M. giganteus*], western grey kangaroo [*M. fuliginosus*], common wallaroo [*M. robustus*]), although other species are killed for the stated purpose of reducing environmental damage and competition with livestock. Since European settlement, six macropods have become extinct and another eleven have had their ranges considerably reduced (Calaby and Grigg 1989). Commercial kill numbers are controlled by an annual quota based on aerial population surveys (annual in most management zones) conducted by state kangaroo management programs and approved by the Australian government

(DSEWPaC 2011). In the decade 2000 to 2010, a combined total of 28.7 million kangaroos of four species were killed from a combined quota of 51.8 million in the commercial zone, while the populations of the four species declined from a combined total of 57.4 million in 2000 to 25.2 million in 2010, in a decade dominated by drought across southern Australia. This kill represents the largest commercial use of terrestrial wildlife for meat and skins anywhere in the world.

Yet there is growing recognition that the need for lethal management of kangaroos has been overstated (Croft 2005; Ben-Ami et al. 2011). A wealth of scientific evidence exists that kangaroo populations do not typically need to be reduced for good conservation outcomes (e.g., Croft, Montague-Drake, and Dowle 2007), that kangaroo populations present a small and variable economic cost to landholders (Arnold et al. 1993), and that any costs can be reduced in other ways (Arnold, Steven, and Weeldenburg 1989). Most governments in Australia now recognize that kangaroos are not widespread pests (i.e., are not a cause of sustained environmental damage or competition with livestock). Nevertheless, many landholders resort to lethal control, either by allowing licensed shooters to kill on their properties for commercial purposes or by utilizing government kill-only licenses. Without clear pest status, one justification for commercial use involves the conservation benefits of use that is sustainable. This approach aims to mitigate habitat degradation in the rangelands of Australia by creating a commercial benefit for landholders to commercially kill kangaroos, and therefore remove some livestock from their lands and reduce total grazing pressure (Grigg 2002). Field studies have tested possible implementation mechanisms (Baumber et al. 2009; Cooney et al. 2009) and alternative benefits, such as greenhouse gas emission abatement, have also been envisioned as macropods emit only very low levels of methane relative to traditional livestock like sheep and cattle (Wilson and Edwards 2008).

Livestock for human consumption endure travel time in trucks, feedlots, and possibly live export before reaching the abattoir. In contrast, the replacement of livestock with kangaroos may reduce the number of livestock suffering this fate. However, the feasibility of replacing livestock with wild kangaroos has been shown to be logistically problematic and with little incentive for landholders (Russell 2008; Ben-Ami et al. 2010), and would incur considerable welfare costs for kangaroos (Ben-Ami et al. 2011; Boom and Ben-Ami 2011). The killing of kangaroos is conducted in remote areas by shooters at night. Intractable welfare concerns exist as not all adults are cleanly shot, despite a code of practice, and dependent young are either left to die or killed inhumanely (RSPCA Australia 2002; Croft 2004; Ben-Ami et al. 2011). Furthermore, there is considerable inequality between kangaroos and livestock in terms of meat production (kangaroos produce far less meat than sheep and cattle because of smaller average body size and slower growth rates [Ben-Ami et al. 2010]), requiring many more kangaroos to produce food for people than livestock. There is no evidence that commercial use of kangaroos (around three million killed annually) has led to a reduction in livestock (Ampt and Baumber 2006) or to any improvement in ecosystem function or biodiversity. Thus the potential conservation benefits, if any, of commercial use are not empirically verified.

Alternative sustainable use mechanisms may achieve better conservation and welfare outcomes than commercial killing for meat and skins. The commercial kill could be curtailed to include only males, thereby minimizing the lethal impact on dependent young. Engagement in nonlethal ecotourism could provide an economic incentive for landholders to promote persistence of kangaroos (Higginbottom et al. 2004) and potentially facilitate a reduction in dependence on livestock. Further research is needed to explore the utility of nonlethal use mechanisms that may promote both welfare and conservation goals. That lethal control has been the dominant paradigm for kangaroo management is a testament to a myriad of factors, yet as we push into the twenty-first century we cannot continue to ignore the needs and well-being of animals. A logical starting point for reviewing the principles of wildlife management must be to examine current law and policy and advocate for appropriate change.

The Influence of Anthrarchy upon Law and Policy

Environmental and animal law in Australasia is heavily influenced by "anthrarchy," which refers to the organization of norms and values that systematically suppress or discount the importance and influence of other species (Ash 2007). In essence, anthrarchy is the institutionalization of speciesism, or the view that humans are superior to other animals (Singer 1975; Spiegel 1997). It relies upon the idea that the universe has been designed just to serve human interests (Wise 1996). Such views are based upon an overemphasis of the competitive and exploitative aspects of existence rather than the cooperative aspects. Anthrarchy assumes that humanity is not a part of nature or biodiversity and as such undermines conservation of biodiversity and the very concept of sustainability (Ash 2005).

There are two ways in which anthrarchy manifests in the laws and policies of Australasia. Firstly, much of the law and policy assumed or assumes that the well-being of other species is not of central concern (Thiriet 2007; White 2009). A common theme through the case studies we have presented is the formulation of laws and policies designed at controlling, eradicating, or introducing species without sufficient regard for the individual impacts upon animals or long-term conservation impacts. More recently, governments have responded to an increasing public interest in animal welfare with codes of practice that set out the "best" methods to kill the animals (Caulfield 2008). With proper law enforcement, these instruments should mitigate some of the damage caused to the well-being of other species. However, the codes are formulated with heavy input from industry, do not have the same legal status as legislation, and are either difficult to enforce or unenforceable (Dale 2009; Ellis 2010).

The second way in which anthrarchy manifests is that nonhuman species are treated as "resources" that can be "harvested," "taken," or otherwise used by humans. Similarly, domesticated animals are treated as property. The designation of animals as property or resources has not led to enhanced consideration for their conservation or well-being (Ash 2007). Domesticated animals have been subject to industrialized production resulting in large-scale cruelty to these sentient beings (e.g., battery hens and sows) (Sharman 2009). Kangaroo species have been classified as "renewable natural resources" that can be "harvested."

The Myth of Management

The word *management* is commonly used throughout Australasia in relation to wild animals. The word *management* implies mastery, or at least a deep understanding for executing control (Ash 2007). Yet as the case studies in this chapter have shown, governments have often failed to show mastery or sufficient understanding of the environment to design, create, or control nature. Furthermore, it is often difficult or impossible to recreate or restore ecosystems to their previous state, and policy makers often fail to understand the delicacy of ecosystems (Bekoff 2006).

Adopting a Compassionate Conservation Approach

Compassionate conservation seeks to transcend the ongoing dilemma of choosing between an environmental and animal approach to wild animals by looking for synergy between these two approaches. Although there is a general belief that compassion and conservation are incompatible (Soulé 1985; Callicott 1989), this need not be the case, and the benefits of combining animal welfare and animal conservation are considerable (Bekoff 2000; Fraser 2010). Adopting a compassionate conservation approach toward wild animals in Australasia would require a paradigm shift in how we think about animals other than ourselves. The shift would involve moving from exploitative-based sustainability to equity-based sustainability (Collin and Collin 1994; Bekoff 2006). It does not necessarily entail the adoption of an animal welfarist or animal rights perspective (see generally Regan 1983; Ellis 2009). Indeed, there is disagreement about what ethical principles should guide conservation policies, yet it is clear that ethics must be considered (Bekoff 2006). What it would entail is the taking of time to find better and more effective solutions to conservation problems rather than immediately turning to a "quick fix" (Bekoff 2006).

There are a number of practical measures that could be taken through legal and policy reform in order to assist in this paradigm shift. Word choice is highly important as it influences human actions (Wittgenstein 1972; Dunayer 2001) and plays a major role in the interpretation and application of law and policy. The term *management* should be avoided because it inappropriately implies mastery and supports anthrarchy (Ash 2007). The terms *harvest, export,* and *import* should not be used to refer to sentient beings. Similarly, the word *resource* should only apply to inanimate, nonliving objects such as money, oil, and coal. The words *it* and *stocks* should not apply to animals as they connote a lack of sentience. Animals should not be viewed as resources or things but instead as nonhuman persons (Wise 1999).

Further measures include the introduction of new legal principles. For example, interspecies equity is an emerging and necessary component of equitable decision making (Earnshaw 1999; Ash 2007). The principle refers to the recognition that the interests of animals are fundamentally similar to our own (Masson and McCarthy 1996) and that animals have an interest in their own lives, liberty, and well-being (Earnshaw 1999). The introduction of such a principle would be the most effective means to ensure that the needs and well-being of other species are considered. Under such an approach, the goal of environmental and animal law would be to facilitate and promote naturally harmonious human behavior toward other animals (Ash 2007). It would also mean that the well-being of individual species and communities of species are to be treated equally.

Another policy measure that would encourage a compassionate conservation approach would be to extend application of the precautionary principle to animal well-being. The precautionary principle requires that practices that may cause harm to the environment or human health should be banned until proven safe (Cameron and Abouchar 1991; Gullett 1997). Compassionate conservation would extend the precautionary principle to situations where there is the potential for harm to animal well-being. This reformulation of the precautionary principle offers a method of elevating animal welfare in a way that integrates conservation and compassion within a fundamental principle of environmental law and policy.

The final method of embracing compassionate conservation is to raise conservation and compassion to the center of the regulatory framework that governs our interactions with wildlife. From this perspective, the central decisions about wild animals must prioritize conservation and animal well-being rather than considering them as incidental to the objectives of land management and the utilization of natural resources.

A Paradigm Shift for Wildlife Management

There is a general reluctance in Australasia to include animal well-being in wildlife management and conservation practice, a fact borne out in the discussion of the numerous case studies provided above. Indeed, the legal and policy frameworks constructed around the management of wild animals are complicit in this lack of concern. There is, however, no convincing argument as to why animal welfare should not be considered alongside animal conservation. Compassion is not an undesirable state, and its adoption does not bring costs and economic hardship. As we have shown, there are many options whereby adopting compassion and reducing welfare costs to animals may have both economic and conservation benefits. Greater transparency of wildlife management policy, from its underlying need to the justification of favored management approaches, is necessary to progress this important field of science in the twenty-first century. As the international movement of compassionate conservation gathers momentum the future for wild animals will improve, but much research is needed to ensure that this movement truly represents a paradigm shift for wild animals in Australasia and elsewhere.

Acknowledgments

We thank members of THINKK, the Think Tank for Kangaroos and the Institute for Sustainable Futures at the University of Technology Sydney for discussions helping to shape our thinking on compassionate conservation. Similarly, we thank Marc Bekoff, Chris Draper, and Freya Mathews for providing further context. Voiceless, the Animal Protection Institute, has provided ongoing support and continuously challenged our perceptions of the treatment of domestic and wild animals.

References

Allen, B. L. 2011. "A Comment on the Distribution of Historical and Contemporary Livestock Grazing across Australia: Implications for Using Dingoes for Biodiversity Conservation." *Ecological Management and Restoration* 12: 26–30.

Ampt, P. and A. Baumber. 2006. "Building Connections between Kangaroos, Commerce, and Conservation in the Rangelands." *Australian Zoologist* 33: 398–409.

Arnold, G. W., D. E. Steven, and J. R Weeldenburg. 1989. "The Use of Surrounding Farmland by Western Grey Kangaroos Living in a Remnant of Wandoo Woodland and Their Impact on Crop Production." *Australian Wildlife Research* 16: 85–93.

Arnold, G. W., D. E. Steven, J. R. Weeldenburg, and E. A Smith. 1993. "Influences of Remnant Size, Spacing Pattern, and Connectivity on Population-Boundaries and Demography in Euros *Macropus robustus* Living in a Fragmented Landscape." *Biological Conservation* 64: 219–30.

Ash, K. 2005. "International Animal Rights: Speciesism and Exclusionary Human Dignity." *Animal Law* 11: 195–213.

———. 2007. "Why 'Managing' Biodiversity Will Fail: An Alternative Approach to Sustainable Exploitation for International Law." *Animal Law* 13: 209–250.

Baumber, A., R. Cooney, P. Ampt, and K. Gepp. 2009. "Kangaroos in the Rangelands: Opportunities for Landholder Collaboration." *Rangeland Journal* 31: 161–67.

Bekoff, M. 2000. "Redecorating Nature: Deep Science, Holism, Feeling, and Heart." *BioScience* 50: 635.

———. 2006. *Animal Passions and Beastly Virtues: Reflections on Redecorating Nature.* Philadelphia: Temple University Press.

Ben-Ami, D., K. Boom, L. Boronyak, D. B. Croft, D. Ramp, and C. Townend. 2011. *The Ends and Means of the Commercial Kangaroo Industry: An Ecological, Legal, and Comparative Analysis.* Sydney: THINKK, the Think Tank for Kangaroos, University of Technology.

Ben-Ami, D., D. B. Croft, D. Ramp, and K. Boom. 2010. *Advocating Kangaroo Meat: Towards Ecological Benefit or Plunder?* Sydney: THINKK, the Think Tank for Kangaroos, University of Technology.

Bennett, E. L., E. Blencowe, K. Brandon, D. Brown, R. W. Burn, G. Cowlishaw, G. Davies, H. Dublin, J. E. Fa, E. J. Milner-Gulland, J. G. Robinson, J. M. Rowcliffe, F. M. Underwood, and, D. S. Wilkie. 2007. "Hunting for Consensus: Reconciling Bushmeat Harvest, Conservation, and Development Policy in West and Central Africa." *Conservation Biology* 21: 884–87.

Boom, K., and D. Ben-Ami. 2011. "Shooting Our Wildlife: An Analysis of the Law and Its Animal Welfare Outcomes for Kangaroos and Wallabies." *Australian Animal Protection Law Journal* 5: 44–76.

Bowman, D. 2012. "Bringing Elephants to Australia?" *Nature* 482: 30.

Brown, G. P., B. L. Phillips, J. K. Webb, and R. Shine. 2006. "Toads on the Road: Use of Roads as Dispersal Corridors by Cane Toads (*Bufo marinus*) at an Invasion Front in Tropical Australia." *Biological Conservation* 133: 88–94.

Cabrera-Guzman, E., M. Crossland, and R. Shine. 2011. "Can We Use the Tadpoles of Australian Frogs to Reduce Recruitment of Invasive Cane Toads?" *Journal of Applied Ecology* 48: 462–70.

Calaby, J. H., and G. C. Grigg. 1989. "Changes in Macropodoid Communities and Populations in the Past 200 Years, and the Future." In *Kangaroos, Wallabies, and Rat-Kangaroos*, vol. 2, edited by G. C. Grigg, P. Jarman, and I. Hume, 813–20. Sydney: Surrey Beatty and Sons.

Callicott, J. B. 1989. "Animal Liberation: A Triangular Affair." In *In Defense of the Land Ethic*, edited by J. B. Callicott, 15–36. Albany: State University of New York Press.

Cameron, J., and J. Abouchar. 1991. "The Precautionary Principle: A Fundamental Principle of Law and Policy for the Protection of the Global Environment." *Boston College International and Comparative Law Review* 14: 1–27.

Caughley, G. 1981. "Overpopulation." In *Problems in the Management of Locally Abundant Wild Animals*, edited by P. A. Jewell, S. Holt, and D. Hart. New York: Academic Press.

Caughley, G., and A. R. E. Sinclair. 1994. *Wildlife Ecology and Management.* Carlton, Victoria: Blackwell Science.

Caulfield, M. 2008. *Handbook of Australian Animal Cruelty Law.* Melbourne: Animals Australia.

Choquenot, D., J. McIlroy, and T. Korn. 1996. *Managing Vertebrate Pests: Feral Pigs.* Canberra: Bureau of Rural Sciences.

Claridge, A. W., and R. Hunt. 2008. "Evaluating the Role of the Dingo as a Trophic Regulator: Additional Practical Suggestions." *Ecological Management and Restoration* 9: 116–19.

Collin, R. W., and R. M. Collin. 1994. "Equity as the Basis of Implementing Sustainability: An Exploratory Essay." *West Virginia Law Review* 96: 1173–90.

Cooney, R. 2008. *Commercial and Sustainable Use of Wildlife: Suggestions to Improve Conservation, Land Management, and Rural Economies.* Canberra: Rural Industries Research and Development Corporation.

Cooney, R., A. Baumber, P. Ampt, and G. Wilson. 2009. "Sharing Skippy: How Can Landholders Be Involved in Kangaroo Production in Australia?" *Rangeland Journal* 31: 283–92.

Croft, D. B. 2004. "Kangaroo Management: Individuals and Communities." *Australian Mammalogy* 26: 101–8.

———. 2005. "Kangaroos Maligned—16 Million Years of Evolution and Two Centuries of Persecution." In *Kangaroos: Myths and Realities*, edited by M. Wilson and D. B. Croft, 17–32. Melbourne: Australian Wildlife Protection Council.

Croft, D. B., R. Montague-Drake, and M. Dowle. 2007. "Biodiversity and Water Point Closure: Is the Grazing Piosphere a Persistent Effect?" In *Animals of Arid Australia: Out There on Their Own?*, edited by C. R. Dickman, D. Lunney, and S. Burgin, 143–71. Mosman: Royal Zoological Society of New South Wales.

Dale, A. 2009. "Animal Welfare Codes and Regulations—The Devil in Disguise?" In *Animal Law in Australasia*, edited by P. Sankoff and S. White, 174–211. Sydney: Federation Press.

Dickman, C. R., R. L. Pressey, L. Lim, and H. E. Parnaby. 1993. "Mammals of Particular Conservation Concern in the Western Division of New South Wales." *Biological Conservation* 65: 219–48.

Dobbie, W. R., D. M. Berman, and M. L. Braysher. 1993. *Managing Vertebrate Pests: Feral Horses.* Canberra: Bureau of Rural Sciences.

DSEWPaC 2011. *Commercial Kangaroo Harvesting Fact Sheet.* Canberra: Department of Sustainability, Environment, Water, Populations, and Communities.

Dunayer, J. 2001. *Animal Equality: Language and Liberation.* Oxford: Ryce Publications.

Dyer, R., P. Jacklyn, J. Russell-Smith, and R. J Williams. 2002. Introduction. In *Savanna Burning: Understanding and Using Fire in Northern Australia*, edited by R. Dyer, P. Jacklyn, I. Partridge, J. Russell-Smith, and R. J. Williams, 1–4. Darwin: Tropical Savannas Management Cooperative Research Centre.

Earnshaw, G. I. 1999. "Equity as a Paradigm for Sustainability: Evolving the Process toward Interspecies Equity." *Animal Law* 5: 113–46.

Edwards, G. P., B. Zeng, W. K. Saalfeld, and P. Vaarzon-Morel. 2010. "Evaluation of the Impacts of Feral Camels." *Rangeland Journal* 32: 43–54.

Eldridge, M. D. B. 2011. "The Changing Nature of Rock-Wallaby (Petrogale) Research 1980–2010." *Australian Mammalogy* 33: i–iv.

Eldridge, M. D. B., T. L. Browning, and R. L. Close. 2001. "Provenance of New Zealand Brush-Tailed Rock-Wallaby (*Petrogale penicillata*) Population Determined by Mitochrondial DNA Sequence Analysis." *Molecular Ecology* 10: 2561–67.

Ellis, E. 2009. "Collaborative Advocacy: Framing the Interests of Animals as a Social Justice Concern." In *Animal Law in Australasia*, edited by P. Sankoff and S. White, 354–75. Sydney: Federation Press.

———. 2010. "Making Sausages and Law: The Failure of Animal Welfare Laws to Protect Both Animals and Fundamental Tenets of Australia's Legal System." *Australian Animal Protection Law Journal* 4: 6–22.

Finch, N. A., and G. S. Baxter. 2007. "Oh Deer, What Can the Matter Be? Landholder Attitudes to Deer Management in Queensland." *Wildlife Research* 34: 211–17.

Fisher, D. O., S. P. Blomberg, and I. P. F. Owens. 2003. "Extrinsic versus Intrinsic Factors in the Decline and Extinction of Australian Marsupials." *Proceedings of the Royal Society B: Biological Sciences* 270: 1801–8.

Flannery, T. F. 1994. *The Future Eaters: An Ecological History of the Australasian Lands and People.* Chatswood: Reed Books.

———. 1998. *Throwim Way Leg.* Melbourne: Text Publishing.

Fleming, P., L. Corbett, R. H. Harden, and P. C. Thomson. 2001. *Managing the Impacts of Dingoes and Other Wild Dogs.* Canberra: Bureau of Rural Sciences.

Fleming, P. J. S., B. L. Allen, and G. Ballard. 2012. "Seven Considerations about Dingoes as Biodiversity Engineers: The Socioecological Niches of Dogs in Australia." *Australian Mammalogy* 34: 119–31.

Florance, D., J. K. Webb, T. Dempster, M. R. Kearney, A. Worthing, and M. Letnic. 2011. "Excluding Access to Invasion Hubs Can Contain the Spread of an Invasive Vertebrate." *Proceedings of the Royal Society B: Biological Sciences* 278: 2900–2908.

Fraser, D. 2010. "Toward a Synthesis of Conservation and Animal Welfare Science." *Animal Welfare* 19: 121–24.

Fraser, D., and A. M. MacRae. 2011. "Four Types of Activities That Affect Animals: Implications for Animal Welfare Science and Animal Ethics Philosophy." *Animal Welfare* 20: 581–90.

Gillies, C. A., and R. J. Pierce. 1999. "Secondary Poisoning of Mammalian Predators during Possum and Rodent Control Operations at Trounson Kauri Park, Northland, New Zealand." *New Zealand Journal of Ecology* 23: 183–92.

Glen, A. S., C. R. Dickman, M. E. Soulé, and B. G. Mackey. 2007. "Evaluating the Role of the Dingo as a Trophic Regulator in Australian Ecosystems." *Austral Ecology* 32: 492–501.

Gorman, J., and C. Kenneally. 2012. "Australia's Changing View of the Dingo." *New York Times*, March 5. http://www.nytimes.com/2012/03/06/science/australias-view-of-the-dingo-evolves .html?pagewanted=all&_r=0.

Grigg, G. C. 2002. "Conservation Benefit from Harvesting Kangaroos: Status Report at the Start of a New Millennium, a Paper to Stimulate Discussion and Research." In *A Zoological Revolution: Using Native Fauna to Assist in Its Own Survival*, edited by D. Lunney and C. Dickman, 53–76. Mosman: Royal Zoological Society of NSW.

Gullett, W. 1997. "Environmental Protection and the Precautionary Principle: A Response to Scientific Uncertainty in Environmental Management." *Environment and Planning Law Journal* 14: 52–69.

Hall, G. P., and K. P. Gill. 2005. "Management of Wild Deer in Australia." *Journal of Wildlife Management* 69: 837–44.

Heinsohn, T. E. 2003. "Animal Translocation: Long-Term Human Influences on the Vertebrate Zoogeography of Australasia (Natural Dispersal versus Ethnophoresy)." *Australian Zoologist* 32: 351–76.

Higginbottom, K., L. C. Northrope, B. D. Croft, B. Hill, and L. Fredline. 2004. "The Role of Kangaroos in Australian Tourism." *Australian Mammalogy* 26: 23–32.

Hornadge, B. 1972. *If It Moves, Shoot It: A Squint at Some Australian Attitudes towards the Kangaroo.* Dubbo: Review Publications.

Howard, T. 2002. "Exotic Grasses and Fire." In *Savanna Burning: Understanding and Using Fire in Northern Australia*, edited by R. Dyer, P. Jacklyn, I. Partridge, J. Russell-Smith, and R. J. Williams, 21–28. Darwin: Tropical Savannas Management Cooperative Research Centre.

Husheer, S. W., D. A. Coomes, and A. W. Robertson. 2003. "Long-Term Influences of Introduced Deer on the Composition and Structure of New Zealand *Nothofagus* Forests." *Forest Ecology and Management* 181: 99–117.

IUCN (2011). IUCN Red List of Threatened Species. Version 2011.2. http://www.iucnredlist.org. Accessed March 15, 2012.

Jarman, P. J., and K. A Johnson. 1977. "Exotic Mammals, Indigenous Mammals, and Land-Use." *Proceedings of the Ecological Society of Australia* 10: 146–66.

Kean, L., and M. O. Price. 2003. "The Extent of Mission Grasses and Gamba Grass in the Darwin Region of Australia's Northern Territory." *Pacific Conservation Biology* 8: 1–10.

King, C. M., and R. A. Powell. 2007. *The Natural History of Weasels and Stoats: Ecology, Behavior and Management.* New York: Oxford University Press.

Kinnear, J. E., M. L. Onus, and N. R. Sumner. 1998. "Fox Control and Rock-Wallaby Population Dynamics—II: An Update." *Wildlife Research* 25: 81–88.

Kirkpatrick, J. B., J. B. Marsden-Smedley, and S. W. J. Leonard. 2011. "Influence of Grazing and Vegetation Type on Post-Fire Flammability." *Journal of Applied Ecology* 48: 642–49.

Kwan, D., H. Marsh, and S. Delean. 2006. "Factors Influencing the Sustainability of Customary Dugong Hunting by a Remote Indigenous Community." *Environmental Conservation* 33: 164–71.

Lever, C. 2001. *The Cane Toad: The History and Ecology of a Successful Colonist.* Otley: Westbury Academic and Scientific Publishing.

Lindenmayer, D., S. Dovers, M. H. Olson, and S. R. Morton, eds. 2008. *Ten Commitments: Reshaping the Lucky Country's Environment.* Melbourne: CSIRO Publishing.

Littin, K. E. 2010. "Animal Welfare and Pest Control Meeting Both Conservation and Animal Welfare Goals." *Animal Welfare* 19: 171–76.

Low, T. 1999. *Feral Future: The Untold Story of Australia's Exotic Invaders.* Melbourne: Penguin.

Ludwig, J., D. Tongway, D. Freudenberger, J. Noble, and K. Hodgkinson, eds. 1997. *Landscape Ecology, Function, and Management: Principles From Australia's Rangelands.* Melbourne: CSIRO Publishing.

Lunney, D. 2010. "A History of the Debate (1948–2009) on the Commercial Harvesting of Kangaroos, with Particular Reference to New South Wales and the Role of Gordon Grigg." *Australian Zoologist* 35: 383–430.

Lunney, D., B. Law, and C. Rummery. 1997. "An Ecological Interpretation of the Historical Decline of the Brush-Tailed Rock-Wallaby *Petrogale penicillata* in New South Wales." *Australian Mammalogy* 19: 281–96.

Marsh, H., H. Penrose, C. Eros, and J. Hugues. 2002. *Dugong: Status Report and Action Plans for Countries and Territories.* Townsville: Early Warning and Assessment Report Series.

Martin, R. 2005. *Tree-Kangaroos of Australia and New Guinea.* Melbourne: CSIRO Publishing.

Masson, J. M., and S. McCarthy. 1996. *When Elephants Weep: The Emotional Lives of Animals.* New York: Delta.

Mathews, F. 2010. "Wild Animal Conservation and Welfare in Agricultural Systems." *Animal Welfare* 19: 159–70.

McKenzie, N. L., A. A. Burbidge, A. Baynes, R. N. Brereton, C. R. Dickman, G. Gordon, L. A. Gibson, P. W. Menkhorst, A. C. Robinson, M. R. Williams, and J. C. Z. Woinarski. 2007. "Analysis of Factors Implicated in the Recent Decline of Australia's Mammal Fauna." *Journal of Biogeography* 34: 597–611.

McLean, I. G., C. Hoelzer, and B. J. S. Studholme. 1999. "Teaching Predator-Recognition to a Naive Bird: Implications for Management." *Biological Conservation* 87: 123–30.

Molyneux, J., D. A. Taggart, A. Corrigan, and S. Frey. 2011. "Home-Range Studies in a Reintroduced Brush-Tailed Rock-Wallaby (*Petrogale penicillata*) Population in the Grampians National Park, Victoria." *Australian Mammalogy* 33: 128–34.

Moriarty, A. 2004. "The Liberation, Distribution, Abundance, and Management of Wild Deer in Australia." *Wildlife Research* 31: 291–99.

Moseby, K. E., B. M. Hill, and J. L. Read. 2009. "Arid Recovery—A Comparison of Reptile and Small Mammal Populations Inside and Outside a Large Rabbit, Cat, and Fox-Proof Exclosure in Arid South Australia." *Austral Ecology* 34: 156–69.

Muhic, J., E. Abbott, and M. J. Ward. 2012. "The Warru (*Petrogale lateralis*) Reintroduction Project on the Anangu Pitjantjatjara Yankunytjatjara Lands, South Australia." *Ecological Management and Restoration* 13: 89–92.

Nimmo, D. G., and K. K. Miller. 2007. "Ecological and Human Dimensions of Management of Feral Horses in Australia: A Review." *Wildlife Research* 34: 408–17.

Nugent, G., and K. W. Fraser. 1993. "Pests or Valued Resources? Conflicts in Management of Deer." *New Zealand Journal of Zoology* 20: 361–66.

O'Connell-Rodwell, C. E., T. Rodwell, M. Rice, and L. A Hart. 2000. "Living with the Modern Conservation Paradigm: Can Agricultural Communities Co-Exist with Elephants? A Five-Year Case Study in East Caprivi, Namibia." *Biological Conservation* 93: 381–91.

O'Donnell, S., J. K. Webb, and R. Shine. 2010. "Conditioned Taste Aversion Enhances the Survival of an Endangered Predator Imperilled by a Toxic Invader." *Journal of Applied Ecology* 47: 558–65.

Parkes, J., R. Henzell, and G. Pickles. 1996. *Managing Vertebrate Pest: Feral Goats.* Canberra: Government Publishing Service.

Regan, T. 1983. *The Case for Animal Rights.* Berkeley: University of California Press.

Reynolds, J. C. 2004. "Trade-Offs between Welfare, Conservation, Utility, and Economics in Wildlife Management—A Review of Conflicts, Compromises, and Regulation." *Animal Welfare* 13: S133–S138.

Rolls, E. C. 1984. *They All Ran Wild: The Animals and Plants That Plague Australia.* Sydney: Angus and Robertson.

Rossiter-Rachor, N. A., S. A. Setterfield, M. M. Douglas, L. B. Hutley, and G. D. Cook. 2008. "*Andropogon gayanus* (Gamba Grass) Invasion Increases Fire-Mediated Nitrogen Losses in the Tropical Savannas of Northern Australia." *Ecosystems* 11: 77–88.

RSPCA Australia 2002. *A Survey of the Extent of Compliance with the Requirements of the Code of Practice for the Humane Shooting of Kangaroos.* RSPCA Australia, Deakin West.

Russell, G. 2008. "Comment of Wilson and Edwards' Proposal for Low-Emission Meat." *Conservation Letters* 1: 244.

Russell-Smith, J., P. Whitehead, and P. Cooke, eds. 2009. *Culture, Ecology, and Economy of Fire Management in North Australian Savannas: Rekindling the* Wurrk *Tradition.* Collingwood: CSIRO Publishing.

Saunders, G. R., B. Coman, J. Kinnear, and M. Braysher. 1995. *Managing Vertebrate Pests: Foxes.* Canberra: Bureau of Rural Sciences.

Sharman, K. 2009. "Farm Animals and Welfare Law: An Unhappy Union." In *Animal Law in Australasia*, edited by P. Sankoff and S. White, 35–56. Sydney: Federation Press.

Short, J., and G. Milkovits. 1990. "Distribution and Status of the Brush-Tailed Rock-Wallaby in South-Eastern Australia." *Australian Wildlife Research* 17: 169–79.

Short, J., and B. Turner. 2000. "Reintroduction of the Burrowing Bettong *Bettongia lesueur* (Marsupialia: Potoroidae) to Mainland Australia." *Biological Conservation* 96: 185–96.

Short, J., B. Turner, and D. Risbey. 2002. "Control of Feral Cats for Nature Conservation. III. Trapping." *Wildlife Research* 29: 475–87.

Short, J., B. Turner, D. A. Risbey, and R. Carnamah. 1997. "Control of Feral Cats for Nature Conservation. II. Population Reduction by Poisoning." *Wildlife Research* 26: 703–14.

Simberloff, D. 2001. "Eradication of Island Invasives: Practical Actions and Results Achieved." *Trends in Ecology and Evolution* 16: 273–74.

Singer, P. 1975. *Animal Liberation.* London: Pimlico.

Soulé, M. E. 1985. "What Is Conservation Biology?" *BioScience* 35: 727–34.

Spiegel, M. 1997. *The Dreaded Comparison: Human and Animal Slavery.* London: Mirror Books.

Sweetapple, P., and G. Nugent. 2009. "Possum Demographics and Distribution after Reduction to Near-Zero Density." *New Zealand Journal of Zoology* 36: 461–71.

Taggart, D. A., D. J. Schultz, C. White, P. J. Whitehead, G. Underwood, and K. Phillips. 2005. "Cross Fostering, Growth, and Reproductive Studies in the Brush-Tailed Rock-Wallaby, *Petrogale penicillata* (Marsupialia: Macropodidae): Efforts to Accelerate Breeding in a Threatened Species." *Australian Journal of Zoology* 53: 313–23.

Taylor, S. S., I. G. Jamieson, and D. P. Armstrong. 2005. "Successful Island Reintroductions of New Zealnd Robins and Saddlebacks with Small Numbers of Founders." *Animal Conservation* 8: 415–20.

Thiriet, D. 2007. "In the Spotlight—The Welfare of Wild Introduced Animals in Australia." *Environment and Planning Law Journal* 24: 417–26.

Vaarzon-Morel, P., and G. Edwards. 2012. "Incorporating Aboriginal People's Perceptions of Introduced Animals in Resource Management: Insights from the Feral Camel Project." *Ecological Managment and Restoration* 13: 65–71.

Wallach, A. D., B. R. Murray, and A. J. O'Neill. 2009. "Can Threatened Species Survive Where the Top Predator Is Absent?" *Biological Conservation* 142: 43–52.

Ward-Fear, G., G. P. Brown, and R. Shine. 2010. "Using a Native Predator (the Meat Ant, *Iridomyrmex reburrus*) to Reduce the Abundance of an Invasive Species (the Cane Toad, *Bufo marinus*) in Tropical Australia." *Journal of Applied Ecology* 47: 273–80.

White, S. 2009. "Animals in the Wild." In *Animal Law in Australasia*, edited by P. Sankoff and S. White, 230–58. Sydney: Federation Press.

Whitehead, P., and C. Wilson. 2000. "Exotic Grasses in Northern Australia: Species That Should Be Sent Home." In "Northern Grassy Landscape Conference," 1–8. Katherine, Northern Territory, Australia, August 29–31.

Wilson, R. G., and J. M. Edwards. 2008. "Native Wildlife on Rangelands to Minimize Methane and Produce Lower-Emission Meat: Kangaroos versus Livestock." *Conservation Letters* 1: 119–28.

Wise, S. M. 1996. "Legal Rights for Nonhuman Animals: The Case for Chimpanzees and Bonobos." *Animal Law* 2: 179–86.

———. 1999. "Animal Thing to Animal Person—Thoughts on Time, Place, and Theories." *Animal Law* 5: 61–68.

Wittgenstein, L. 1972. *Philosophical Investigations.* Oxford: Blackwell.

Woinarski, J., C. Pavey, R. Kerrigan, I. Cowie, and S. Ward, eds. 2007. *Lost from Our Landscape: Threatened Species of the Northern Territory.* Palmerston: Department of Natural Resources, Environment and the Arts.

22

Explaining China's Wildlife Crisis

Cultural Tradition or Politics of Development

Peter J. Li

WILDLIFE PROTECTION is a global challenge. Climate change is threatening, for example, the very survival of polar bears and Canadian harp seals, species depending on ice for survival (Stirling and Parkinson 2006; Palmer 2006). On top of the natural disaster, human greed adds to their misery. Trophy hunters and the sealing industry are the most imminent threats. Arctic habitat protection is certainly the direct responsibility of the range countries. Yet countries like China have an unshakable responsibility as well (Aldworth 2010; CCTV-9 2010).[1] Their markets, or more precisely, the closure of them, could spell the end of human-made tragedy currently still befalling the polar bears and harp seals. China today occupies a decisive position in worldwide wildlife protection. Chinese actions or inactions in this regard have global impacts.

This chapter is about China's wildlife crisis. As the following sections attempt to demonstrate, abuse of and assault on wildlife in captivity and in the wild have reached an unprecedented level on the Chinese mainland in the reform era (1978–present). Shocking brutality against wildlife animals has been frequently exposed by Chinese and international media. To readers outside East Asia, they ask if the Chinese are culturally indifferent to animal suffering. Indeed, does the Chinese culture sanction cruelty to animals? Or is it the contemporary politics of economic development that is more directly responsible for the crisis?

Ignoring Nature and Wildlife Devastation

Like the rest of the world, China has witnessed a serious deterioration of the natural environment. In terms of wildlife devastation, China's contemporary history can be divided into two phases, that is the pre-reform years (1949–78) and the reform era (1978–present). The two distinct eras share a strikingly similar modernist attitude towards nature despite their opposing development models. While the pre-reform era, politically totalitarian and economically autarkic, saw nature as object of conquest and wildlife resources for human benefit, the Stalinist command economy with a low level of productivity had by and large limited the Leninist Party–state's ability to wreak havoc on nature in ways that the modern capitalist system is better equipped to. China's reform era saw the rise of an authoritarian and developmentalist regime obsessed with growth and modern production technology. The end of the command economy in the mid-1980s opened the Pandora's box. Wildlife animals, as resources for economic development, have been thrown into the biggest survival and welfare crisis.

The State's War against Nature: 1949–78

China adopted the Soviet development model in the 1950s. Agriculture was placed under a collective farming system called "the People's Commune," the Chinese equivalent of the Soviet state farms. To Eurasian communist leaders, collective farming served to eradicate bourgeois petty producers and place rural production under strict state command. Under state control, the rural sector was to serve the state's industrialization objective. Agricultural production, resource allocation, and labor input were strictly planned by the state. Farmers were therefore tied to collective farm work. No rural households were capable of engaging in wildlife farming. Wildlife domestication, started in 1954, was a state monopolized production. The malfunctioning Stalinist economic system ensured a loss-making wildlife farming operation. In the pre-reform era, wildlife animals impacted by the state farming operation were significantly smaller in number.

Assault on nature and wildlife was largely a state action. This included the massive land reclamation campaign in the border and mountainous regions in the early 1950s, frenzied deforestation during the Great Leap Forward campaign of 1958, and the nationwide mass campaign to exterminate sparrows as a result of the "great leader's" appeal (Shapiro 2001). Land reclamation destroyed the natural habitat of a large number of wildlife species such as Siberian tigers, South China tigers, Chinese river dolphins, and Chinese alligators to name some of the most impacted. Deforestation caused by the "Mass Steel Production Campaign" led to the disappearance of some of the nation's pristine forests. The nationwide sparrow killing spree instilled in the minds of the young that small animals like sparrows did not deserve human compassion. Sparrow killing resulted in insects growing out of control. Labor diversion to nonproductive political campaigns led to humanity's biggest man-made famine of 1960–62 killing over thirty million Chinese peasants. To tide over during the nationwide hunger, people began to take to the mountains. Government officials joined the hunting expedition. Sichuan alone reportedly decimated more than 62,000 deer in the wild in 1960 (State Council 1962). Mongolian gazelle was hunted near extinction (Geng 1998, 162–67).

In China's pre-reform era, there were only a few reactive national decisions related to wildlife protection. These included the 1950 Measures on Protecting Rare Wildlife Animals placing wildlife under state control, the 1956 People's Congress decision to create China's first nature reserve, the 1957 regulations on hunting, the 1961 Forestry Ministry notice on strengthening wildlife management, and the 1962 State Council Instructions on Wildlife Conservation (Jin 2002, 314; Cai 2000, 121). These policies had some effect on limiting assault on wildlife. The Chinese government had however contradicted its own policies by sanctioning a host of shortsighted wildlife use operations. The State Council Instructions on Wildlife Conservation also included tigers in the protection list. Yet the state trade companies had continued to purchase tiger pelts until 1974 (Zheng 2001, 232–33).

In the pre-reform era, wildlife conservation, like environmental protection in general, was not a concern to the Chinese government. What occupied their efforts were ideological campaigns indoctrinating the masses with the Maoist version of a socialist society. Viewing animals as objects of compassion was bourgeois and therefore ideologically questionable. Private production was outlawed. Prohibited were also productive activities utilizing wildlife animals as resources. In human-nature relations, the Party extolled the concept of human conquest of nature (Shapiro 2001, 67–94). Wildlife devastation was by and large state behavior in the pre-reform era.

An All-Out War on Nature in the Reform Era

With only 6.5 percent of the world's territory, China is home to more than 6,347 species of vertebrates, 14 percent of the world's total. Among these vertebrates, there are 711 mammals, 3,862 fish, 1,294 birds, 412 reptiles, and 295 species of amphibians (Ma, Zou, and Zheng 2003). The entire world knows that giant pandas are native to China and are endangered. Not known to the outside world are other even more critically endangered animals such as South China tigers, ibis, Chinese river dolphins, and Chinese alligators to name the most famous. China's vast territory is temporary home to a vast number of migratory species. In 1980 China joined CITES (Convention on International Trade in Endangered Species of Wild Fauna and Flora) as a way to attract foreign know-how and capital for wildlife conservation, ignored and mismanaged in the past. In 1989 the Chinese government adopted the Wildlife Protection Law (WPL), the nation's first national law for animal protection.

China since 1989 has ironically witnessed a greater wildlife crisis. Globally, more than 593 species of birds, 400 mammals, and 209 amphibians are believed to be in endangered status. In China, rapid industrialization and increased human activities have threatened the survival of 398 species of vertebrates (Zheng 2001, 236). More than 130 of the world's 400 endangered mammals are in China. Inside China, there are fewer than fifty Siberian tigers in the wild. The Yangtze River dolphin, a species that had lived in the waters of central China for more than twenty million years and was referred to as a living fossil, is reportedly extinct (Lovgren 2006). South China tigers are all in captivity. Animals whose extinction has been confirmed include Przewalski's horse, stubby-nosed antelope, Donc langur, hog deer, Taiwan cloud leopard, sarus crane, Hebei rhesus macaque, and white-headed gibbon. Countless other species are being relentlessly exploited, poached, and farmed to the brink of extinction (Yang and Yi 2004).

Chinese wildlife traders have engaged in a worldwide sourcing expedition. According to the International Union for Conservation of Nature (IUCN), some 30 percent of the sixty-four species of shark and ray are threatened with extinction while 24 percent of them are near extinction. Shark finning to supply Chinese catering businesses is a major contributing factor (Gill 2009; Foreign Policy Journal 2008). A so-called tiger trail linking India and Nepal with China allowed traffickers to smuggle tiger body parts over to the Chinese side (WWF 2009). Since the early 1990s Chinese traders have left no stones untouched in the pristine forests of Southeast Asia and Siberia. Along the Sino-Indochinese borders, live and dead wild animals are shipped in great quantities into China on a daily basis (Li et al. 1995, 112–58). In the Russian Far East region, those involved in wildlife trafficking uncovered by Russian customs between 1999 and 2006 were mostly Chinese (Lyapustin, Vaisman, and Fomenko 2007). In 2002, 512 bear paws and four tiger pelts bound for China were intercepted by Russian customs (BBC 2002; Wildlife Alliance, 2007). Live North American freshwater turtles and African elephant tusks have found buyers in Mainland China (Voice of America 2002; Xinhua News Network 2009).

Cruelty to wild animals has reached an unprecedented level on the Chinese mainland. Shark finning is another act of humans' gross inhumanity to nonhuman animals. Fishermen cut fins off and then toss the traumatized body back into the salty sea. Bear farming, tiger farming and fur animal farming are perhaps the most brutal farming operations. Bear farming victimizes more than 10,000 Asiatic black bears in cruel bile extraction from an open wound cut in their stomachs. Incarcerated bears suffer from intentional food deprivation, endure physical abuse, and languish physically and mentally in extreme boredom and discomfort. Tiger farming, raising some 5,000 tigers, has also been condemned for its ulterior motives and shocking welfare conditions. The entire world has been shocked to learn the brutal farming and slaughter conditions of Chinese fur animals when the *Fun Fur* investigative report was published (Hsieh et al. 2005). Outdated housing, poor management, cruel practices such as animal performance, photo ops (tiger de-toothed, de-clawed, drugged, and tied to a podium for young visitors to take pictures with), live feeding, and visitor abuse illustrate severe welfare problems in China's zoological gardens.

What explains China's wildlife crisis? Are the Chinese culturally incapable of compassion for nonhuman animals?

Human-Animal Relations in Chinese Culture

Cultural perspective is no stranger in China studies. It was first used to explain China's failure in modernization in the nineteenth and early twentieth centuries. To scholars of this approach, China's "static unchanging civilization" is to blame for the nation's backwardness. With the burden of a stagnant culture, China could never modernize (Cohen 1984, 57–96). Relevant to our study of animal cruelty is the perceived Chinese culture of cruelty (Nathan 1990). For example, Jonathan Spence, a renowned historian, sees Chinese history in the last four hundred years as one of political repression, territorial conquest, brutal violence, and intramural brutality (Spence 1990). The cultural assumption has led many to ask "why does a culture that condemns violence, that plays down the glory of military exploits, awards its highest prestige to literary, rather than martial figures, and seeks harmony over all other values, in fact display such frequency and variety of violent behavior?" (Nathan 1990, 30).

Cultural study of Chinese contemporary politics has continued. In his study of political violence in contemporary China, Barend J. ter Haar argues that violence during the Cultural Revolution (1966–76) "was by no means an innovation of the Maoist era but had important roots in traditional Chinese religious culture" (ter Haar 2002, 27–68). To him, China has never truly moved away from the tradition of "martial violence" despite the claim made by other scholars. The cultural evolution from "martial violence" to "refinement," in his opinion, is an ongoing process never yet completed (ter Haar 2000, 131). China is still in the grips of a violent ideology since its past predetermines Chinese attitudes and behaviors.

Are the Chinese culturally predestined to act cruelly to one another and to animals?

Daoism, Buddhism, and Confucianism

Daoism, Buddhism and Confucianism are three ancient thought systems that are still shaping Chinese world outlook. They have in the past 2,000 years promoted attitudes, values and mores that condemn violence. They offer insights into what constitutes socially acceptable behavior in China and other East Asian nations.

Daoism is a native Chinese philosophical concept of the right "way" for dealing with the cosmos, with one another, and with fate. Daoism stands for the unity between humanity and nature. Daoists do not see humans as being superior to or more intelligent than nonhuman individuals. Humans therefore do not have the right, still less the privilege, to treat other members of the universe as subordinates. To reduce the harm humans can do to nature, Daoists advocate frugality in consumption and restraint in behavior. Daoists propagate the much-hailed "three treasures": compassion, frugality, and modesty.

Daoism stands firmly against killing and cruelty to animals. It calls for rescuing animals from danger and exploitation (Sun, He, and Huang 2009, 182). In many texts on Daoist prohibitions, acts such as kicking and whipping farm animals, capturing animals in hibernation, destroying nests, harvesting eggs, and terrorizing animals are prohibited cruel acts (Sun et al., 183–84). Daoism has had great impact on people's attitude and behavior through the popularization of folk Daoism, a more pragmatic and earthly form of Daoist teachings easier for the commoners to understand. Through the belief in karma, folk Daoism has succeeded in guiding people to act in ethical ways that can benefit them in return. The Daoist concept of species equality perhaps puts China as one of the first animal rights advocating countries in the world.

Buddhism was introduced into China between the first and fourth centuries. It advocates harmony and a peaceful mind. Personal spiritual cultivation is therefore the means to that end. More than Daoism, its impact on Chinese society has been tremendous. Like Daoism, Buddhism is against the human-centered perspectives regarding nonhuman lives as inferior to humans. The spread of Buddhism in China can be attributed to many factors. Yet its idea of karma has certainly played a unique role in attracting Chinese followers. The belief that killing, including brutality against animals, would invite misfortune resonates well with the Chinese (Lin 1988, 126–27). For more than one thousand years, Buddhist practice of mercy release, the setting free of captured animals out of compassion, was a gesture exercised by the general public and endorsed by Chinese emperors. Some Chinese emperors even called on the nation to practice vegetarianism (Sun, He, and Huang 2009, 187). During parts of the Ming Dynasty (AD 1368–1644), there was a societal movement for prohibiting killing. Mercy release and suspension of slaughter were practiced at important celebrations such as the Chinese New Year, birthdays, weddings, business openings, and national and family events. According to a Beijing college professor, prohibition of killing and mercy release were not simply religious acts, they were also state policies (Mang 2009, chaps. 4 and 5).

Confucianism is a native thought system created more than 2,000 years ago. It contains a complex set of social, political and moral values. Of the three ancient thought systems, the impact of Confucianism on China has been most fundamental. Many Chinese may profess to be Taoists, Buddhists, or Christians, but in the final analysis, they are Confucianist. Confucianism is pragmatic and anthropocentric. Not only does Confucianism believe that humans are the most precious, it also sees that animals exist for human purposes (Zhang 209). As a state orthodox system, Confucianism was concerned more with the interest of a benevolent government, a harmonious society and cordial relations among the humans.

Confucianist anthropocentric outlook, however, was no obstacle to the development of ideas of compassion for nonhuman animals (Sun, He, and Huang 2009, 177–78). In the Confucian text *Analects*, the word *ren* (benevolence) was mentioned more than two hundred times (Mang 2009, 96–105). Not only did Confucius argue for benevolence of governance, he also called for the rulers to extend benevolence over every aspect of the society. He placed importance on protecting life and opposing violence. Mencius, arguably the most famous Confucian next to Confucius, saw a sense of pity as one important addition to the "four virtues" (benevolence, righteousness, rites, and intelligence) of Confucianism (Sun, He, and Huang 2009, 177–78).

The Confucian idea of "reasonable use" and the "Doctrine of the Golden Mean" served to neutralize Confucian pragmatism and anthropocentrism. Confucianists do not oppose use of wildlife. Yet, they call for measured use with a reason. The Doctrine of the Golden Mean stresses moderation and restraint in attitude and behavior. It cautions against extremism, excesses, and indulgence. The doctrine is therefore inadvertently conducive to ecological protection. The doctrine stresses the virtue of "nothing too much." Use of wildlife should therefore be limited and not excessive. Confucian scholars in later dynasties even proposed that benevolence or love should be extended beyond the human race (Sun, He, and Huang 2009, 179).

The preceding discussion is far from a complete coverage of China's past tradition related to human-nature relations. Yet it has tried to demonstrate that China has a more complex legacy that includes ideas and practices for compassion and protection of nonhuman animals. Blaming China's past for the contemporary flaws is misleading and misses the real target.

The Politics of Reform and Economic Modernization

China is one member of the anthropocentric world community. Admittedly, the enormity of wildlife-related industry and the magnitude of cruelty befalling wild animals in China overshadow those in the rest of the world. If culture is no explanation for China's runaway wildlife crisis, we need to examine China's contemporary politics for an answer.

Reform, Poverty Reduction, and Local Growth

When the Chinese government initiated the reform program in 1978, the Chinese economy was on the verge of collapse. A majority of the 900 million Chinese eked out a living on starvation rations determined by the Leninist Party–state. A Stalinist command economy had made Mainland China one of the poorest nations on earth. In the final days of Mao's rule, societal discontent was such that even the staunchest Maoist supporters in the party acknowledged that the communist regime was facing a legitimacy crisis. The reform was to salvage the established Leninist order by jump-starting economic growth to end the food crisis that had gripped the nation for more than two decades. Liberalization of the rural sector by de-collectivizing rural production and lifting the ban on private rural production were the first reform measures. These measures were expected to generate quick results in poverty reduction.

Poverty reduction was a greater challenge in the vast inland and mountainous areas. Tapping into local resources and converting local conditions into comparative advantages were first recommended as a way out for the underdeveloped regions. In the 1980s, two catch phrases reflected the government's eagerness for a quick fix of the Chinese economy. "Those living on a mountain live off the mountain while those living near the water live off the water." The other slogan was that "it is a good cat, be it white or black, as long as it catches mice." Color, ideological persuasion, or political correctness is to be secondary to productivity or economic results. Achieving fast growth has since been the national consensus. To the reformist leaders, a good economic performance that can end poverty was the only way to restore people's trust in the party. This official obsession with growth has allowed local authorities and private individuals to delve into all kinds of business activities including those that are increasingly challenged by society because of their adverse ethical implications.

In the reform era, China has seen an expanding wildlife-related industry. This industry has also undergone noticeable production realignments. First, the formerly state monopolized production became private enterprise. In the early 1980s millions of peasant households began to engage in wildlife-related business activities. In southeast provinces such as Jiangxi, Zhejiang, and Anhui, peasant households began to farm wild animals or to conduct interprovincial wildlife transport. Second, the erstwhile small-scale sideline production in wildlife products, by the end of the 1990s, had become increasingly industrialized. A bear farmer in Heilongjiang's Mudanjiang started his bear farm with only three bear cubs. By 2009, his farming operation had become a conglomerate type business raising more than 1,000 bears for bile extraction. Third, regional specialization has emerged for greater production coordination, better use of raw materials, proximity to processing centers, and access to markets. Wildlife farming for the catering business has concentrated in south, southwest, and southeast parts of the country, areas closer to Guangdong, the world's capital of wildlife eating. Bear farming has shrunk to some fourteen provinces in the northeast, southwest, and northwest regions. Fur animal farming concentrates in north and northeast China.[2]

Wildlife-Related Industry and Productivity Increase

In the reform era, Chinese officials are evaluated by their record in facilitating local growth. Production growth is evaluated by both quantity increase and the value of the output. By the mid-1980s, Chinese farmers had moved to the production of value-added products. Growing cash crops or converting the limited farmland into wildlife farming can generate greater production value. Converting fishing ponds into turtle farms brings a higher profit margin. There is little doubt that farming or dealing in bears, tigers, foxes, quails, snakes, peacocks, and other wild animals is significantly more productive in terms of sales revenue and profit than growing food grain or other agricultural produce. Wildlife farming suggests greater income for the producers and greater revenue potential for the local state coffer.

Profit awareness underlies decisions for profit-maximizing production models. Maximal use of production space is one result of the efficiency consideration. The smaller the space allowance to caged wildlife animals, for example, the greater the profit margin a farm can obtain. This explains the prevalent space deprivation on Chinese wildlife farms. Similarly, denial of proper food to farm bears, practiced by many bear farm owners, is more an act based on a crude efficiency calculation than sheer cruelty imposed on the bears. The bear farming community believes that good food for the bears reduces bile production. To extract more bile, the so-called liquid gold, bears cannot be fed too well. This practice is similar to forced molting of laying hens through feed deprivation to induce a new laying cycle. Therefore, a reduced food ration cuts the input cost and maximizes bile extraction.

Maximization of profits has also driven the reckless to target wild animals in the wild. Wildlife farming requires starting capital and operating input. Yet, poaching wild animals is a zero-investment gain. Despite the Wildlife Protection Law, enforcement failures have allowed assault on the nation's wild animals to continue unabated. The currently abundant supply of farmed wildlife products has reinforced the belief that wild animal parts have greater values in nutrition or medicinal effect. Wild-caught turtles are sold at a significantly higher price than domesticated turtles. This mind-set of the consumers suggests that the claim that wildlife farming is conducive to conservation is dubious.

Wildlife and Local Tax Income

The attraction of wildlife-related production to local authorities can never be underestimated. In 2004, the value of China's wildlife farming was estimated at 100 billion yuan (Zhang, Zhou, and Wang 2004, 27). The bear farming industry contributed some ¥8 billion to that total. Heibao in Heilongjiang Province, the world's biggest bear farm, alone produced 5,100 kilograms of bile powder in 2008 (Zhang 2009). Its CEO, a deputy to the local Provincial People's Congress, came up with a statistic of the value of China's farm bears. According to his calculation, the value of one farm bear in a ten-year-period amounts to 10 million yuan in economic results (Heilongjiang People's Congress 2005). Using his estimate, China's bear farming industry of 10,000 bears can generate 100 billion yuan within ten years. Its revenue potential is attractive to local governments. Revenue potentials have often overshadowed any other considerations such as animal welfare.

Fur animal farming is perhaps the only wildlife production that holds a hegemonic position in the local economy of the farming regions. Suning, Hebei Province, has seen a rising economic power due to fur production. In 2004, fur farming generated sales revenues of 2.8 billion yuan, the biggest gain of all fur-producing counties in the country. In Suning, fur animal farming made up 80 percent of the local GDP. Its revenue contribution to the local state coffer reached 65 million yuan, 32 percent of the local tax revenue (China Fur Information Net 2009). It is therefore not a surprise that the 2005 *Fun Fur* report condemning live skinning elicited a knee-jerk reaction from the local government. One year after the report was released, Suning's farming industry handed a staggering 90 million yuan in tax to the local government. The same year's sales revenue hit $3 billion (Suning County People's Government 2009).

Other fur animal farming regions also zealously protect this industry. Liaoyang, Liaoning Province, produces 300,000 fox pelts and over 70 million yuan of sales revenue a year. In 2008, the city produced sales revenue of 1.45 billion yuan, accounting for 26 percent of the total rural output and 37 percent of the peasants' total income (Liaoyang Animal Inspection Bureau 2009). In China, three biggest fur animal farming provinces hold some 70 percent of the country's 100 million fur animals. Fur animal farming is an important part of their rural economy. Fur processing and product manufacturing are important productions of Zhejiang, Fujian, Guangdong, Hebei, and Liaoning (China Special Farming Information Net 2008). These provinces extract enormous value from the postfarming productions of the raw materials.

The contribution of fur animal farming to local economies is most vividly demonstrated by Suning's rapid ascendency in the economic ranking of Hebei Province. With a population of 330,000 and a predominantly agricultural economy, Suning in 2002 was one of the poorest counties (ranked eighty-sixth). By the end of 2004 and with the rapid take-off of its fur animal farming productions, it jumped to the fortieth position (Suning County People's Government 2009). This change in economic standing can reflect very positively on the performance of the local leaders whose career mobility hinges on their ability to generate local growth.

Conclusions

There is no denial that China poses a major threat to wildlife within its own borders and in the world at large. Never in its 5,000-year history did China ever raise and keep hundreds of millions of wildlife species in captivity as it does today. Are the Chinese culturally predestined to be indifferent to animal suffering? Or is the contemporary politics of economic reform more directly responsible for wildlife devastation and animal suffering? The preceding sections have attempted to answer these two questions. Yes, China has a wildlife eating culinary subculture in parts of the country. Yet the mainstream diet in the country has never been dominated by wildlife, not even by domesticated farm animal products. China was and still is influenced by the ancient thought systems of Daoism, Buddhism, and Confucianism. While the first two philosophical ideas reject killing and animal abuse, Confucianism, given its anthropocentric outlook, calls for moderation and measured use of natural resources. None of these three thought systems stands for human exploitation of the natural world in ways that have taken place in the last three decades.

China's postsocialist developmental state has pursued a market-oriented reform program. Catch-up and GDP growth are the obsessions of the Chinese reformist authorities. The aim is to restore the legitimacy claim of the ruling Chinese Communist Party and ensure stability of the established ruling order. For this purpose, the government sees poverty reduction through fast economic growth a top priority. The performance of local officials has therefore been linked to local economic performance. And local growth determines their upward career mobility. Economic liberalization has sanctioned a host of productive activities at the expense of the wildlife species and also profit-seeking measures in nonprofit organizations. To subsidize daily operation, zoos have engaged in for-profit activities such as animal performance, live feeding, and other welfare compromising programs. Government officials, particularly local leaders, have paid scant, if any, attention to the animal welfare or ecological consequences of fast economic growth.

It is therefore not Chinese culture but the current "development first" mindset that is behind the nation's wildlife crisis. Encouragingly, Chinese national authorities have realized that the current mode of development is not sustainable. A consensus among the national leaders has been reached: China's development should in the long run be eco-friendly. We expect this new acknowledgment to have a positive impact on China's efforts in wildlife protection.

Notes

1. Rebecca Aldworth (executive director of Humane Society–Canada) has led campaigns against Canadian sealing industry in the last twelve years.

2. Interview with a ranking official of China Association of Wildlife Conservation, Beijing, China. March 24, 2008.

References

Aldworth, Rebecca. 2010. Interview with China national TV, CCTV-9, Beijing, November 30.

BBC. 2002. "Bear Paw Poachers Caught Red-Handed." BBC News, May 8. http://news.bbc.co.uk/1/hi/world/europe/1975844.stm. Accessed May 8, 2009.

Cai, Shouqiu. 2000. Huanjing ziyuan faxue jiaocheng (Textbook on environmental and resource laws). Wuhan: Wuhan University Press.

CCTV-9. 2010. "Say No to Seal Products." CCTV-9 special report. http://english.cntv.cn/program/cultureexpress/20101206/103281.shtml. Accessed December 18, 2010.

China Biodiversity Office. 2004. "First State Report on China's Execution of the 'Convention on Biological Diversity.'" http://www.biodiv.gov.cn/swdyx/145241087982698496/20040301/1046429.shtml. Accessed July 2, 2006.

China Fur Information Net. 2009. "Suning Holds a Press Forum on the Development of Fur Industry." http://www.fur.com.cn/jiaoyishichang/scdt/scdt/sc/1934.html. Accessed October 1, 2009.

China Special Farming Information Net. 2008. "The Current State and Future Prospects of China's Fur Animal Farming Industry." http://www.zt868.com/newshtml/2008-11-10/20081110220626.htm. Accessed November 30, 2008.

Cohen, P. A. 1984. Discovering History in China: American Historical Writing on the Recent Chinese Past. New York: Columbia University Press.

Foreign Policy. 2008. "Shark Near Extinction as China Gets Richer." Foreign Policy (February 20). http://blog.foreignpolicy.com/posts/2008/02/20/shark_nears_extinction_as_china_gets_richer. Accessed May 1, 2008.

Geng, Biao. 1998. Geng Biao Huiyi Lu (Geng Biao remembers: 1949–1992). Nanjing: Jiangsu People's Press.

Gill, Victoria. 2009. "Many Sharks "Facing Extinction.'" BBC News report. June 25. http://news.bbc.co.uk/2/hi/8117378.stm. Accessed May 1, 2010.

Heilongjiang People's Congress. 2005. "Deputies Proposed Economic Development Plans by Referring to the Value of Farm Bears, a Heilongjiang People's Congress Report" (January 28). http://www.hljrd.gov.cn/rdyw/200507260044.htm. Accessed September 15, 2006.

Hsieh-Yi, Yi-Chiao, Y. Fu, M. Rissi, and Barbara Mass. 2005. *Fun Fur? A Report on the Chinese Fur Industry.* A Swiss Animal Protection, Care for the Wild and East International investigative report.

Jin, Ruilin. 2002. *Huanjin faxue* (The study of environmental laws). Beijing: Beijing University Press.

Li, Yimin, and D. M. Li. 1995. *An Investigation of the Wildlife Trade in Guangxi and at the Sino-Vietnamese Border Areas.* Beijing: China Science and Technology Press.

Liaoyang Animal Inspection Bureau. 2009. "Liaoyang's Fox and Deer Dominated Wildlife Farming Industry." An information brochure of Liaoning Provincial Animal Husbandry and Veterinary Bureau. http://www.lnah.gov.cn. Accessed September 30, 2009.

Lin, Yutang. 1988. *My Country and My People.* Taipei: Mei Ya Publications, 126–27.

Lovgren, Stefan. 2006. "China's Rare River Dolphin Now Extinct, Experts Announce." *National Geographic News,* December 14. http://news.nationalgeographic.com/news/2006/12/061214-dolphin-extinct.html. Accessed October 24, 2007.

Lyapustin, Sergey N., A. L. Vaisman, and P. V. Fomenko. 2007. *Wildlife Trade in the Russian Far East: An Overview.* A TRAFFIC (wildlife trade monitoring network) report.

Ma, Jianzhong, H. F. Zou, and G. G. Zheng. 2003. "The Current State and Future Directions of China's Wildlife Animals and Habitat Protection." *Journal of China Agriculture Science* (zhong guo nong ye ke ji dao bao) 5 (4).

Mang, Ping. 2009. *The World of The Interrelated Self and Other: Chinese Belief, Lives, and Views of Animals.* Beijing: China University of Politics and Laws.

Nathan, Andrew J. 1990. "Why Does China Eat Its Own? A Culture of Cruelty." *New Republic* (July 30 and August 6).

Palmer, Nick. 2006. "The Killing Has to Stop." *New Scientist* 189 (2544).

Shapiro, Judith. 2001. *Mao's War against Nature: Politics and the Environment in Revolutionary China.* Cambridge: Cambridge University Press.

Spence, Jonathan. 1990. *The Search for Modern China.* New York: W. W. Norton.

State Council of the People's Republic of China, 1962. "Guowuyuan guanyu jiji baohu he heli liyong yesheng dongwu zhiyuan de zhishi" (State Council instructions on actively protecting and reasonably using animal resources). http://www.people.com.cn/GB/33831/33836/34143/34234/2551386.html. Accessed April 20, 2006.

Stirling, Ian, and C. L. Parkinson. 2006. "Possible Effects of Climate Warming on Selected Populations of Polar Bears (*Ursus maritimus*) in the Canadian Arctic." *Arctic* 59 (3): 261–75.

Suning County People's Government. 2009. "The Basic Information about Suning County." http://www.clii.com.cn/web/xiaoning/02.htm. Accessed March 1, 2009.

Sun, Jiang, L. He, and H. Zheng. 2009. *An Introduction to the Theory of Animal Protection Legislation.* Beijing: Law Press.

ter Haar, Barend J. 2000. "Rethinking 'Violence' in Chinese Culture." In *Meanings of Violence: A Cross-Cultural Perspective,* edited by Goran Aijmer and Jon Abbink. Oxford and New York: Berg.

———. 2002. "China's Inner Demons: The Political Impact of the Demonological Paradigm." In *China's Great Proletarian Cultural Revolution: Master Narratives and Post-Mao Counternarratives,* edited by Lien Chong. Lanham, MD: Rowman and Littlefield.

Voice of America. 2002. "Hong Kong Sent Back to the US Smuggled Turtles." http://www.voanews.com/chinese/archive/2002-01/a-2002-01-11-22-1.cfm. Accessed February 2, 2009.

Wildlife Alliance, 2007. "Smugglers Apprehended in Russia with Bear Paws, Tiger Pelt." Wildlife Alliance website, http://wildlifealliance.org/news/press-releases/smugglers-apprehended-in-russia.html. Accessed March 3, 2009.

WWF. 2009. "Curbing the Wildlife Trade: Nepal." A WWF report. http://www.panda.org/who_we_are/wwf_offices/nepal/our_solutions/projects/index.cfm?uProjectID=NP0902. Accessed March 28, 2009.

Xinhua News Network, 2009. "Fang Cheng Gang Police Checkpoint in Guangxi Uncovered Suspected Elephant Trunks." Xinhua News Network, January 7.

Yang, Xiaohong, and L. Yi. 2004. "China's State Forestry Bureau Released the Results of a Nationwide Wildlife Animal and Plant Resource Survey." *Xinhua News Network,* June 10. http://news.beelink.com.cn/20040610/1599518.shtml. Accessed September 18, 2005.

Zhang, Ming. 2009. "Report on the Development of Mudanjiang's Pharmaceutical Productions." A speech made by the director of Mudanjiang Food and Drug Inspection Bureau. http://www .mdj.hljda.gov.cn/news.aspx?id=672. Accessed July 5, 2009.

Zhang, Songhui. 2009. "On the Daoist Idea of All Lives Being Equal and the Confucian Anthropocentric Ideology." *Hunan Daojiao*, 2006. http://www.hunandj.com/tjld/36/2006227104524 .htm. Accessed March 28, 2009.

Zhang, Wei, X. H. Zhou., and W. Li. 2004. "On the Standardized Management and Industry Construction of China's Wildlife Farming Operation." *Chinese Wildlife* 25 (3): 27.

Zheng, Yi. 2001. *Zhongguo zi huimie: Zhongguo shengtai bengkui jingji baogao* (China's ecological winter). Hong Kong: The Mirror Books.

Zhou, Shifen. 2006. *Chuang shi ji di 20 zhang: Chong du 20 shi ji* (The 20th chapter after the birth of Jesus Christ: Revisiting the 20th century), particularly chap. 11. Shanghai: East China Normal University.

23

A Triangular Playing Field

The Social, Economic, and Ethical Context
of Conserving India's Natural Heritage

Vivek Menon

Introduction

INDIA PACKS in 8.1 percent of the global biodiversity, 7 percent of
the world's recorded fauna, and 11 percent of its recorded flora in
2.4 percent of the global surface, making it one of the twelve mega-
diverse countries of the world (Anon 2005). It also has a 2,300-year
recorded history of nature conservation.[1] Nature conservation has,
down the ages, been influenced by the prevalent religion, ethic, and
social objective, in conformity with global trends (Menon and Lavigne
2006). It is postulated here that the current conservation policies of
India are shaped by the limitations imposed by three sides of a triangle.
On one side are the realities of being a nation mired in poverty. India
is home to about a third of the world's poor people, which, according to various indexes, comprises between 27 and 41 percent of its
1.2 billion human population (Anon 2007). It ranks 119 among 169
countries in the Human Development Indexes of the United Nations
Development Programme (UNDP), 2010. The social, political, and
economic ramifications of these statistics shape the conservation of
Indian nature in a particular mold. The second boundary is that set,
ironically, by the developed nation "aspirational" mind-set of the Indian government and certain sections of its society. In other words,
by the economic growth engine that is currently chugging at 8.75
percent and is expected to cross 9 percent by the end of 2010. This
adds a second dimension of pragmatism to conservation. The baseline
of the triangle is, however, the strong ethical and value-based origins
of the Indian conservation mind-set. This is the only side that has
idealistic influences. These three realities, at once stark and impres-

sive, contradictory yet compelling, drawn from the recesses of a long history and periscoping into futuristic eras, lay a unique framework around nature conservation in India. Both policies and on-ground implementation cannot but be influenced by these forces, creating a unique mix that causes commentators such as Oates (2006, 278) to say: "Although it was argued in the 1980s and 1990s that traditional conservation, involving the enforcement of hunting laws and the establishment of strictly protected areas, is incompatible with the realities of life for people in poor rural areas in developing countries, I noted that the example of India shows that this is not so. While having one of the densest human populations in the world and persistent human poverty, India managed to create and maintain effective nature reserves and to stave off the extinction of endangered species like the tiger."

Of late, such statements have been questioned by social and environmental commentators (Gadgil and Guha 1995; Shahabuddin 2010), their arguments buttressed by spiraling human wildlife conflict and by the perilous state of those much-endangered species and ecosystems that Oates commends India at having conserved. Yet the answers to such challenges also seem to be governed by the triangular playing field.

The Creation of the Ethical Baseline

Preservation of nature in ancient India was based, as was much else of its prevalent social norms, on the principles of ahimsa and dharma. In the Vedas the concept of ahimsa was first chiefly espoused in the Tittreya Samhita of the Yajurveda,[2] although it finds reference in the Atharvaveda, Manusmriti,[3] and other such works of the times. Of the four great aims of humanity prescribed in Hinduism (and adapted to Buddhist and Jain doctrines), dharma, artha, kama, and moksha, the latter was always considered the ultimate goal, but the first the means to it in life.[4] Nature conservation, too, in the same vein was shaped by the spiritual, ethical, temporal, and moral code of dharma. This Vedic, Hindu, and later Buddhist and Jain tradition was central to Indian nature conservation but was also influential in the immediate vicinity of the country.[5]

Around 320 BC, Kautilya, the Machiavellian premier of Chandragupta Maurya, ostensibly records in the Arthashastra[6] thus: "The following animals shall be declared protected species a) Sea fish that have strange or unusual forms, b) fresh water fish from lakes, rivers, tanks or canals, c) game birds or birds for pleasure such as curlew, osprey, datyuha, swan, chakravakha, pheasant, bringharaja, partridge, mattakokila, cock, parrot and mynah and d) all auspicious birds and animals." Here Kautilya displays the beginnings of pragmatism that is embodied in clause c), for example, mingling with those ordained by religion in clause d). The beginnings of pragmatic conservation also starts finding mention in the rules of the Chief Protector of Animals (and controller of Animal Slaughter) in which he is exhorted, "to not allow the slaughter of animals in reserved parks and sanctuaries,"[7] while pragmatically "if any of the animals (wild or tame) in the sanctuaries turn out to be dangerous, they shall be taken out of the sanctuaries and then killed."[8] In a clear map drawn in the Arthashastra of a Hypothetical Kautilyan state are three kinds of forest: a productive forest, a recreational forest, and an elephant forest (Rangarajan 1987)

Fifty-odd years later, his grandson Asoka embraced Buddhism and commissioned thirty-three inscriptions between 269 and 231 BC. Many of them contained epigraphically his unique mix of Buddhist learning and personal moral code and a derived wish to protect animals and conserve nature. For example, he put on stone a list of protected animals in his realm, something mirrored later in the 1972 version of the Indian Wildlife Protection Act's Schedule I. Asoka, who was referred to by his Buddhist name in the edicts, says in Edict 5, "Beloved-of-the-Gods, King Piyadasi, speaks thus: Twenty-six years after my coronation various animals were declared to be protected—parrots, mainas, //aruna//, ruddy geese, wild ducks, //nandimukhas, gelatas//, bats, queen ants, terrapins, boneless fish, //vedareyaka//, //gangapuputaka//, //sankiya// fish, tortoises, porcupines, squirrels, deer, bulls, //okapinda//, wild asses, wild pigeons, domestic pigeons and all four-footed creatures that are neither useful nor edible" (Dhammika 2010). Wildlife conservation and animal welfare codes are inserted in a parallel fashion as if one cannot be separated from the other. "Cocks are not to be caponized, husks hiding living beings are not to be burnt and forests are not to be burnt either without reason or to kill creatures. One animal is not to be fed to another."

During the long period of Mughal and then later British rule, these concepts were forsaken in part, and Indian nobility joined both their external sovereigns as well as the tribal population of India in hunting wild animals. Forests were protected in large measure only for the pragmatic purposes of production and recreation. However, during this time, several legislations such as the Wild Birds Protection Act of 1887 and the Wild Birds and Animals Protection Act of 1912 came into being. Also, interestingly, this period saw the emergence of Gandhiji, who took the principles of ahimsa into the social and political spheres of the country on a scale not seen before. His views on the environment and animal welfare are well documented. Writings such as "our sense of right and wrong had not become blunt, we would recognize that animals had rights, no less than men" (Gandhi in Hingorani and Hingorani 1985); "I do believe that all God's creatures have the right to live as much as we have" (Harijan, January 19, 1937); and "we should feel a more living bond between ourselves and the rest of the animate world" (Patel and Sykes 1987, 50) speak for themselves

(Weber 1999). In many ways this was a reinvention of this Vedic tradition, and the nascent nature conservation movement of independent India leaned heavily on these precepts. After independence, this ethic of conservation got enshrined in the constitution of India, and later in 1972 when the Wildlife Protection Act was promulgated, carrying with it the spirit of Asoka, Kautilya, Buddha, and Gandhi, as a value-based yet pragmatic law that fitted the cultural and social milieu and realities of the newly formed nation state. Several authors have documented this, including Rangarajan (2001) and Kumar[9] (2002). However, Ranjitsinh (personal communication), one of the people who drew up the act, attributes it to ground realities and neither to ancient principles within India nor to the influence of European or American conservation dictates.

Conserving Nature Amid Poverty

India is estimated to have a third of the world's poor, and more than 80 percent of them live on less than two US dollars per day. In 1981, 60 percent of India's population was termed poor, which in 2005 had marginally improved to 41 percent. This poverty line marks an earning of US$1.25 a day.[10] Other studies, especially national ones by the Planning Commission, show only 27.5 percent of India as poor. A study by the Oxford Poverty and Human Development Initiative using a Multi-Dimensional Poverty Index (MPI) found that there were 421 million poor living under the MPI in eight north India states of Bihar, Chattisgarh, Jharkhand, Madhya Pradesh, Orissa, Rajasthan, Uttar Pradesh, and West Bengal. This number is higher than the 410 million poor living in the twenty-six poorest African nations.

This has a great bearing on the social programs of the government. But what does this mean for the conservation of nature? Let us examine this with two quotes from formidable thinkers, the environmentalist Suzuki and the economist Eckholm. "The question arises as to whether truly poor countries can afford the philosophies of saving animals and wildlife, because in any resource crunch, 'human needs must come first.' And since there are so many countries like India . . . where there are so very many needy humans, it doesn't seem likely that wild animals and plants stand much of a chance" (Suzuki and Dressel 2002). "Without a rapid reversal of prevailing trends [of food shortage and deforestation], in fact, India will find itself with a billion people to support and a countryside that is little more than a moonscape" (Eckholm 1976). When Eckholm wrote these words, India had 16.89 percent of its geographical area under forest cover (Anon 1987).

India tackled the poverty syndrome by pragmatically realizing that values were not enough to save forests and wildlife. It promulgated the Wildlife Protection Act in 1972, the Forest Conservation Act in 1980, the Environment

(Protection) Act in 1986, banned all killing of wild animals (except for the rarest of rare exceptions) in 1991, banned all timber exports in 1995, and put in place the Biological Diversity Act in 2002. The new millennium is as good a time in history as any to check on the two prophesies. India reached Eckholm's first prophesy of a billion people in 2000. In doing so, had it transformed itself into a lunar-scape? Although there were worrying signs in east-central India and parts of western India that this could become true in a lifetime, shy of the millennium, India had 19.39 percent of its land classified as forests, an *increase* of 3 percent (Anon 1999). This is an incredible statistic. While human numbers had doubled in the intervening three decades,[11] forest cover had gone up by 3 percent.

What of wild animals? Protected area for wildlife had gone up by 2000, to 5 percent of the country's surface, again a tribute to the strong yet visionary policies and laws of the land. If one looks at two charismatic megafauna at two ends of the country a similar pattern plays itself out.

In the early part of the twentieth century there were fifteen or twenty rhinoceroses reported from Kaziranga National Park in Assam. Gee (1964) estimates 625 rhinos in all between India and Nepal. By the turn of the century, Kaziranga National Park in India alone had 1,552 rhinos. This was a remarkable conservation achievement. At the other extremity of the country, the lion was showing similar resilience. From an oft-quoted record of the *Kathiawar Gazetteer* of 1884, "there are probably no more than ten or a dozen lions and lionesses left in the whole Gir forests," to a remarkable recovery under the patronage of successive nawabs of Junagadh, the Asiatic lion was fighting for its survival in Asia. From the time of Eckholm's prophesy to that of Suzuki and Dressel's, lion populations stabilized and rose incrementally. "Between 1974 and 1995 five censuses were held and the results showed a steady increase in the population of lions from 180 to 304. . . . In 2001, the pug mark method employed in the previous year's rapid survey was refined and a result showing a lion population of between 327 and 332 was declared" (Divyabhanusinh 2005). There is no doubt that India set aside land and protected its wild animals in the face of poverty, often ignoring the needs of its voting populace for a more esoteric, even if intrinsic, need to preserve nature.

A large part of the credit to this goes to the leader of the nation for most of this period, Indira Gandhi. Unlike her namesake, the Mahatma, India's first woman premier did not act with only a value system to guide her. Cold political calculations and swift decision makings were the hallmarks of her tenure at the top. She applied both to conserve nature and wildlife to the extent that poverty was essentially disregarded when the big picture decisions were made. Poverty did make a difference only in that despite her unqualified support at the top, forest cover could not go up more than 3 percent in thirty years, as the inexorable need for fuel wood, grazing land, and timber took their toll on Indian nature. But as Rangarajan (2001) documents in his concise history of the wildlife movement of the country, "It was perhaps typical of the model of preservation pursued in this country that villagers with no place else to go for their cattle were calmly asked to make do with alternatives."

Suzuki and Dressel quote a former Indian Forest Service officer, Manoj Misra, as summarizing the phenomenal success of India conservation in the face of poverty: "Even if we are poor we have to save our animals. That's the basis of our life, that's our identity, that's morality, that's our heritage and we know it." The ethical baseline had met its pragmatic hypotenuse.

Conserving Nature Despite a Growth Fixation

India, after growing by a robust 9 percent for three consecutive years, slipped to 6.7 percent in 2008–9 after being hit by the global financial crisis. The country's economy picked up on government stimulus and expanded to 7.4 percent in 2009–10 For the current year, the growth forecast of 8.75 ± 0.35 percent in the midyear analysis is higher than all predictions by government agencies. GDP grew 8.8 percent and 8.9 percent in the first and second quarters, respectively. Both in the times of Asoka or even when Indira Gandhi was speed-tracking environmental conservation, India was not an economic superpower. It was also, in its own words, a developing nation. Today, the official government position is that India is a developed country, although writers such as Bożyk (2006) refer to India as a newly industrialized country, a status between developing and developed. Aspirationally, however, India is a developed nation, pushing for a seat in the UN Security Council, subtly demonstrating its relatively new nuclear capability as well as flexing its technological and economic muscle.

The challenges of this new reality on environment can be easily visualized by following the two-year term of the current Minister of State (Independent Charge) for Environment and Forests, Shri Jairam Ramesh. An educated, erudite, and pragmatic green minister, Ramesh has had to seesaw between ideology and pragmatism, environment and growth. Soon after taking over as minister he felt the need to clarify: "We have been getting feedback from several states and industry organizations that very often environment was emerging as a stumbling block to faster growth. The Prime Minister has told me to clear this impression that the Environment Ministry was a regulatory hurdle in the process of economic growth."[12] Later, citing violations and environmental loss, he held up or denied environmental clearances to some of the largest developmental projects in the country. He was instantly hailed by the environmental community as the new messiah for the green movement in India. On growth, Ramesh said unequivocally that "if you are reporting a 9 per cent GDP growth . . . the real GDP growth in terms of accounting for ecological degradation, loss of natural resources, loss of bio-diversity would probably be somewhere closer to five and a half to six percent."[13] The prime minister seemed quick on the uptake: "Environment clearances have become a new form of Licence Raj,"[14] he said, addressing the National Conference of State Ministers of Environment and Forests. "This is a matter that needs to be addressed. There are trade-offs that have to be made while balancing developmental and environmental concerns." Speaking in 2010 of a work ethos that had passed away two decades ago seemed to imply that environmental clearance as practiced by the ministry was retrogressive. Soon thereafter, the ministry cleared two large mining projects worth billions of dollars that were previously withheld, although the link between the statement and the clearances can only be implied.

The situation is similar if not more worrying at the state government level, with a large majority of state governments clearing or attempting to clear developmental projects that are considered important to economic growth to the detriment of wildlife. In a recent meeting of a State's Advisory Board for Wildlife meeting, of the twenty-seven points on the agenda, fourteen were for diversion of forest lands from protected areas for roads, four for irrigation and pipeline-related forest clearance, and one for mining clearance. Apart from agenda items put forward by individual members of the board, the state itself had not put up a single point concerning wildlife management or conservation.

While the realities of poverty and the value systems of ecosystem conservation seem to have been balanced, however forcibly, by Indian policy, the target of a certain economic growth is proving a stiffer test for conservation

A Triangular Plot and a Trio of Implementers

We have seen Indian conservation and animal welfare policies to be shaped by a triangular plot. This in turn spawns conservation projects and programs that are implemented on the ground by a triumvirate of implementers: the government, communities, and not-for-profit charitable organizations. Through most of India, the government maintains legal and management control over natural resources.

However, in the fifth and sixth Schedule Areas (in the northeast and the Andaman and Nicobar islands) large tracts of forestland are owned by, conserved, or destroyed by communities. The Wildlife Protection Act, 1972, still does not confer a right to these communities for conserving wild flora and fauna, but some interpretations give the community a quasi-legal right to manage wildlife as well in their territories (Dutta 2008, 2010). In many pockets of the country, community-led or community-managed conservation has started to spread.

Recording this, Shahabuddin (2010) notes, "specific animal species that are endangered are being protected [by communities]: the golden langur in the Chakrashila Sanctuary in Assam and the spotted-billed pelican in the village of Kokkrebellur in Karnataka." But this is only the continuation of an old tradition: "The practice of exercising restraint with regard to fauna and flora for cultural and religious reasons provides an insight into a kind of control very different from that attempted via official fiat," notes Rangarajan (2001, 61), while documenting the protection of "storks, groves and antelopes."

Facilitating the conservation of nature in areas owned by the government and that owned by the community are civil society institutions, imprecisely referred to as NGOs.[15] They are, in many cases, implementers of conservation projects, in others, donors to such projects, and at grassroots level often the conscience of the conservation movement. The multiplicity of these institutions with their array of biases, mandates, and approaches makes conservation of nature uniquely chaotic.

It is interesting to note that the two nongovernmental members of the triumvirate of implementers have different weightages of these realities influencing them. The community is heavily influenced by the grinding poverty that it is often immersed in, and while value systems allow it to indulge in conservation despite these realities, the third factor of national economic growth rarely influences conservation decisions. Examples where the two conflicting factors can be best viewed are during times of extreme man-animal conflict. It has been seen even in the personal experiences of the author that villagers chase away marauding elephants pelting stones while still chanting Jai Ganesh or hail to Thee O Elephant God. Similarly, even after repeated man killing and eating cases, villagers implore conservationists to "take the tiger away" and not to kill it, in many parts of India.

Civil society institutions on the other hand have other external realities that influence their agendas. Most of this is to do with resources. A significant chunk of nature conservation finances for the NGO sector is raised outside the country, and donor bias that is inherent in such resources shape NGO policies to a large extent. The other reality that shapes their policies is interinstitutional competition for these resources as well as for turf. These two external influences add on to the three that largely influence government decision making. A large number of civil society organizations have value-based missions, and their vision is largely idealistic. In nature conservation in India there is of recent vintage a split between animal-centric groups and those that are anthropocentric. While both groups will claim to be based by their respective value judgments, one is more driven by human interests than the other. Both interestingly are relatively united in opposing economic-development-related threats to nature.

While the gulf separating the two may be exaggerated, it does exist, and this leads to the implementation of conflicting projects on the ground and the active and at times aggressive lobbying by both groups in the setting of policy.

Conclusion

It is understood that the Indian government conserves forest and wildlife through a strict rendering of the law of the land shaped by the three dominant factors, which has been discussed in the previous sections. However, most of these laws have been set in place by a previous regime. Currently, economic growth and poverty alleviation are certainly the mantras of the political class. Protecting ecological wealth for inherent value is therefore less relevant in governmental circles today. The ethical baseline has been kept alive, even in the government, by civil activism and community beliefs. The values of such groups are also influenced by poverty and the growth stratagem, and all this forms a unique mix in the cauldron of conservation. This unique potion, almost as inexplicably as the idea of India as a nation-state, works. India has today more than half the world's tigers, more than 60 percent of its Asian elephants, more than 85 percent of the rhinos in Asia, and all of the lions in the continent. These four large animals together cause the loss of more than five hundred lives annually. It also has a fifth of its land under forests, 1.5 percent of the world's forests. For a country with a billion human inhabitants this is a rather unique conservation achievement.

Notes

1. Based on Asokan edicts on animal welfare and conservation erected in the third century BC.

2. TS 5.2.8.7.

3. One should not use their God-given body for killing God's creatures, whether these creatures are human, animal, or whatever. Yajurveda 12.32. "O human! animals are Aghnya—not to be killed. Protect the animals." Atharvaveda 6.140.2. By not harming any living being, one becomes fit for salvation. Manu-Smriti 6.60.

4. The two middle human endeavors have had lesser moral standing.

5. "Arahat Mahinda's message of Ahimsa or kindness to all beings, brought with it the first message of conservation to this country (Sri Lanka)" (Jayawardene 2002) and "The Buddhist theory (of conservation) has been deeply anchored in the Khmer heart" (Thon 2002).

6. A treatise on Statecraft.

7. What is critical to note is that both the Asokan edicts and Kautilya's Arthashastra prescribe punishments and pecuniary penalties on those who dissent from their "laws," thus validating them as the first legislations on wildlife conservation and animal welfare in India.

8. From 2.26, 1, 2, 4, 10, 14.

9. Kumar, writing of the reluctance of modern conservation to legalize hunting in India says, "The development of wildlife laws in India . . . has been inspired by that tradition and thus India has largely forsaken the route of using wildlife, when it involves taking a life" (Kumar 2002).

10. PPP, in nominal terms ₹21.6 a day in urban areas and ₹14.3 in rural areas.

11. From 554.91 in 1970 to 1,021.08 in 2001.

12. Express News Service, May 30, 2009.

13. Speaking at the Stakeholders Consultations on the Economics of Ecosystems and Biodiversity in India.

14. Wikipedia (s.v. Licence Raj) defines Licence Raj, the Permit Rajas: "the elaborate licenses, regulations, and accompanying red tape that were required to set up and run businesses in India between 1947 and 1990."

15. Their essentially charitable nature (at least legally) does not come out in this abbreviation of the term *nongovernmental organization* when a not-for-profit organization would be a more precise manner of definition of these bodies.

References

Anon. 1987. "State of Forests Report." Forest Survey of India, Ministry of Environment and Forests, Government of India.

———. 1999. "State of Forests Report." Forest Survey of India, Ministry of Environment and Forests, Government of India.

———. 2005. "Securing India's Future: Final Technical Report of the National Biodiversity Strategy and Action Plan." Delhi/Pune: TPCG and Kalpavriksh.

———. 2007. "Poverty Estimates for 2004–5." Planning Commission, Government of India.

Bozyk, P. 2006. "Newly Industrialized Countries." In *Globalization and the Transformation of Foreign Economic Policy*. Aldershot: Ashgate Publishing.

Dhammika, Ven S. 2010. "The Edicts of King Asoka." An English rendering of *Access to Insight*.

Divyabhanusinh. 2005. *The Story of Asia's Lions*. Mumbai: Marg Publications.

Dutta, R. 2008. "Forest and Wildlife Conservation in Bodoland Territorial Council: A Policy Analysis." In *Bringing Back Manas: Conserving the Forest and Wildlife of Bodoland Territorial Council*, edited by V. Menon et al. New Delhi: Wildlife Trust of India.

———. 2010. "Forest and Wildlife Conservation in Garo Hills Autonomous District Council: A Policy Analysis." In *Canopies and Corridors: Conserving the Forests of Garo Hills with Elephants and Gibbons as Flagships*, edited by Kaul et al. New Delhi: Wildlife Trust of India.

Eckholm, Erik P. 1976. *Losing Ground: Environmental Stress and World Food Prospects*. New York: Norton.

Gadgil, M., and R. Guha. 1995. *Ecology and Equity: The Use and Abuse of Nature in Contemporary India*. Delhi: Penguin India.

Gee, E. P. 1964. *The Wild Life of India*. New York: Dutton.

Hingorani, A. T., and G. A. Hingorani, eds. 1985. *Encyclopaedia of Gandhian Thoughts*. New Delhi: All India Congress Committee.

Jayawardene, Jayantha. 2002. "The Rediscovery of Ahimsa: Sri Lanka's Conservation Philosophy through the Ages." In *Heaven and Earth and I: Ethics of Nature Conservation in Asia*, edited by V. Menon and M. Sakamoto. New Delhi: WTI and Penguin Books.

Kumar, A. 2002. "A History of Compassion: Indian Ethics and Conservation." In *Heaven and Earth and I: Ethics of Nature Conservation in Asia*, edited by V. Menon and M. Sakamoto. New Delhi: WTI and Penguin Books.

Menon, V., and D. Lavigne. 2006. "Attitudes, Values, and Objectives: The Real Basis of Wildlife Conservation." In *Gaining Ground: In Pursuit of Ecological Sustainability*, edited by D. Lavigne and S. Fink. Guelph, ON: International Fund for Animal Welfare.

Oates, J. F. 2006. "Conservation, Development, and Poverty Alleviation: Time for a Change in Attitudes. In *Gaining Ground: In Pursuit of Ecological Sustainability*, edited by D. Lavigne and S. Fink. Guelph, ON: International Fund for Animal Welfare.

Patel, Jehangir P., and Marjorie Sykes. 1987. *Gandhi: His Gift of the Fight*. Rasulia: Friends Rural Centre.

Pearce, D., and D. Moran. 2008. *New Global Poverty Estimates: What It Means for India*. Washington, DC: World Bank.

Rangarajan, L. N. 1987. *The Arthashastra / Kautilya*. Edited, rearranged, translated, and introduced by L. N. Rangarajan. New Delhi: Penguin Books India.

Rangarajan, M. 2001. *India's Wildlife History, an Introduction*. New Delhi: Permanent Black.

Shahabuddin, G. 2010. *Conservation at the Crossroads: Science, Society, and the Future of India's Wildlife*. Ranikhet: Permanent Black.

Suzuki, D., and H. Dressel. 2002. *Good News for a Change: Hope for a Troubled Planet*. Toronto: Stoddart.

Thon, Yi. 2002. "The Role of Buddhist Wats in Environmental Preservation in Cambodia." In *Heaven and Earth and I: Ethics of Nature Conservation in Asia*, edited by V. Menon and M. Sakamoto. New Delhi: WTI and Penguin Books.

Weber, T. 1999. "Gandhi, Deep Ecology, Peace Research, and Buddhist Economics." *Journal of Peace Research* 36 (3).

24

Conservation and Its Challenges in Kenya

Josphat Ngonyo and Mariam Wanjala

Introduction

CONSERVATION IN KENYA is not an alien phenomenon. It dates back to time immemorial where communities living in or around protected areas were able to coexist with the resource that met their daily requirements without compromising the ability of future generations to meet their own needs. That was basic conservation.

However, with the era of development and the urge to industrialize, the pressure on natural ecosystems became quite intense and both animals (wild and domestic) and plants became threatened with extinction.

In recent years, governments have developed policies and set up strategies that would contribute to the conservation of both flora and fauna in their natural habitats due to the repercussions of nature neglect.

Conservation

Conservation projects have been embarked on to ensure protection or restoration from loss, damage, or neglect of a natural resource. Specifically, these projects have been set up to achieve various objectives:

- Protection of *endangered species* with a limited territorial range. Since it is not always possible to relocate animals or breed them in captivity, protecting their natural habitat can be very important.
- Maintaining *biodiversity*, or preserving a nation's unique (keystone) natural environment.
- Wildlife *rehabilitation*. In these instances, the project focuses on injured and abandoned wildlife and nurses it back to health before re-

leasing it into its habitat or sending it to another location that simulates the original habitat.

Some of these projects that have taken precedence in most of Kenya's game parks and reserves, sanctuaries, and wildlife rehabilitation centers include:

Rhinos: Rhinos have been a high conservation priority since they are highly endangered. Establishment was a joint effort of the KWS, David Sheldrick Wildlife Trust, World Wide Fund, and Eden Wildlife trust, to ensure translocation and safe haven in sanctuaries like Lake Nakuru Rhino Sanctuary. This rhino sanctuary is particularly an important sanctuary for both Kenya's indigenous black rhino and the white rhino, introduced to the country from South Africa. The sanctuary carries out breeding programs and plans on translocation when the rhino population builds up.

Elephants: The elephant program was triggered by the alarming decline in the elephant population by 85 percent in a span of twenty years. The main reason for the setup of this program was the fact that the interaction of elephants and human beings was the chief source of their endangered status. As a result the elephant became the number one wildlife killer of humans. The elephant program now concentrates on research that will help minimize and contain conflict with humans. This initiative has been carried out in areas that are prone to elephant invasions, including Mwea, Trans Mara, and Voi (see also Moss, Croze, and Lee [2011] for detailed discussions of the Amboseli elephant studies).

Wetlands: the wetlands program was set up due to the fragility of the marine ecosystem and other wetlands. These areas are cradles of immense biodiversity that may become extinct if not conserved. Some of the activities undertaken in this project include a sea turtle survey, mangrove mapping, effective coastal-zone planning, and management. Collaborative studies have also been carried out in major lakes, including Naivasha, Victoria, Nakuru, and Bogoria, under the National Wetlands Committee.

Forests: Kenya's forest ecosystems are managed not only for their trees—a source of wood products—but also as a biodiverse habitats for plant and animal life and, more importantly, for the role they play in protecting the country's water and soil. Loss of forest cover in Kenya has contributed to diminishing livelihoods of many Kenyans caused by reduced land productivity, famine and, drought. In the recent past the Kenyan government has sought to conserve the country's "water towers" (the Mau Complex, Mt. Kenya, the Aberdare Range, Mt. Elgon, and the Cherengani Hills) due to the direct or indirect impact of deforestation on all aspects of economy and growth. Kenya's ten forest conservancies now have community representation to the Forest Conservation Committees (FCC).

Education and Youth: Sensitization of individuals is aimed at increasing people's knowledge and awareness about the environment and associated challenges; developing the necessary skills and expertise to address the challenges; and fostering attitudes, motivations, and commitments to make informed decisions and take responsible action (UNESCO, Tbilisi Declaration, 1978). The Kenya Wildlife Society has put in place education centers in Nairobi, Nakuru, Tsavo east, and Meru National Parks and information centers in Saiwa Swamp, Kakamega, Hells Gate, Tsavo west, Malindi, Watamu, Kisite, Kiunga, and Arabuko Sokoke, that offer conservation education programs.

Conservation Challenges

There are several challenges that face most of these conservation projects, which can be conveniently categorized as follows:

Economic Challenges

With an increase in population in the country, the demand for wood products has increased tremendously. Timber is needed in large quantities to construct houses in both rural and mostly urban centers. Large forests are being excised and cleared to make room for housing estates and other development-oriented projects. Environmentalists are being tasked to monetize most of these ecosystems so that governments can visualize the value of the "asset" in question. A case study of Mabira Forest, Uganda, is a clear indication of the pressure development projects geared to improve the country's economic welfare put on existing environmental ecosystems. Following a proposed plan for clearing a third of the reserve for agricultural use, local researchers calculated the values of the forest. This economic evaluation of the forest shows that from a short-term perspective, growing sugarcane would lead to more economic benefits than maintaining the forest reserve, with a return of 3.6 million US$/year in contrast with 1.1 million US$/year for conservation. However, sugarcane production is only optimal during a short time span—five years. When comparing both land use alternatives over the lifetime of the timber stock—60 years—the benefits from the forest and the ecosystem services it provides exceed those of the sugarcane planting.

Sociocultural Challenges

Land Use Changes: Land is important in Kenya as it is the base upon which activities like agriculture, wildlife conservation, urban development, human settlement, and infrastructure are carried out. There have been remarkable land use changes over the years. These land use changes, particularly agriculture and rural and urban development, have negatively affected conservation across the landscape. Furthermore, these changes have been exacerbated by lack of a national land use policy. Thus land that ought to be in a reserve or protected area is encroached upon by individuals due to increase of population and the pressure to obtain more to consume agriculturally.

Destruction of Wildlife Habitats: Wildlife habitats provide an important resource base for rural people's livelihoods. However, rapidly increasing populations and other complex socioeconomic factors have put enormous pressure on the limited productive land, forcing the rural poor to resort to poor land use practices for subsistence. Poor cultivation methods, deforestation, charcoal burning, and overgrazing are four main factors causing severe wildlife habitat degradation.

Insecurity: Insecurity in most of the wildlife areas is a serious threat and challenge to wildlife conservation and management efforts. The security relates to the protection of wildlife, communities living in those areas, and visitors. This situation has been exacerbated by the state of insecurity in the neighboring countries, which has led to the proliferation of small arms in the region.

Inadequate Incentives: Wildlife is found both within and outside protected areas. Whereas protected areas have been set aside for purposes of wildlife conservation, areas outside protected areas that serve as dispersal areas are communally or individually owned. Currently there are inadequate incentives to motivate communities and landowners to adopt land use practices that are compatible with wildlife conservation and management.

Management Effectiveness Assessment and Prioritization: Given the enormous and competing social challenges such as poverty, health care, and education, wildlife conservation and management receives much less in budgetary allocation, yet its scope is wide. Efficient and effective wildlife conservation and management requires regular assessments and strategic actions aimed at addressing wildlife priority issues.

Human Wildlife Conflict and Compensation: Increasing human-wildlife conflict (HWC) is a major problem in wildlife areas. Acute water shortage and inadequate dry season pasture has severely affected wildlife, livestock, and humans. As competition for the available ecosystems continues, there have been rising levels of human-wildlife conflicts. In addition to climate variability, increased HWCs have been attributed to extending human activities in areas originally preserved for wildlife. Currently, compensation is paid by the government. The amounts payable, which relate only to human injury and death, are very low, and there is no compensation for wildlife damage to crops, livestock, and property. In addition, the bureaucratic process followed before the payment of compensation disadvantages a large majority of the rural poor.

Conservation of Shared Wildlife Ecosystems: Habitat requirements for wildlife species are critical for their survival and reproduction. Most wildlife species have evolved and adapted to large home ranges, some of which straddle the boundaries of two or more countries. This affects their life cycle and migration, raising the need to promote a harmonized approach to the conservation and management of shared wildlife ecosystems.

Climate Change: Globally, the climate is changing, resulting in direct physiological impacts on individual species, changes in abiotic factors, changed opportunities for reproduction and recruitment, and altered interactions among species. Climate change may also produce more conducive conditions for the establishment and spread of invasive species, as well as change the suitability of microclimates for native species and the nature of interactions among native communities. There is inadequate data on the impact of climate change on biodiversity.

Political Challenges

A new study shows that political corruption and bad governance, rather than human population pressures and poverty, may present the greatest threat to wildlife in developing countries. The researchers found that high levels of corruption in African countries strongly correlate with declining elephant and black rhinoceros populations. Most of the practices that threaten the ecosystems and natural heritage have been linked to failure of a good governance system. Most of the blame has been apportioned on the relevant departments and their failure to change with times, but above all, on the institutional capacity of the state to change the policy and legislative framework for managing natural areas. Corruption and governance failure also manifest locally. They reflect the failure by the state and its organs like the Forest Department to implement their mandates and act transparently for sustainable management of the forest ecosystem. There has consequently been governance-related practices that have threatened the country's biodiversity, with the most acute being corruption and inefficiency of the officers. The Forest Department bore the blunt of the blame from the local people, and most environmental organizations have largely attributed the nonproductivity of Kenya's forests to nonperformance and abetting the corrupt practices of forest officers.

The Way Forward

The problems facing conservation in Africa must be addressed not only through technical solutions but also through better taking into account the relationship between the immovable heritage and its relevant communities and overall environment.

Integrated Conservation and Development Projects (ICDPs)

Community involvement in resource management is now regarded as an essential component of conservation projects. This is an approach that aspires to combine social development with conservation goals (Hughes and Flintan 2001; Ngonyo 2010; Moss, Croze, and Lee 2011). These projects look to deal with biodiversity conservation objectives through the use of socioeconomic investment tools. Biodiversity conservation is the primary goal, but ICDPs also like to deal with those social and economic requirements of communities that might threaten biodiversity. They wish to improve the relationships between state-managed protected areas and their neighbors, but do not inevitably seek to delegate ownership of protected area ecosystems to local communities. They usually receive funding from external sources and are externally motivated and initiated by conservation organizations and development agencies. ICDPs are normally linked to a protected area (Hughes and Flintan 2001). ICDPs, through benefit sharing, are believed to discourage poaching and promote economic development, hence they try to benefit indigenous populations in several ways: through the transfer of money from tourism, the creation of jobs, and the stimulation of productivity in agriculture. ICDPs that have yielded fruit from their establishment include Bwindi Impenetrable Forest (Uganda), Lake Mburo National Park (Uganda), Mount Elgon, and Ngorongoro Conservation Area, Tanzania.

Control of Ecosystem Use

The country should form Natural Ecosystem Management Committees to control and manage the use to ensure protection of its interests. Kenyan NGOs, states, and scholars should reject and resist current attempts to globalize biological ecosystems through the so-called common property rights. (Common property rights imply no exclusive use, no right of transfer, and, in the limit case, no net gain of income.)

Resource Valuation Systems

Although some dislike the thought of trying to put an economic value on biodiversity (some things are just priceless), there have been attempts to do so in order for people to understand the magnitude of the issue: how important the environment is to humanity and what costs and benefits there can be in doing (or not doing) something. The Kenyan government, NGOs, economists, and ecologists should develop a value system for Africa's natural ecosystems that integrates their cultural, ecological, and economic values. They should protect the integrity of Africa's ecosystems against pervasive and exploitative international profit markets.

Management of Transboundary Ecosystems

The Kenyan government, NGOs, and scientific institutions should form regional bodies and scientific panels to ensure the conservation and rational management of transboundary ecological ecosystems such as the Lake Victoria basin, the Nile River, and tropical forests. These areas should be exploited judiciously for the benefit of all the people, while being conserved for future generations.

Review of Policies and Laws

Laws and regulations governing the conservation, utilization, and management of natural ecosystems in Kenya must be reexamined with a view to:

- discarding those that inhibit and prohibit the control and participation of local communities in the conservation, management, and utilization of natural ecosystems;
- protecting the cultural norms of ethnic communities that pertain to the conservation and management of natural ecosystems in the region.

The country should initiate studies and compile national umbrella laws on the environment to include substantive enforcement procedures on all environmental matters. Such laws should also include general procedures for the implementation of relevant treaties to which the state is a party. Those who are involved in the development or enforcement of national laws should work in collaboration with human ecologists to understand the relationship between human communities and their environment. This will help elucidate how those human communities perceive their own relationship to specific legal provisions. In the process, the laws and their enforcement would incorporate that understanding.

For the purpose of enhancing the efficacy of the laws on the environment and natural ecosystems, it is imperative that the government initiates studies of human ecology in relation to the legal culture of the people. This initiative should be conducted within national institutions, and on a comparative basis, so that the experiences of different communities can provide a comprehensive background for discussion.

Local Management of Ecosystems

NGOs and development agencies working in Kenya must identify and promote community-based strategies that integrate local indigenous knowledge into natural ecosystems conservation and management.

Local people, the ultimate owners and guardians of natural ecosystems, must be the direct beneficiaries of the income that accrues from the use of ecosystems by:

- sharing collected revenues from wildlife reserves through tourism;
- integrating them into resource management and control committees at local and national levels;
- eliminating middlemen and processes that reduce income from the use of natural ecosystems.

A practical example is the Northern Rangeland Trust (NRT). NRT is a registered Kenyan trust with a board of trustees and with constituent communities as members. Central to the NRT structure is the Council of Elders made up of individuals nominated by each member organization, which is representative of the diversity of ethnic groups and stakeholders that exist in the area and provides a platform for exchanging ideas, experiences, and addressing common issues affecting pastoralists. Within the council is a selected Conflict Resolution Team, which has been instrumental in dealing with historical ethnic rivalries that predominantly surround issues of access to natural ecosystems. This is done through establishment and support of a collective, community-led conflict resolution mechanism that builds upon traditional systems and includes members from representative ethnic groups to mediate potential natural resource conflicts.

Alternative Approaches to Tourism

Promotion of tourist activities must be reoriented to integrate information on the cultural and ecological values of natural ecosystems to the Kenyan people. Folklore, myths, taboos, and totems based on flora, fauna, lakes, and rivers should be included in tourist information packages. Domestic tourism must be promoted.

Ecotourism should also be promoted in reserves and parks, seeing that it has all-around benefits to both the people and the environment (see tables).

Environmental Education

Kenyan scientists and teachers should evolve a new approach to the study of natural sciences that integrates indigenous principles of natural resource conservation and management. A holistic approach to the study of natural organisms and systems should be adopted. Deliberate programs of training should be mounted to produce top-level experts in the various fields of natural ecosystems and the environment. The objective should be to create a pyramidal structure

Environmental benefits of ecotourism

Direct benefits	*Indirect benefits*
• Incentive to protect natural environments	• Exposure to ecotourism fosters environmentalism
• Incentive to rehabilitate modified environments	• Areas protected for ecotourism provide environmental benefits
• Provide funds to manage and expand protected areas	
• Ecotourists assist with habitat maintenance and enhancement	
• Ecotourists serve as environmental watchdogs	

Economic benefits of ecotourism

Direct benefits	*Indirect benefits*
• Generate revenue and employment	• High multiplier effect and indirect revenue and employment
• Provide economic opportunities for peripheral regions	• Stimulation of mass tourism
	• Supports cultural and heritage tourism
	• Areas protected for ecotourism provide economic benefits

Sociocultural benefits of ecotourism

Direct benefits

• Fosters community stability and well-being through economic benefits and local participation
• Aesthetic and spiritual benefits and enjoyment for residents and tourists
• Accessible to a broad spectrum of the population

of expertise, with those in the top echelon constituting the critical mass of training for innovation in the sustainable management of natural ecosystems.

Relevant institutions should establish at least one research center in each of the various environmental sectors. Such centers should earn their excellence through competitive research and establish their own innovative capabilities. Specific experts should organize themselves into "think tanks" and face the challenge of open debate.

The public must challenge researchers to conduct competitive and innovative research. To reinforce this challenge, the government should remunerate the researchers in a manner that permits them standards of living reasonably comparable to those of their international counterparts. Locally designed educational programs on the value of natural ecosystems must be integrated into educational curricula at all levels.

References

Hughes, Ross, and Fiona Flintan. 2001. "Integrating Conservation and Development Experience." International Institute of Development.

Moss, Cynthia, Harvey Croze, and Phyllis C. Lee, eds. 2011. *The Amboseli Elephants: A Long-Term Perspective on a Long-Lived Mammal.* Chicago: University of Chicago Press.

Ngonyo, Josphat. 2011. "Kenya: Conservation and Ethics." In *Encyclopedia of Animal Rights and Animal Welfare*, 2nd ed., edited by Marc Bekoff, 343–46. Santa Barbara, CA: Greenwood Press.

Relevant websites:

http://www.oceandocs.net/bitstream/1834/805/1/HUGHES,%20R1-24.pdf

http://www.easternarc.org

http://www.elephanttrust.org

http://www.savetheelephants.org

http://gdrc.org

http://kenyaforests.wildlifedirect.org

25

Is Green Religion an Oxymoron?

Biocultural Evolution and Earthly Spirituality

Bron Taylor

Conservation is getting nowhere because it is incompatible with our Abrahamic concept of land. We abuse land because we regard it as a commodity
belonging to us. When we see land as a community to which we belong, we
may begin to use it with love and respect. [Aldo Leopold, 1949, xviii]

Man-nature dualism is deep-rooted in us. . . . Until it is eradicated not only
from our minds but also from our emotions, we shall doubtless be unable
to make fundamental changes in our attitudes and actions affecting ecology. The religious problem is to find a viable equivalent to animism. [Lynn
White, 1973, 62]

Introduction: A Sensory Spirituality of Belonging to Nature

OUR SPECIES HAS long sought to understand our place in the universe and our relationships to other living things. Curious creatures
that we are, we see moving things and wonder where they come from
and what gave rise to them. The world's diverse religious traditions
have sought to pull back the veil to illuminate what the ultimate un-
seen causes might be. Such curiosity is understandable but it may be
our undoing. Religions often lead us to ignore nature, distracting us
from what we *can* see and know—namely, our absolute dependence
on the biosphere and that our own flourishing is mutually dependent
the flourishing of the rest of the living community.

My criticism here is nothing new. Since Charles Darwin published
On the Origin of Species in 1859, the greatest environmental thinkers
in the Western world have often criticized the dominant religions of
their day, viewing them as promoting beliefs and priorities that lead
inexorably to nature's destruction.

Henry David Thoreau proclaimed that he had nothing to learn from missionaries, but much to learn from American Indians, whom he believed had greater understanding of nature than most European Americans. Moreover, because he had spent a lifetime closely observing nature, when he first read Darwin's theory of natural selection (shortly after its publication), he immediately recognized that it provided a compelling explanation for the planet's stunning biological diversity.

A few years later, in 1867, Wisconsin native John Muir had wandered his way to the small town of Cedar Keys on Florida's gulf coast, where he contracted and eventually recovered from malaria. Writing about what he learned from his experiences in nature during that long walk as well as his recent and severe illness, and referring sardonically to missionaries and other "profound expositors of God's Intentions," Muir humorously ridiculed the Christianity dominant in his time for promoting the anthropocentric conceit that all of nature was made for human beings:

> The world, we are told, was made especially for man—a presumption not supported by all the facts. A numerous class of men are painfully astonished whenever they find anything, living or dead, in all God's universe, which they cannot eat or render in some way what they call useful to themselves. (1997 [1916], 825)

He added that "venomous beasts, thorny plants, and deadly diseases of certain parts of the earth prove that the whole world was not made for him [humans]," and concluded that he was "glad to leave these ecclesiastical fires and blunders" in order to "joyfully return to the immortal truth and immortal beauty of Nature" (Muir 1997 [1916], 827). For Muir, who would later found the Sierra Club and help to spur the emergence of global environmentalism, close attention to nature was the pathway to spiritual truth.

Aldo Leopold, the forester, wildlife manager, and ecologist, who is considered by many to be the most important environmental philosopher in the twentieth century, like Muir, contributed to the emergence of on-the-ground conservation by cofounding an environmental organization (in his case, the Wilderness Society in 1935). Like Henry David Thoreau and Muir before him, Leopold believed that there was something fundamentally wrong and environmentally destructive about anthropocentric Western religions. "Conservation is getting nowhere because it is incompatible with our Abrahamic concept of land," he wrote. "We abuse land because we regard it as a commodity belonging to us" (1949, xviii).

Three decades later, in 1967, the historian Lynn White Jr. advanced similar criticisms to those earlier made by Thoreau, Muir, and Leopold. Although Christian himself, he charged that Western Christianity was "the most anthropocentric religion the world has seen," and averred, "we shall continue to have a worsening ecologic crisis until we reject the Christian axiom that nature has no reason for existence save to serve man" (1967, 1207). White linked this belief to the idea that with Christianity, humankind is not understood to be "part of nature" but instead "made in God's image" (1967, 1205). Writing at a time of growing environmental alarm, and publishing in the widely read journal *Science*, helped to make White's thesis famous; and it convinced many of the culpability of Abrahamic religions in general, and Christianity in particular, for the widespread decline of environmental systems (Whitney 2005).

The religious worldviews and perceptions criticized by Thoreau, Muir, Leopold, and White, and by scores of environmentalists during and since they wrote their influential works, are rooted in a profound ignorance of nature, including the human place in it. And they all offered a similar remedy, beginning with a sensory exploration of the living world; none of them prioritized speculation about supernatural realms or divine beings.

Their own understandings led them neither to reject religion altogether nor to a life devoid of spiritual meaning, however, and they all considered a spiritually meaningful existence as critical to the reformation of environment-related behaviors. With regard to the environmental crisis, White famously argued, for example, that "since the roots of our trouble are so largely religious, the remedy must also be essentially religious, whether we call it that or not" (White 1967, 1207). Noting that his own culture remained strongly Christian, in his famous *Science* article he suggested pragmatically that perhaps Saint Francis of Assisi could become a model of connection to nature that could overcome human alienation from her.

The other three environmentalist luminaries also advanced what I have called "spiritualities of belonging and connection to nature," viewing these as essential to fostering environmentally sustainable behaviors (Taylor 2001a, 2001b). Thoreau, for example, in his famous essay "Walking," not only argued that "in Wildness is the preservation of the World," but he also expressed a deep loyalty to nature and an appreciation of the interconnectedness of nature, his conviction that human beings are properly understood to belong to nature, as well as antipathy for the dominant religion of his day:

> I wish to speak a word for Nature, for absolute freedom and wildness, as contrasted with a freedom and culture merely Civil—to regard man as an inhabitant, or a part and parcel of Nature, rather than a member of society. I wish to make an extreme statement, if so I may make an emphatic one, for there are enough champions of civilization: the minister and the school committee and every one of you will take care of that. (1862, 1, or 2001, 225)

It is by close attention to nature that one arrives at such perceptions, as Thoreau noted in his most famous work, *Walden*:

> Fishermen, hunters, woodchoppers, and others, spending their lives in the fields and woods, in a peculiar sense a part of Nature themselves, are often in a more favorable mood for observing her, in the intervals of their pursuits, than philosophers or poets even, who approach her with expectation. She is not afraid to exhibit herself to them. (1970 [1854], 339)

Muir likewise expressed a deep sense of belonging, connection, and loyalty to nature. He spoke of feeling "Nature's love" from mountains, waterfalls, plants, birds, and other animals, and he articulated many now-famous aphorisms of metaphysical and ecological interdependence, including "When we try to pick out anything by itself, we find it hitched to everything else in the universe" (1997 [1911], 245). Muir's central epistemological premise was that one must seek spiritual experience and wisdom directly in wild nature (Fox 1981, 43). This sensory epistemology he shared with Thoreau, Leopold, and the majority of environmentalists today.

The quotation from Leopold criticizing Abrahamic religions continued by asserting that they promote a belief that nature is "a commodity belonging to us," explicitly contrasting this anthropocentric view with a spirituality of "plain citizenship" and belonging to nature. "When we see land as a community to which we belong, we may begin to use it with love and respect (Leopold 1949, xviii).

An Organicist Spirituality, Animistic Kinship Ethics, and Reverence for Life

The sensory and sensual spiritualities of these environmental thinkers reflect an approach that is the opposite of ignoring nature, for they all depended on the close observation of it. Moreover, in their own ways, they not only shared a deep sense of belonging and connection to nature, they also advanced an understanding that everything in the biosphere, and universe, is related and mutually dependent. Although they used different metaphors they all articulated organicist worldviews, by which I mean the various living things in the biosphere function as do the organs in a body, working together to maintain the functioning of the entire living system. And they also expressed and advanced animistic perceptions, which I am defining here as the perception that understanding, communication, and perhaps even communion is possible between human and nonhuman organisms.

Interestingly, in a little known article published a few years after his famous one in *Science*, Lynn White asserted that to overcome the separation between humans and nature (which all of these figures considered to be a critical part of the essential problem), we must "find a viable equivalent to animism" (see the epigraph for the full quotation).

I think this is precisely what Thoreau, Muir, Leopold, and thousands of scientists, artists, poets, musicians, and nature lovers of diverse kinds and cultures have been doing ever since Darwin published his gestalt-changing book. These same actors have also been expressing and promoting Gaian spiritualities and organicist understandings of ecological interdependence and mutual dependence.

I do not mean to imply that they are doing something new within the human experience, for as historians and anthropologists have long shown, organicist and animistic perceptions have long been a part of the human experience. What I have been arguing, including in *Dark Green Religion: Nature Spirituality and the Planetary Future* (Taylor 2010), is that such perceptions are taking new, entirely sensory and naturalistic forms, divested of the supernaturalistic beliefs that often have typically accompanied such perceptions in the past. Organicism (and its contemporary expression in Gaia hypothesis), and its animistic forms (often grounded in ethology, the study of the consciousness and behavior of nonhuman animals) are often today grounded in the sciences. Consequently, they are fully compatible with modern epistemologies and they can, I think, provide what White was seeking: environmentally beneficent perceptions that are viable in a modern context where, generally speaking, greater levels of education correlate positively with secular and scientific understandings, and negatively with religious perceptions and practices related to supernatural divine beings or forces.

I have also been arguing, however, that there is no good reason to say that any one trait typically associated with religion, such as supernaturalism, is essential to this sort of phenomenon. Even if one thinks belief in nonmaterial divine beings is essential to religion, there is still explanatory power in examining the *religion-resembling* beliefs and practices of individuals and groups who do not profess beliefs in supernatural divine beings or forces, as, for example, when people use terminology typically associated with religions (such as metaphors of the sacred or its opposite), or engage in performances and ceremonies, or participate in other activities, that are reminiscent of traditional religions. Indeed, when we use the lenses of the social sciences, without worrying about where the boundaries of religion lie, we can see a dramatic increase of religion-resembling beliefs and practices in which people consider nature to be sacred in some way, and time focused in the close observation of nature to be a critical pathway to spiritual truth. Through such an approach we can see how close observation of nature can become a spiritual epistemology revealing the interconnectedness, mutual dependence, and kinship of life, leading to a reverence for life and a desire to protect biotic diversity.

Biocultural Evolution and Planetary Spirituality and Identities

There have long been human cultures that have coevolved within environmental systems in ways that have not reduced the diversity and fecundity of the living world, where knowledge of nature and lifeways well adapted to natural systems were integrated into the cultural and religious life of the people. Such cultures have usually been small scale and sometimes characterized by animistic, polytheistic, or pantheistic worldviews, which expressed beliefs in divine beings or forces, and that sometimes enjoined practices that regulated ecosystems or protected them, as has been the case with designated sacred groves and ritual sites.

But with the global spread of agricultures beginning some 10,000 years ago, and for the past 150 years increasingly intensified by fossil-fuel-driven industrial machines and calorie inputs, religious restrictions on environmental behavior have increasingly lost their power. Nevertheless, we can learn from the ecological wisdom that is often embedded in traditional cultures, which at their best arose and maintained themselves as the result of long observation and experimentation within environmental systems, reflecting a form of science. Such "traditional ecological knowledge" as it is labeled today by anthropologists and others, can be hybridized with modern scientific knowledge. What we learn from nature when we explore it, learning from both traditional ecological learning and modern scientific methods is that the flourishing of all species depends on the well-being of the entire environmental systems that all life forms are embedded in and partially constitute.

With a clear understanding that we are a part of nature we will also recognize that we are subject to natural laws—except not by some divine fiat or privilege. Like other organisms, if we exceed the carrying capacity of the habitats we depend on, our numbers will decline, often rapidly, and sometimes in ways that will involve great suffering. And with the contribution of an evolutionary perspective, we will realize that all living things came to be what they are through exactly the same process of natural selection, and from this, we can deduce our continuity and kinship with all other species. Through such lenses we can also see that our capacity for empathy and cooperation evolved to improve our chances for survival. Through empathetic sentiments, combined with our knowledge that all organisms strive to exist and survive, we can surmise an ethical obligation to respect the struggle for existence of other organisms, and as much as possible, express with our lives a live-and-let-live ethic.

Today evolution is a biocultural process, not just a biological one. Biocultural evolution is today giving rise to and shaping naturalistic spiritualities and biocentric (life-centered) environmental ethics. Thoreau, who died soon after reading Darwin's opus and unfortunately had little time to assimilate all of its implications, nevertheless envisioned just such a cultural shift in his writings, recognizing that wildness in both human culture and environmental systems was essential to the well-being of both. Muir believed similarly, and Leopold explicitly argued that cultural evolution toward a deeper-felt affection and kinship with all life is critically important to the flourishing of environmental systems. As Leopold put it, "I have purposely presented the land ethic as a product of social evolution because nothing so important as an ethic is ever 'written.'" For Leopold, the "land ethic" was shorthand for biocentric ethics that value all terrestrial and aquatic ecosystems. He concluded this passage as such, "The evolution of a land ethic is an intellectual as well as emotional process" (1949, 225).

I believe that a worldview and ethic can be deduced from our long and increasingly sophisticated observation of nature. While such an ecological and evolutionary ethics can be grafted onto the world's long-standing religious traditions, for those who find supernatural religious beliefs implausible, worldviews based entirely on the senses, both in everyday observation of the world and as aided through scientific methodologies and technologies, can provide ethical guidance and ethical meaning. I believe and have attempted to document that such ecospiritualties are growing globally and presage a long-term and environmentally salutary trend (Taylor 2010).

Indeed, the long-term trend *must* be a closer understanding of our place in nature, for only when we stop ignoring nature will we be able to develop lifeways, livelihoods, spiritualities, and cultures that will be adaptive and resilient. Only time will tell whether cultural evolution can accelerate rapidly enough to significantly slow down the precipitous decline of the earth's biocultural complexity, a process in which the world's currently predominant belief systems have been deeply complicit. The hard work of cultural transformation has never been more urgent, and it begins with a clear understanding of nature, our belonging to it, that our social and cultural lives are embedded in it. As the bioregionalist Freeman House, one of the many 1960s countercultural figures to go back to the land to work on cultural transformation and environmental restoration, once put it, "its time to get serious about this business of co-evolution" (House 1990, 113; cf. House 1999; Taylor 2000).

References

Chidester, David. 2005. "Animism." In *Encyclopedia of Religion and Nature*, edited by B. Taylor, 1:78–81. London and New York: Continuum International.

Fox, Stephen. 1981. *The American Conservation Movement: John Muir and His Legacy.* Madison: University of Wisconsin Press.

Harvey, G. 2006. *Animism.* New York: Columbia University Press.

House, Freeman. 1990. "To Learn the Things We Need to Know: Engaging the Particulars of the Planet's Recovery." In *Home!: A Bioregional Reader,* edited by V. Andruss, C. Plant, J. Plant, and E. Wright, 111–20. Philadelphia: New Society.

———. 1999. *Totem Salmon: Life Lessons from Another Species.* Boston: Beacon Press.

Leopold, A. 1949. *A Sand County Almanac.* New York: Oxford University Press.

Muir, J. 1997 [1916]. "Cedar Keys." In *Muir: Nature Writings,* edited by William Cronon, 818–27. New York: Library of America. Originally published in J. Muir, *A Thousand Mile Walk to the Gulf* (New York: Houghton Mifflin, 1916).

Taylor, Bron. 2000. "Bioregionalism: An Ethics of Loyalty to Place." *Landscape Journal* 19 (1 and 2): 50–72.

———. 2001a. "Earth and Nature-Based Spirituality (part I): From Deep Ecology to Radical Environmentalism." *Religion* 31 (2): 175–93.

———. 2001b. "Earth- and Nature-Based Spirituality (part II): From Deep Ecology and Bioregionalism to Scientific Paganism and the New Age." *Religion* 31 (3): 225–45.

———. 2010. *Dark Green Religion: Nature Spirituality and the Planetary Future.* Berkeley: University of California Press.

Thoreau, H. D. 1862. "Walking." *The Atlantic* online. http://www.theatlantic.com/doc/186206/thoreau-walking.

———. 2001. *Henry David Thoreau: Collected Essays and Poems.* New York: Library of America/Penguin.

———, ed. 1970 [1854]. *The Annotated Walden: Walden; or, Life in the Woods, Together with "Civil Disobedience."* New York: Barnes and Noble.

White, L. 1967. "The Historic Roots of Our Ecologic Crisis." *Science* 155: 1203–7.

———. 1973. "Continuing the Conversation." In *Western Man and Environmental Ethics,* edited by I. G. Barbour, 55–64. Menlo Park, CA: Addison-Wesley.

Whitney, E. 2005. "White, Lynn—Thesis of." In *Encyclopedia of Religion and Nature,* edited by B. Taylor, 2:1735–37. London and New York: Continuum International.

26

Avatar

The Search for Biosynergy and Compassion

Anthony L. Rose and A. Gabriela Rose

JAKE SULLY: Well, if I'm like a child, then maybe you should teach me.
NEYTIRI: Sky People cannot learn, you do not see.
JAKE SULLY: Then teach me how to see.
NEYTIRI: No one can teach you to see.
[Cameron 2009]

A Persistent Blindness

THE MOST SEEN FILM of all time to date—*Avatar*—exposes the
moviegoer to a deep flaw in modern human life. We have become
blind to the personal and universal soul of nature. Animal nature.
Plant nature. Human nature. All Nature. We look at other beings, at
the multitude of living forms, at the myriad landscapes and life spaces,
and at ourselves—and we do not *see*. Myopia and rapid judgment result in stereotypes and knee-jerk reactions that diminish our insight
into the deep inner meaning, emotion, beauty, and spirit of the countless beings in our ego-centered worlds. James Cameron's masterpiece
about the awakening of the sleeping human soul by the love-energy
of synergistic life on a far-off planet instructs us that while we cannot
teach others to *see*, we must encourage one another to open our eyes.

This essay explores antidotes to the modern human folly of "Ignoring Nature" by building on *Avatar*'s messages of planetary harmony,
spiritual and species communion, and global conservation. We shall
examine the wisdom implied in key quotes, events, and images from
the film, and we shall interpret these ideas in light of biosynergy research and theory (Rose 1994 to 2011), ancient myth and folklore, and
our own personal experiences. The first section of the essay emphasizes the benefits of seeing and treating all earth-life and ourselves as

kindred spirits in a harmonic universe and includes personal experiences of the first author. The second section focuses on the transformative and healing power of the human-horse connection, with experiences and research of the second author. Together these broad and focused inquiries result in a clarion call to see the world through one another's eyes.

Alas, to truly see, we must acknowledge and lament the attitudes, actions, and social systems that drive our persistent blindness. This author's long lifetime of observations as an applied social psychologist, working in schools and hospitals, military and religious institutions, businesses, governments, and charities, is most ardently expressed by turning lament into rant!

> We humans, by the billions, have vision so narrow and myopic that our senses of self and of place are clichés and our imaginations are rusty nails buried in the detritus of lost and failing civilizations. Urban-homo-sapiens are blindfolded, boxed, lined up on treadmills, and heaped into trailers and jet planes that propel us on linear journeys from delivery room to morgue. We begin dying with the first cry of abandonment and start going blind when our eyes are trained to focus on twirling overhead figurines to distract us while mom and dad watch TV or drive off to the work cubicle. Our minds focus on surfaces, eye candy, makeup and bling, muscles and curves, grins and frowns, feathers and fur, petals and thorns, signs of possible pleasure and pain to lose or gain from the myriad objects and individuals that come into view. In the end what we see is all about Me, rarely about You.

> Sadly, nature is ignored most by the high achievers; those who heed the cry to "focus child, focus worker, focus leader!" Focus on the algebra equation, the production output, the homeland security, the corporate resources—human, natural, technological, organizational, economic, opportunistic, ballistic resources. Focus on your task and I'll focus on mine and we'll get it all done fast, so we can accomplish the mission, whatever it might be. Save the neonate conceived in a test tube to live in brain-dead oblivion. Catch the teenager raised in a ghetto to lock-up in emotional fury. Hire the graduate trained in an autocracy to be yoked in lifelong ennui. Hamstring the scientist, physician, executive, visionary, endowed with genius, to follow the common path. Shallow inarticulate unexamined lives emerge as artifacts of blind and artificial societies: those who escape from the senseless cacophony are obliged to report what they see, what they feel, what they imagine our natural world might be when experienced in its unfettered glory. Escape, report, reinvent! (Anthony Rose)

As the dominant species on earth, we must recognize our shortcomings and address them head on. But all the remedial work will go nowhere, without the vision to see how glorious life on this planet is meant to be.

Envisioning Another Reality

In producing the movie *Avatar*, James Cameron and his colleagues escaped the persistent blindness and reported with supreme brilliance. They constructed a planet to contrast our own, and called it Pandora: a place where living beings are united in loving spirit and entwined in common ancestral history and karmic fate. A place endowed with clans of great and powerful beasts that coexist in harmony, roam the jungles, and soar the skies sensitive to one another's natures, commune through mutual service to the greater good, mingle spirits in the glistening ever-present synergy of eternal life.

And then, they threw a wrench in the works. Cameron and company invented a flock of self-righteous human invaders and sent them to fly their space ships to Pandora. The planet's indigenous humanlike beings called these interlopers "the Sky People."

The Sky People traveled in interplanetary warships from an ecologically shattered earth to conquer and pillage the mineral resources of Pandora—the lush and life-rich moon of a far-off giant planet. Jake Sully, a paraplegic ex-marine, joined the mission to be freed from the confines of his wheelchair and to experience the vigorous life of an avatar. His solo adventure in a holomorphic body designed to emulate the planet's humanoid inhabitants—the Na'vi—begins with exhilaration but soon turns tragic. Jake stumbles into the forest den of giant rhinoceros-like creatures that recognize him as an invader and roar their warning to him. Rather than kneel down submissively and lower his eyes to show he means no harm, he turns and runs. The creatures pursue him.

The Na'vi princess, Neytiri, emerges from her hiding place in time to kill the lead creature just before it crushes Jake. This causes the other behemoths to run away. Jake Sully possessed the body of a Na'vi warrior, but was blind to nature's ways and had put himself and nature at mortal risk. The human proclivity to flee in terror was insane behavior for the human-Na'vi avatar who had invaded the lair of great beasts of the forest. When Mo'at, the mother-queen of the Na'vi people, learned of his transgression, she told Jake how he might acquire the Na'vi capacity to *see*.

> Mo'at: Do as we do, and learn it well. Then we will see if your insanity can be cured. (Cameron 2009)

The insane terrors that cause humans to fear, flee, or fight those other beings who are strange to us have been imbedded in our psyches and our societies for millennia. They emerged when we shifted from collaborating with earth-life to conquering and controlling it: from gathering and hunting to farming and industrializing. Entwined with our mad sense of god-given dominion is a projected negativity about the natural world. We diminish wild animals to the status of mindless hostile beasts and consider humans who commune with and protect wildlife to be fools standing in the way of our righteous hegemony. Still, there are many women and men who have set aside their terror to follow experienced guides in search of the beast in his lair.

> In 1984 I traveled from urban Los Angeles through Paris to Kigali and trekked up Mt. Karisoke to visit with mountain gorillas. Two men who, like the Na'vi, were attuned to nature in these primeval rain forests led us on the trek to find the Sumi group. Back home people thought me insane to attempt this adventure. When I expressed my concerns, the Rwandan guides merely shrugged and told me to follow their lead. Six hours later we came upon a large troop of shaggy black-haired apes of all sizes. Exhaustion turned to exhilaration! When a male gorilla stood to face us and pound his chest, a surge of terror almost compelled me to turn and run. Obedient to my guides, I dropped to my knees and lowered my head. The great silverback huffed and the gorillas returned to their repast. I didn't possess the body of a gorilla tracker—far from it. But I did as my guides did, honored the gorilla's preeminence, and was privileged to SEE our kindred great ape cousins in their peaceable world. (Anthony Rose)

We who have faced the beasts and been accepted by them have begun to restore our sanity. Since that event in the Ruwenzori Mountains I've traveled the world to *see* and commune with other great apes, as well as elephants and rhinos, dolphins and sharks, tigers and lions, wolves and bears, and scores of other animals. Discussing these experiences with hundreds of fellow travelers, we've found that profound interspecies epiphanies turn fear of nature into fondness for it (Rose [1998] 2006). This discovery has informed our work with colleagues across equatorial Africa to initiate programs in conservation education that promote compassion and biosynergy (Rose 1996a; Rose et al. 2003; Rose et al. 2008). While our contribution to the ethos of conservation has caused modest positive changes, John Cameron's impact with *Avatar* upped the ante for us all. By exposing the merciless madness of Eurocentric manifest destiny and helping us *see* compassionate nobility in indigenous cultures, endangered species, and natural ecosystems, *Avatar* has challenged millions of people to enable local and global biosynergy of all life on this planet. *Avatar* and its profound messages compel us to give back to earth what she bequeaths to us.

Biosynergy: Giving Back

> Neytiri: She says that all energy is only borrowed; at some point you have to return it. (Cameron 2009)

Jake has explained to the Na'vi leaders that the invaders from earth, the Sky People, have come to harvest minerals on Pandora and take them back to earth to fuel the remains of human civilization. In the above quote, Neytiri translates her mother's declaration of the megaprincipal of biosynergy. For life to survive in the universe, from the most Spartan natural ecosystem to the most diverse planetary biosphere, biological energy must function in synergy. No individual, no species, no society, no nation, no continent, no planet can take, hoard, and consume the energy of others for long. Ultimately, consumers must return the resource to those from whom they have taken it, or they will burn themselves out and die while those from whom the resources were taken diminish and disappear. Only through a continuous cycle of give and take will living systems function synergistically and life thrive sustainably (Corning 2003; Rose 2010).

My first adventure into tropical rain forest occurred in 1982 at Ketembe research station on Gunung Leuser in Aceh, northern Sumatra. After days of searching, we came upon the first wild-born orangutans I had ever seen: a mother with infant in arms and her female daughter were feasting in the lower branches of a huge fig tree. The three red apes stared into my eyes, acknowledged my awe, shifted and settled into full view, and continued foraging. I was profoundly impacted by this encounter, which set the stage for the next three decades of devotion to great ape conservation. Yet it was a different interaction that was the peak epiphany of those life-changing travels in Indonesia:

I stopped at the bend in a narrow trail, studied the thin lines that crisscrossed my soiled map, and was about to tilt a water bottle toward my lips when I saw the wormy creatures. At least 20 leeches were wiggling out of the leaf litter and heading my way. The prior evening I had extracted thirteen of these bloodsuckers from my toes and ankles. Now a company of annelids was streaking toward the scuffle and heat of my feet, gnashing their teeth in the excitement of a probable feast. I deftly hopped over the encroaching circle and stood fast. In an instant they had whirled 'round and were heading my way. I leapt back over them again. They turned again. I bent down, offered my forefinger to the leader of the pack, and watched him crawl on board and attach his jaws to my skin. In a minute or two he had doubled in size, filled with my blood. I removed him and looked closely at the small red dot on my finger. In thanks for my liquid offering, he had injected anticoagulant. "Tit for tat—you feed me; I clean up after," he seemed to be saying.

Biosynergy! The realization spiraled through me. I was part of an ecosystem that was in a state of synergy, with all life forms engaged in mutual service. More than service, in mutual attraction, fascination, interdependence, harmony. I had entered to explore and learn, and the biosynergy of the place had transformed me from observer to participant, from interloper to inhabitant, from utilizer to synergizer. (Rose 2007, 124)

The profound discovery of biosynergy in that rain forest changed my life goals and redirected my professional energies; in the three decades since that incident I've been giving back to nature constantly. Still, the years of immersion in tropical forests and in the struggle to save them have not yet transformed me into one who *sees* the infinite soul of nature at every turn. Like Jake, the human hero in Cameron's cinematic world, I cannot help but fall back into seeing nature from a human-centered perspective. The blinders that narrow my vision, attached at the start of my civilized urban life, are still in place. For example, while composing this treatise on *seeing*, I stopped to describe the scene in my own backyard in the vernacular of the ethologist.

> Outside the window a young male kestrel sits in still silence on the top branch of the coral tree, facing north. To his right he sees our half-acre garden reaching uphill to the house, and in our long string of picture windows he notes reflections of the trees, rooftops, and cliff sides that roll down to the ocean around him. With the double vision of his habitat provided by our mirror walls, the kestrel appears safe from predators. He can spend more than the usual time searching for mice, lizards, and birds in the brush and bushes beneath him. (Anthony Rose)

This report of what the kestrel sees is entirely about material surfaces, with the exception of the suggestion that he "appears safe from predators." Ethologists and naturalists, like myself, are trained to stick to these external observations and simplistic assumptions. We do so in part to assure that our status as "scientists" will be safe from predatory professionals who might hammer us if we stray into immaterial, ethereal, and mythological arenas.

Can we risk setting aside our fixation on human constructs and imagine what the kestrel himself sees, feels, dreams? Suppose we become kestrel avatars! Enter this garden by the sea as raptor. Perch atop the coral tree and *see*.

> I see spaces, white airways rising empty to the clouds, safe places high above the crowds, corridors of clear blue light that streak straight to the sea, zigzag flight paths from tree to tree to tree: familiar trails and clearings in the sky where my ancestors and I can hover, dive, and fly. All's well for now atop this isolated glade. I inhale the Glory winged Isis made, and look for quarry hiding in the shade. (Anthony Rose)

Poetic fancy or perceptive intuition? No student of falconry will challenge the premise that the kestrel is a creature of the airways whose first concern is finding room to fly. What's on the ground is important, but secondary, to birds of prey. Few would deny this. On the other hand, to impute raptor awareness of the centuries-old Egyptian goddess of the heavens, as done in the last sentence of this kestrel rapture, is anathema to most human-centered minds.

Still, if we can create deities in our own image, why do we deny the same ingenious capacity to other creatures? While objectivists and speciesists may consider other animals as inferior faithless beings, much of humanity past and present has felt otherwise. In countless contemporary and ancient cultures humans *see* other animals and nature as James Cameron's *Avatar* depicts them: as sacred, numinous beings living in harmonious clans, devoting their bodies and souls to enriching the biosynergy of all life on their planet. These visions have been recorded as myth and folklore for millennia.

The Na'vi/Pandora Tale and Interspecies Myth and Bonding

Of the allegoric elements from the film that parallel human mythology, three are antidotes to our human blindness to nature: the unifying power of Eywa—the Mother Tree; the synergy of Na'vi and other animal clans; and the bonding of Na'vi warriors with their great flying partners, the Ikran. These cinematic constructions are similar, respectively, to icons of most human religions, to animistic human-animal relationships, and to special cases of human biosynergy with animals that have evolved in conjunction with our species.

The forms of the "Tree of Life" in Judeo-Christian-Islamic and in Vedic-Buddhist myth are honored by billions of people. Eywa—the "Mother Tree of All Souls" on Pandora presents this icon for the first time in human history as a dynamic and gloriously depicted three-dimensional video graphic phenomenon. To watch the mystical concept of life's eternal connections perform as the central force in this heroic tale reinforces our archetypal dreams of paradise. The Hebrew Etz Chaim from the Book of Proverbs, the Christian healing tree from the Book of Revelations, the Hindu tree of immortality from the Rig-Veda, the Buddhist Bodhi Tree from tales of Siddhartha Gautama—all these sacred religious ideas come to life in *Avatar*. No wonder hundreds of millions of people worldwide were uplifted spiritually by this film.

Depictions of synergistic relationships among the Na'vi and the other animal clans on Pandora reflect not only the research and theories of ecologists but are also representative of a plethora of mythic interspecies connections believed and asserted by people indigenous to natural ecosystems, as well as by animists worldwide. *Avatar* presents vivid examples of animistic beliefs about human kinship and mutuality with other species, akin to folklore expressed by aboriginal peoples from the Congo Basin to the Australian outback to the plains and tundra of North America. Many millions of people of the predominant faiths (Christianity, Islam, Hinduism, and Buddhism) hold animistic beliefs, and social movements within these institutions are calling for increased reverence for nature. Perhaps a billion people worldwide practice religions that consider other species as sacred kindred beings and heed the call to care for the Creation.

Thus far, this essay has focused on illuminating the harmonic soul of nature on a large scale, reflecting our preeminent concern for the conservation of global biodiversity (Rose 2011). The two important *Avatar* allegories discussed above reflect this big-picture outlook. People who acknowledge the workings of the Tree of Life and of animistic interspecies synergy in the primeval world are on the path to *seeing*. But the ultimate key to this quest for communion with nature is within the seeker's individual grasp. It's the animals that live in our midst that are best positioned to open our eyes, and our hearts and souls. The *Avatar* depiction of the bonding of Na'vi warrior and loyal steed—the four-winged beast called Ikran—harkens humanity's most vital historic, mythic, and personal interspecies bond: the bond between human and horse.

It's long been recognized that biosynergy with animals that have evolved in conjunction with our species has had enormous influence on the human condition. The evolution of wolves into dogs, for example, has produced a plethora of treasured service animals, and has led to the presence of cherished canine companions in many millions of human homes. Similarly, cats and other animals kept as companions and pets advance interspecies awareness and bonding. But it's the synergy of humans and horses that has influenced human civilization and inspired human mythology and worldview most profoundly.

Horses enabled human pursuits such as herding and plowing, trading of goods, migrating to more fertile territories, and the conquest of distant lands and societies for millennia (Chamberlin 2006). In fact the concept of "horse-power" affirms our equine partners as the antecedent icons of our modern modes of transportation: our synergy with automobiles grew out of our biosynergy with horses. Clearly, horsepower has given advantage to those who harness it. European hegemony on horseback enslaved and trampled the peoples and landscapes of the Americas, much as cars, trains, airplanes, and their by-products contribute to the destruction of biodiversity today. Nonetheless, our bonds with horses throughout time have enabled humanity to experience and appreciate earth life in superhuman ways. In *Avatar*, when Jake bonded with his Ikran and they charged across the Pandora skies, both human Na'vi-avatar and his faithful steed merged their bodies, hearts, minds, and souls. Thus unified, they could soar, *see*, and *be* as one.

The balance of this essay will focus on human-equine bonds that transcend most other human interspecies connections. Like the mythic centaur, humans astride horses embody a union wherein two creatures merge into one. Shifting our search from nature's biosynergy to the compassionate bonding of human and horse moves us from I SEE YOU, to WE ARE ONE! In effect we will attempt to take the I-Thou relationship postulated long ago in theology (Buber [1923] 1970) and embraced in psychology (Rogers 1980; Rose and Auw 1974) beyond the limits of humanism and apply it to the connection of all beings on our planet.

The Human-Horse Connection

Limitations live only in our minds. But if we use our imaginations, our possibilities become limitless. [Jamie Paolinetti, 2003]

In contrast to modern man's proclivity to ignore nature, Cameron's depiction of the synergy of Na'vi and Ikran offers a vivid allegory similar to one of humanity's most profound interspecies relationships—the human-horse bond. Human history and myth are filled with tales of the connection between people and horses. The horse has carried us across plains and over mountains to explore, exploit, commune, and conquer. Horses have enabled our communities and civilizations to expand; they've helped us plow our fields, herd our livestock, transport our goods, carry our cavalries. Horsepower in its many forms has enabled the global conquest of earth by humankind.

In *Avatar*, the union of Na'vi warriors and their flying steeds, the Ikran, blocked the human conquest of Pandora. The Sky People's warships, with all their gun power, were still no more than mechanical extensions of nature-blind human invaders. They were no match for squadrons of interspecies fighters, inspired and guided by their vision of an eternally unified biosphere. While the political message of *Avatar* has inspired moviegoers with renewed hope for the preeminence of global biosynergy, the personal message of communion among individuals of different species touches an equally great audience. Hundreds of millions of people have watched and extolled the love that allowed Jake and Neytiri, human and Na'vi, to truly *see* each other. We identify most easily with interpersonal communion when both parties are similar: the human-humanoid bonds in *Avatar* are accepted by film audiences, much as human-ape bonds are acknowledged by natural scientists (Rose 1996b). But when the bond is between two beings of different evolutionary lineage, like horses and humans, it's importance requires explanation to be genuinely understood.

The original connection between human and horse was that of hunter and prey. The earliest record of human concordance with horses dates to the mid-Paleolithic (100,000 to 35,000 BP) in Western Europe where cave paintings, engravings, and portable effigies have been discovered. Monumental Upper Paleolithic cliff drawings and cave frescoes have been found in Siberia and Western Mongolia as well (Kelekna 2009). "Early in the history of human civilization, on the prairie grasslands and boreal forests of central Asia, horses became like the buffalo to the Blackfoot—both suppliers of goods and sacred gods" (Chamberlin 2006, 62). The human-equine connection changed to one of cohabitants when we began to herd them, but it was their initiation as beasts of burden for carrying, pulling, driving, and riding that raised their position from resource providers to personal allies and enablers. Only then did they emerge in our minds and myths as transcendent beings, as earthlings and as deities with extraordinary elegance and mystic powers that matched and sometimes surpassed our own.

Arabian scribes have celebrated the horse as "a drinker of the wind, a dancer of fire" and declared: "The sum total greatness in the Arabian [horse] is not just her incredible beauty—it is her ability to bond with humans" (Hausman and Hausman 2003, 4). American novelist John Steinbeck (1939) put it quite simply: "A man on a horse is spiritually as well as physically bigger than a man on foot." Across centuries and continents writers of all ilk have extolled the horse as a paradoxical compatriot—wild and tame, dangerous and devoted, mundane and magical. From the divine winged-horse Pegasus to the horse-man centaur, myths of human relations with magic horses reinforce research findings that humankind's most *profound interspecies events* (PIEs) tend to emerge in relationships that are fraught with challenges (Rose [1998] 2006). Human bonds with horses, as with other animals, become indelible when confronted by the seven elements of the "ultimate PIE" (Rose 1994). Those elements are:

1. Initial extreme difficulty for the human to gain access to the animal.
2. Attraction and perseverance by the human in pursuit of a connection.
3. Reversal of mistrust by the animal with regard to the pursuing human.
4. A striking first contact, followed by closer and longer interactions.
5. Intervening events that separate the pair, leaving the bond unfulfilled.
6. Heroic acts by the pair to reach/protect/reconnect with each other.
7. Profound shifts in perception of self/other by both members of the pair.

While these crucial elements were factored from data on human bonding with wild animals, they trace precisely the development of the bond between Jake and his Ikran as portrayed in *Avatar*. The difficult beginning, human perseverance, reversal of mistrust, and a striking first contact are all in Cameron's screenplay. Similarly, these events transpire in the bonding of horse and rider. Profound shifts in worldview occur on the silver screen, as they do on the riding trail. Separation and heroic reconnection also take place in the film, as they often do in the life of horse and rider. Once humans and horses share these experiences, their bonds remain no matter how long the separation, like a flame waiting to be fanned and refueled.

> I returned to horse riding after eight years and it was just like coming home. My love of horses reawakened in an instant. What I had lost in riding skills was offset by the gain of life experience. In those years my capacity to love had grown, expanding my passion for riding. Getting back in the saddle hasn't been easy, but that has driven me to connect with each horse I mount. Scores of people have told me, "when you become a horse person, you are a horse person forever." My hiatus was less than a decade, but I've seen connections that span a lifetime.

My aunt, Gail, rode her black stallion almost daily from age twelve till she went off to college at seventeen. Upon returning to a stable for the first time in sixty years, she lit up when a young horse looked in her eyes, walked over, and nuzzled her. In their own ways they had said, "glad to *see* you again." (Gabriela Rose)

The Compassionate Bond of Horse and Rider

There is a fire that burns inside every horse lover—a flame that lasts a lifetime. Once refueled, the bond gains strength through adversity. Separation in time and space can be matched in its bonding power by recovery from a sudden fall from grace. Human and horse confirm their mutuality during smooth riding and expand it by meeting challenges—some intended, others accidental.

> Ironically, my love of horses and riding skyrocketed when I fell off Nellik. Luckily I cracked my helmet, not my head! I stood up at once, climbed back on, and realized that my fear of falling was gone. She seemed to sense my newfound calmness, shivered relief, and snuffled. This was my first fall since returning to riding; it felt good to be done with it. We slipped into a trot, this time with no resistance from Nellik or tension in me. My seat was better, my core tighter, and my legs were looser. The fall had bounced my attention out of my head: left me thoughtless, mindful only of our connected bodies and intentions. Once the technicality of the movement is learned there is no reason to get caught in your head. You have to let go of the gripping going on in your mind and realize that your body will do what it needs to do. It's easy to love something that is always pleasant. True love compels one to get beyond a negative experience. My love for horses strengthened after I fell. My need to ride increased. I no longer fear falling. I can relax and sense the horse; allow feelings to guide us. Nellik and I now experience our connection more naturally; riding has shifted from a fear-driven cognitive process to a physical and emotional—and spiritual—connection. (Gabriela Rose)

Native American horse trainer Gawani Pony Boy (Horse Follow Closely) says that many riders use various methods of "telling" to instruct their horse. This, he says, conditions the animal to do what is expected of it—to obey. "The problem with demanding obedience is that there are only two possible outcomes for the horse: to obey and be rewarded or disobey and be punished" (Hausman and Hausman 2006, 113). The limits of telling another animal what to do are confounded by memory. Pony Boy explains that if you raise your hand in anger as if you are about to hit your horse when she disobeys, she may shy away whenever you raise your hand, fearing punishment. Animal trainers of all types are aware of the adverse side effects of punishment. With horses, evolved as prey animals that instinctively run from threat, commands given from a position of taskmaster are bound to backfire eventually. Looking back on my accident with Nellik, it's clear that I had been pulling her reins, trying to turn her head physically in a direction I had chosen, rather than leading her gently, asking her to affirm my intentions with my own body movement, and hers.

Most Native Americans understood that the horse is a herd animal and that its movements are directed subtly by the cooperative nature of the herd as a whole. The horse is an imitative being that learns by watching, looking, and thinking—but mostly by following or emulating the movements of the leader. Pony Boy calls this the relationship of Itancan (leader) and Waunca (imitator). He says that when, as a human, you establish leadership in the proper way your horse will want to follow your lead naturally.

Once the leader-imitator relationship is established, human and horse can become the best of friends. The gentleness of an animal so large is astounding. The way he nuzzles your arm and hand to show affection. The way she exhibits her comfort simply by letting her tongue hang out. The whisper of their breath, asking for contact. Their stillness and solidity in preparation for the mount. All these small behaviors exhibit the solidarity of the bond between horse and rider.

THE HORSE SONG OF THE NAVAJO

Holy wind blows through his mane,
His mane of rainbows.
My horse's ears are of round corn,
His eyes of stars.
I am wealthy because of him.
I am eternally peaceful.
I stand for my horse.
[Hausman and Hausman 2003, 85]

What the Navajo knew as part of their ongoing way of life, industrialized people strive for on weekends, vacations, and holidays—if we dare to take the risk. Communing with horses is not just another road trip in the convertible or minivan. Not a motorcycle or bike ride. Freedom, speed, and mobility become visceral body/mind experiences while on horseback.

The challenges of the workweek melt away when I'm on a horse. I live in the now. What is happening at each moment: the sensing of the horse's great body, the scenery moving around us at a pace governed by two beings, all give me the feeling I could do anything, go anywhere, be anything. The horse and I have six legs and one intention. We feel as one, see as one, move as one. He knows what I am thinking; I know what he's feeling; we do what we want. There are no limits, only those we believe we have. Together we fly! Fortunately, my horse and I are whole, healthy, and equally able to gain from our relationship. Such is not the case with some riders I've known. To participate in the human-equine relationship can give even more to people with disabilities. (Gabriela Rose)

The Healing Power of the Horse

When a person is unable to access what is going on in his or her own body and mind, the horse will express it behaviorally or even become locked up in his own body. [Kohanov 2001, 202]

The hero of the film *Avatar*, Jake Sully, is a man crippled in war, unable to use his legs, giving up his will to live. He escapes his wheelchair and his depression by volunteering to be transported to another galaxy and serve as the avatar of an indigenous tribesman in a place called Pandora. Initially it was the power of technology that gave Jake back his legs. But it is his bonding with the Na'vi people, his Ikran steed, and ultimately with the giant flying dragon Toruk, king of all Ikrans, that healed him of his earthly human maladies. By becoming uni-fied with these otherworldly beings, Jake transformed into a healthy, loving, and spiritually fulfilled person.

Like *Avatar*'s earthling hero, children and adults with physical and psycho-logical limitations can find tremendous healing from bonds with other beings, and with horses in particular. Whether born on the autism spectrum or dis-abled by illness or accident, those with special needs are in many ways impris-oned by their maladies. But as soon as they get on that horse they begin to set themselves free. Over time, many can go on horseback anywhere the healthy human goes, but with an even greater appreciation. The invalid's imbalance is corrected by the secure upstanding steed. Four strong and powerful legs that carry the person with speed and agility replace two wobbly limbs. They can ex-perience nature as never before—on a moving, living, feeling platform sixteen hands in the air. There is a fluidity of movement available to them that they find no other place. Not the mobility enabled by wheelchair, automobile, or airplane—cold mechanical movement. On horseback, they become enabled by a warm organic living being. As infants, kindred grownups carried them where they couldn't otherwise go. Now the horse has become the carrier and the kin: a kin of a very different kind. This author's experiences working with special needs riders affirm the truth of remarkable events such as this one, reported about the observations of Barbara Rector, cofounder of the Equine Facilitated Mental Health Association (EFMHA):

Barbara found that instinct alone couldn't explain the sophisticated reactions horses exhibited in the presence of people dealing with all types of physical, mental, emotional, and spiritual challenges. Josh, a student attending one of her early workshops, happened to be blind from birth, and he did surprisingly well in getting an Arabian gelding named Dundee to cooperate with him in the round pen after some initial adjustments. When his session was over, however, Josh asked if he could have a few minutes to himself before the next person took her turn. He just wanted to wander around the ring alone because he'd never experienced the sensation of moving at liberty in a new environment without his cane. The horse, who had never encountered a person with a visual impairment, watched Josh's tentative steps for a moment before walking directly to the young man's side and putting his withers in the same position a guide assistant would take. Each time Josh lost his footing, Dundee moved in close, providing his body as balance. Together, they walked, jogged, and ran. Even Barbara was amazed as they began to move in figure eights. No one has ever been able to explain how Dundee knew Josh was blind, or why the horse was motivated to act as the boy's guide. (Kohanov 2001, 202–3)

As we study human interspecies bonds in earnest, hypotheses abound to explain why the horse moves close to help the person in need. The mother mare helps her newborn colt to stand in the pasture. The therapy horse helps the blind man to make his way around the ring. Horses have evolved for millennia as herding helpers of humankind. For decades, beginning in Europe and spreading around the world, equine therapy programs have developed to place disabled people on the backs of sensitively trained horses and enable them to go where they can't go on foot.

To see people, young and old, become transformed by this healing experience is astounding. Like the crippled Jake Sully whose Na'vi-avatar bonded with his Ikran mount, they are no longer defined by their disability, but by the conjoint spirit and desire they share with the horse. To witness the joy on the faces of these riders as they break through their limitations is exhilarating. To watch them attain the eternal love of horses is heartwarming. To know that a horse can heal a human is inspiring. Still, there is more to the horse-human bond than merely mounting and riding, loving and healing, showing and going.

Ultimately, as we've discussed earlier in relation to bonding among all beings, human and horse must "*see*" each other—peer into their bonded hominid and equine souls—and become internally connected. Once this is done rider and steed can take the next step: they can look together from inside outward and truly *see* the world through each other's eyes.

We Must *See* and *Be* as *One*

When Cortez came to the new world the Aztecs had never seen a horse and rider before and assumed it was one being of magic. [Hausman and Hausman 2006, 147]

For the Aztecs to mistake Cortez and his horse for one being, there must have been a strong connection between the two. Envisioning horse and rider as one magical creature signifies complete melding of the two beings in their feeling and vision, as well as their movement. Ask any professional rider who has performed with horses in equestrian competitions; they will confirm that to excel in dressage, jumping, slalom, cross country, and circuit racing requires that human and horse feel and see as one united being. For Cortez to take his cavalry across the sea and ride into a new world, every horse and rider had to be strongly united. The Aztecs were right.

In Greek mythology centaurs were beings with a human upper body and the lower body of a horse. They have been treated as possessing of special wisdom, and as dangerous, uncanny characters. Cheiron, the foremost of centaurs, was of a special, more civilized lineage and had been schooled by Apollo (god of light, medicine, and music) and Artemis (goddess of wild animals and healer of young women). Cheiron was admired for his visionary wisdom, beneficence, and healing powers, and for his ability as a teacher and mentor (Tripp 2007). It is reasonable to assume that the integrity of horse and human in Cheiron reflected Greek experience with well-bred and trained horses; the Greeks knew that horse riding uplifted people and expanded their capacity. From the Olympian tales of ancient Greece to the *Avatar* stories of modern times, people have depicted the human-equine bond as an epitome of interspecies integrity.

The interplay of human and horse takes *seeing* to another level, beyond that of any other interspecies connections we can make: even those between people and dogs. Herding dogs manage other animals as they see them, in accord with human training and instruction. Guide dogs become the eyes for blind humans. Sled dogs pull the weight and follow the voice and line commands of seeing humans. These canine-human connections approach, but don't quite reach, that of horse and rider. People simply don't ride dogs. For human and horse to reach the full potential of their bond—to walk, trot, gallop, run, and jump together safely and swiftly over trails and barriers—they must function in unity. Ultimately, horse and human rider must synergize their mutual devotion, shared intentions, and physical capacities so as to *see* through each other's eyes and move through the world as one being.

Turning a blind eye to nature robs humanity of the capacity to see the biosynergy and compassion that vitalizes life on earth. Without that vision, unable to *see* one another, we are lost and all life is at risk of destruction and extinction. "*I see you*" in the form expressed between Jake and Neytiri in the film *Avatar* was extended to all living beings on the imaginary planet Pandora. The Na'vi saw into the hearts and souls of all nature and knew that the vision was reciprocal. It is thus that they united to defeat the human invaders. But in the end it was the higher-level bond between Na'vi warriors and their faithful chargers, the Ikran, that would finish the fight. Na'vi and Ikran could *see* through one another's eyes, and soar as *one* through the embattled skies to achieve the final victory for themselves and all life on their planet.

What we learn from the unity of horse and rider on Earth and of Na'vi and Ikran on Pandora is a lesson we must expand to our relationships with all of nature. Yes, we must *see* through the eyes of every living organism, and seek the biosynergy and compassion that will save life on this planet, but there is another step to take. Somehow, through imagination, intuition, and insight, we must find ways to unite with each living being as we do with our kindred horses.

Scientist and conservationist, filmmaker and moviegoer, farmer and forester, therapist and theologian, lay person and professional—all humans who care about nature must commune in partnership with all the diverse flora and fauna of this earth. We must do more than stop ignoring nature. We must do more than see through the surfaces into the soul of nature. We must envision ourselves astride all life in the biosphere, entwined with all spirits, seeing with all eyes, riding with all beings into a future that will restore our biosynergy, energize our compassion, and save us from the global catastrophe that will surely befall this planet, if we humans continue to blindly go our own self-centered way.

I See You must expand to *We Are One*, if biosynergy and compassion are to enable life to prevail on Earth.

References

Buber, M. [1923] 1970. *I and Thou.* Translated by W. Kaufmann. New York: Simon and Schuster.

Cameron, J. 2009. Director, *Avatar.* Santa Monica, CA: Lightstorm Entertainment.

Chamberlin, J. E. 2006. *Horse: How the Horse Shaped Civilization.* New York: Bluebridge Books.

Corning, P. A. 2003. *Nature's Magic: Synergy in Evolution and the Fate of Humankind.* Cambridge: Cambridge University Press.

Dines, L. 2005. *Horse Tails and Trails.* Minocqua, WI: Willow Creek Press.

Hausman, G., and L. Hausman. 2003. *The Mythology of Horses: Horse Legend and Lore throughout the Ages.* New York: Three Rivers Press.

Kelekna, P. 2009. *The Horse in Human History.* New York: Cambridge University Press.

Kohanov, L. 2001. *The Tao of Equus: A Woman's Journey of Healing and Transformation through the Way of the Horse.* Novato, CA: New World Library.

Paolinetti, J. 2003. Director, *The Hard Road.* Santa Monica, CA: Native Productions.

Rogers, C. R. 1980. *A Way of Being.* Boston: Houghton Mifflin.

Rose, A. L. 1994. "Description and Analysis of Profound Interspecies Events." Proceedings of the 15th Congress of International Primatological Society, Bali, Indonesia.

———. 1996a. "Commercial Exploitation of Great Ape Bushmeat." In *Seminaire sur l'impact de l'exploitation forestiere sur la faune sauvage,* edited by R. Ngoufo, J. Pearce, B. Yadji, D. Guele, and L. Lima, 18–20. London: WSPA and Cameroon MINEF.

———. 1996b. "Orangutan, Science, and Collective Reality." In *Orangutan: The Neglected Ape,* edited by R. Nadler, B. Galdikas, N. Rosen, and L. Sheeran, 29–40. New York: Plenum Press.

———. [1998] 2006. "On Tortoises, Monkeys, and Men." In *Kinship with the Animals,* updated ed., edited by K. Solisti and M. Tobias, 15–32. San Francisco: Council Oak Books.

———. 2007. "Biosynergy: The Synergy of Life." In *Encyclopedia of Human-Animal Relationships,* vol. 1, edited by M. Bekoff, 123–29. Westport, CT: Greenwood Publishing.

———. 2010. "Biosynergy: The Synergy of Life." In *Making the Invisible Visible: Essays by Fellows of the International Leadership Forum, Western Behavioral Sciences Institute*, edited by R. Farson, 109–17. Norcross, GA: Greenway Communications.

———. 2011. "Bonding, Biophilia, Biosynergy, and the Future of Primates in the Wild." *American Journal of Primatology* 73: 245–52.

Rose, A. L., and A. Auw. 1974. *Growing Up Human*. New York: Harper and Row.

Rose, A. L., P. J. Fraser, and D. E. Ndeloh. 2008. "Holistic Conservation Values Education." Talk at IPS Conservation Education Symposium, 22nd Congress of International Primatological Society, Edinburgh, Scotland, August.

Rose, A. L, R. A. Mittermeier, O. Langrand, O. Ampadu-Agyei, and T. Butynski, photography by Karl Ammann. 2003. *Consuming Nature: A Photo Essay on African Rain Forest Exploitation*. Los Angeles: Altisima Press.

Steinbeck, J. 1939. *The Red Pony*. New York: Heinemann.

Tripp, E. 2007. *Classical Mythology: An Alphabetical Guide*. New York: Plume Press.

Some Closing Words

Moving Ahead with Heart, Peace, and Compassion

Marc Bekoff

Our world is becoming global. Many of us are in daily contact with people from other parts of the world, other cultures, and language groups. As globalization proceeds, we will need, urgently, every ounce of empathy, tolerance, and communication skill we can muster. This is what our connection with animals has given us; this is what we need so badly for the future. [Pat Shipman 2011, 280]

THE LATE THEOLOGIAN Thomas Berry (1999) stressed that our relationship with Nature should be one of *awe*, not one of *use*. Individuals of all species have inherent or intrinsic value because they exist, and this alone mandates that we coexist. They have no less right than we do to live their lives without our intrusions, they deserve dignity and respect, and we need to accept them for who they are.

As I read and reread the essays in this book my brain got filled to the brim over and over again. Taken together they make a strong case for compassionate conservation. Clearly, there are many people who truly care about healing our wounded planet, and we can only hope that current and future students will follow in their wake and that everyone who can do something does whatever it is they can to make the planet a better place for all beings. We must expand our compassion footprint and step lightly, if at all, into the lives of other animals. As much as I prefer a hands-off policy, we really are all over the place and there are situations that demand that we do something, often having to make various trade-offs we wish we didn't have to do.

Tigers in Bangladesh: Moving Me out of My Comfort Zone

On March 30, 2011, I received an e-mail from Christina Greenwood, project manager of the Sundarbans Tiger Project (Sundarbans

Tiger Project 2011), about a conservation project in Bangladesh dealing with tiger-human conflict, overseen by the Wildlife Trust of Bangladesh and the Zoological Society of London. The e-mail arrived at the best and worst of times. On the one hand it couldn't have come at a better time, as I was reading some of these essays and struggling with how to reconcile my own inability to come to terms with what I would like the world to be like and what it really is. On the other hand it came as I was rushing about to leave town and I didn't need more on my plate. But it served as an important wake-up call that moved me out of my comfort zone, and the net effect was that it came to be a most welcomed e-mail.

The reality of the situation in Bangladesh made it a perfect case study because it brought into focus many of the issues with which my colleagues and I wrestle each and every day that all center on the daunting question, "Who lives and who dies?" Greenwood wrote that she and her colleagues were working on a tiger conservation project in Bangladesh, and one of the issues they face is tiger-human conflict. Around fifty people per year are killed by tigers when they enter the forest for fishing or wood collection, and when tigers stray into villages adjacent to the forest, villagers beat them to death or put out poisoned bait.

Greenwood wondered if tigers who come into villages should be euthanized, especially the injured ones. Would this really be the most humane choice since there were no obvious options? Of course, Greenwood and her team want to do the most ethical thing for both the tigers and the people. However, it's a very tricky situation in that the villagers are very poor, veterinary care for injured tigers is extremely limited, there aren't any rehabilitation centers, and zoos are in bad condition and can't take in any additional animals. Collaring animals and learning more about their movement patterns might help in the future (Siddique 2011), but decisions had to be made immediately.

I struggled with what I would likely do in this real world (or as my colleague Lori Gruen calls it, this nonideal world) scenario. Here is what I hurriedly wrote: "I'm leaving for amsterdam tomorrow so let me give you a brief summary of some of my ideas about this because there are no 'easy' or 'fast' answers—in the best of all possible worlds, which this is not, such conflicts would be avoided or there would be enough money to take care of the problem tigers—good zoos or rehab centers with proper veterinary care—clearly this isn't the case—so the way I see it, cutting through the chase, is that the tigers are ultimately likely to suffer different sorts of deaths because there's no money to care for them in ways that could give them a 'good' life—life in a zoo would be horrible, there are no rehab centers, they can't get the proper veterinary care, and if they're left to their own they'll be poisoned—so it's a matter of our having to make life and death decisions for them, something we do for billions of other animals . . .

while I know some might/would disagree with me, and I wish I had more time right now to write more, I see that in some/many instances in the situations with which you're faced, euthanasia—mercy killing to avoid prolonged suffering and likely/sure death—would be the more ethical option—of course each situation should be considered on its own—case by case—but it *seems* to me that in the dire situations and options you describe, euthanasia would be more humane and ethical than putting a tiger(s) in a zoo or leave him or her on their own to die in the wild or be poisoned—I deeply wish it weren't so."

After I sent my e-mail I sat at my desk for more than an hour wondering if I did the best I could, and every day since I wonder what other options might be available in this situation, one that is mirrored in many other places around the world. What if this were about less charismatic species or individuals, snakes or rats, for example, or about animals we thought to be insentient, or about an abundant species in which there were "disposable" individuals and the loss of some wouldn't endanger them? We often play the numbers game, ignoring the well-being of individuals (Bekoff 2010). Would we still struggle with the decision to euthanize, kill, the injured tigers? Would Greenwood and her team even be working on this project?

Speciesism in the real world directly informs the practical strategies we choose to use. For example, we need to be able to identify those characteristics of an individual or species that warrant keeping them alive or allowing them to suffer or die, and when we factor in behavioral and ecological variables this becomes a difficult practice. Nonetheless, we treat "higher/more valuable" species different from and better than "lower/less valuable" species although these designations, based on our attempts to draw lines separating species, are fraught with error and ignore nature, including evolutionary continuity (Bekoff 2010). Speciesism leads to specious conclusions, because while some of the comparisons that are made seem to make sense and are superficially pleasing, they actually lack real merit on close inspection (Bekoff 2011). *Specious speciesism undermines our collective efforts to make the world a better place for all beings.*

Questions and issues such as these keep me awake at night, and I'm sure I'm not alone, tossing and turning and wrestling with an ever-changing and elusive reality. What is the most compassionate thing we can do? Is killing in the name of conservation a compassionate option? Should individuals be traded off for the good of their own or other species or for ecosystem integrity? Can we really put a price on an animal to reflect their value? What does compassionate conservation entail? Can we reconcile the differences between those people interested in individuals and those more interested in species, populations, and ecosystems? How can we rewild our hearts and bring animals back into the picture? (Bekoff in press). How can we build and maintain corridors of compassion and coexistence? How can we cross disciplines and work together? How can we walk the talk so that something actually gets done? Can we do better?

These questions were dealt with at the first international meeting on compassionate conservation held in Oxford, England (Compassionate Conservation 2010) and followed up with a session at the 2011 Asia for Animals meeting held in Chengdu, China. The demands of the real world require us to expand our personal and professional comfort zones and think and act out of the box because we simply cannot do everything that needs to be done. Compassionate conservation is the perfect catalyst and guide for such a move.

We've arrived at this crossroads because we've chosen to ignore nature for a myriad of reasons. Patience may not be a virtue in these troubled times. There can be long time lags between our taking action and seeing their effects. Invoking the precautionary principle, it's safe to say that even if we think we need more data because of a lack of consensus among researchers, for most of the situations we're facing we know enough right now to take action. We all agree that our troubled and wounded world needs a lot of compassionate healing, right now, not when it's convenient. There is a compelling sense of urgency.

Given our overproducing and overconsuming ways, there's no doubt that the problems with which we're faced in the future will increase, and the relative number of people who are able to work on them will decrease. So, those who *can* do something are mandated to do something. In the future there likely will be fewer people who will actually be able to make a positive difference in our relationships with animals and ecosystems. Joel Cohen (2009), head of the Laboratory of Populations at the Rockefeller University and Columbia University, offers the sobering fact that the difference in the population numbers between less developed areas of the world (the have nots) and more developed regions of the world (the haves) will have increased from twofold in the 1950s to about sixfold by 2050. This means that it is imperative—perhaps it is truly a moral imperative—that those who can do something good for animals and the Earth do it because the division between those who can and those are can't is rapidly growing, and this will be challenging to humanity as the ratio shifts. Of course, because not all "the haves" choose to do much if anything at all, it is even more essential that those who choose to do something do it for as long as they can and not succumb to the inevitable disappointments, frustrations, and burnout that are associated with animal and environmental activism.

Our unique contribution to the wanton decimation of the planet and its many life forms is an insult to other animal beings and demeans us. Michael Tobias (2011) notes, "Estimates vary, but nearly half-million people are born per day, approximately 150 per minutes of which 50 to 65% will die before reaching their reproductive years. But that leaves at least 225–250,000 new people to feed, clothe and house, every day: a city about the size of Indiana's second largest metropolitan region of Fort Wayne."

Academics, Advocacy, and Activism Go Hand in Hand

It is impossible to be neutral on issues of animal or environmental protection. Many scientists like to think science is objective and that they don't have an agenda and that they have no obligation to interact with nonresearchers, but this is rarely the case (Bekoff 2010). Indeed, it shouldn't be. In his wonderful and bold book, *A World of Wounds*, renowned biologist Paul Ehrlich (1997, 15) wrote: "Many of the students who have crossed my path in the last decade or so have wanted to do much, much more. They were drawn to ecology because they were brought up in a 'world of wounds,' and want to help heal it. But the current structure of ecology tends to dissuade them. . . . Now we need to incorporate the idea that it is every scientist's obligation to communicate pertinent portions of her or his results to decision-makers and the general public." I could not agree more.

Animal suffering continues in all corners of the world. However, there are also "good" things happening, and these can be recalled to keep us inspired and engaged when it looks like there is little or no hope. From time to time people ask me about animal activism, burnout, and other matters associated with working for animals, so I've penned some short "one-liners" that I've found helpful over the years. Whether one agrees or disagree with some of them, I know we all agree that we must keep on working for animals and earth and peace and justice for all. So here are some thoughts that keep me going, in no particular order.

- Think positively. Don't let people get you down. I'm not a blind optimist but along with all the "bad" things there are "good" things happening, and that's what kindles and rekindles me, at least. Negativity is a time and energy suck and all of you good people need to keep doing what you are for as long as you can, and this means, at least for me, rekindling from time to time and taking deep breaths and enjoying whatever it is I enjoy. The bottom line is take care of yourself so you can do what you do for as long as possible.
- Concentrate on successes, what works, and put the failures aside.
- We are not the radicals or the "bad guys" who are trying to impede human "progress." We are caring people and we don't have to apologize for feeling. We should be unapologetic and compassionate activists working for a better world. In fact, those who care about animals and the Earth should be seen as heroes who are not only fighting for animals but also for humanity. Biodiversity is what enables human life as well as enriches it. It is imperative that all of humanity

reconnects with what sustains the ability of our species to persist, and that we will act as a unified collective while coexisting with other species and retaining the integrity of ecosystems. There are no quick fixes and we need to realize that when animals die, we die too.

- We need to encourage scientists to act as concerned citizens as well as get citizens to act as responsible stewards.
- Be proactive. We need to look at what's happening and prevent further abuse and not always be "putting out the fires" that have started.
- Be nice and kind to those with whom you disagree and move on. Sometimes it's just better to let something go, so pick your "battles" carefully and don't waste time and energy. Don't waste time "fighting" people who won't change, and don't let them deflect attention from the important work that needs to be done. Don't get in "pissing matches" with people who want you to waste precious time and energy fighting them, time and energy that must go into working for animals and earth and peace and justice.
- If we let those who do horrible things get us down or deflect us from the work we must do, they "win" and animals, earth, and we lose. While this may be obvious I thought it worth saying again because it's a common ploy to get people to get into tangential discussions and arguments that take them away from the important work that must be done.
- Teach the children well, for they are the ambassadors for a more harmonious, peaceful, compassionate, and gentle world.

Reasons for Inspiration and Hope:
There's No Going Back to the Way Things Were

It's difficult to be optimistic given the challenges with which we're faced. However, in April 2011, Bolivia announced it would grant all nature equal rights to humans. Among the eleven rights are the right to life and to exist, the right to continue vital cycles and processes free from human alteration, the right to pure water and clean air, the right to balance, the right not to be polluted, and the right to not have cellular structure modified or genetically altered (Vidal 2011).

Erle Ellis (2011a) notes that while it's true we've transformed Earth beyond recovery, rather than looking back in despair we should look ahead to what we can achieve. He writes, "There will be no returning to our comfortable cradle. The global patterns of the Holocene have receded and their return is no longer possible, sustainable, or even desirable. It is no longer Mother Nature who will care for us, but us who must care for her. This raises an important but often neglected question: can we create a good Anthropocene? In the future will we be able to look back with pride? . . . Clearly it is possible to look at all we have created and see only what we have destroyed. But that, in my view, would be our mistake. We most certainly can create a better Anthropocene. We have really only just begun, and our knowledge and power have never been greater. We will need to work together with each other and the planet in novel ways. The first step will be in our own minds. The Holocene is gone. In the Anthropocene we are the creators, engineers and permanent global stewards of a sustainable human nature." Ellis (2011b) aptly refers to Earth as "the planet of no return."

In addition to being proactive we need to be positive and exit the stifling vortex of negativity once and for all. Negativity is a time and energy bandit and depletes us of the energy we need to move on. We don't get anywhere dwelling in anguish, sorrow, and despair. As Elin Kelsey (2012) puts it, "pessimism is a turn-off." In a taxicab in Vancouver, British Columbia, I saw a saying in front of a church that really resonated with me: "Make the most of the best and the least of the worst." Amen.

Let's Make This the Century of Compassion: We Need a Revolution as We Wrestle with Reality

Compassionate conservation is not an oxymoron and is here to stay. Two things are for sure. First, if we develop compassion and empathy there is hope for our planet, all of its inhabitants and landscapes. Second, if we continue to ignore nature we will make little progress dealing with the problems at hand.

We don't own the world and we suffer the indignities we impose on other beings. Ethics needs to be firmly implanted in conservation biology even if it moves us outside of our personal and professional comfort zones, and some projects have to be put on hold or terminated. After all is said and done, more is usually said than done. This is not a criticism but rather reflects the numerous global challenges we face. We must confront the realities of the world and move on, considering each situation on a case-by-case basis. What obtains for tigers in Bangladesh might not for gray wolves in western North America or elephants in Kenya. Should we save ants rather than charismatic species?

We need to tap into our basic goodness to get and keep the momentum going. Accepting there will always be inevitable differences in this vast and diverse world, those who care about animals and who are ready to do something to make the world a better place for all beings should be able to make substantial progress on righting the wrongs, working as a tightly knit community. Being practical and facing the realities of a given situation is essential. Being proactive also is mandatory. We really know enough right now in most situations to warrant preemptive action before a situation gets out of hand. There are ample warning flags, canaries in the coal mine, telling us that time is not on our side.

"Who lives and who dies?" is an incredibly difficult question. Difficult choices will always have to be made, as we won't be able to do everything that needs to be done. That's just the way it is. There aren't enough resources, person-power, and money to right all the wrongs.

What would a global moral imperative look like? Guiding principles would be (1) do no intentional harm, (2) respect all life, (3) treat all individuals with respect and dignity, and (4) tread lightly when stepping into the lives of other animals. Let's give it a try.

There are numerous animals and habitats worth protecting and saving so let's get on with it. The animals are constantly telling us what they want and need. Their manifesto, simply put, is treat us better or leave us alone (Bekoff 2010). As we save other animals, we'll be saving ourselves.

Compassion Rocks

Compassion begets compassion and there's actually a synergistic relationship, not a trade-off, when we show compassion for animals and their homes. There are indeed many reasons for hope (Goodall 1999, 2009). As I mentioned in the preface, there's also compelling evidence that we're born to be good (Keltner 2009; Bekoff 2010; see also Rifkin 2009 and de Waal 2010) and that we're natural-born optimists (Sharot 2011). Therein lie many reasons for hope that in the future we will harness our basic goodness and optimism and all work together, cooperatively, as a united community (see also Williams 2008; Sennett 2011; and Barringer's [2007] discussion of the "spiritual handshake" that stresses service to the common good). We can look to the animals for inspiration and evolutionary momentum (Bekoff and Pierce 2009). So, let's tap into our empathic, compassionate, and moral inclinations to make the world a better place for all beings. As a six-year-old told me at a kid's event I was leading, "compassion rocks."

Michael Soulé (2002), founder of the field of conservation biology, perhaps said it best: "We're certainly a dominant species, but that's not the same as a keystone species. A keystone species is one that, when you remove it, the diversity collapses; we're a species that when you add us, the diversity collapses. We can change everything, dictate everything and destroy everything."

People who care about animals and nature should not be considered "the radicals" or "bad guys" who are trying to impede "human progress"; in fact, they could be seen as heroes who are not only fighting for animals but also for humanity. Biodiversity is what enables human life as well as enriches it. It is imperative that all of humanity reconnects with what sustains the ability of our species to persist and that we act as a unified collective while coexisting with other species and retaining the integrity of ecosystems. There are no quick fixes, and we need to realize that when animals die, we die too.

We need to retain hope and try as hard as we can to realize our dreams for a better future for all animals and our planet as a whole. If we present a negative picture to youngsters and those working to make the planet a better place for all beings then we can hardly expect them to do much at all. As we learn more about what we can and cannot do I truly believe that we will be able to succeed in many of our attempts to right the numerous wrongs and bring more compassion, peace, and harmony to our world. As Nelson Mandela (played by Morgan Freeman in the movie *Invictus*) told the South African soccer team, we must exceed our own expectations (http://www.lariat.org/AtTheMovies/nora/invictus.html).

We must work on long-term solutions of different scales and realize that Earth is really our one and only home. And we need to think "out of the box." What we've been doing hasn't worked very well. Now with seven billion of us and counting upward, the problems are likely to get bigger and more difficult to solve. As we ignore, fragment, and fracture nature we do the same to ourselves, often without even realizing how we're self-destructing, because we need nature just as we need oxygen. Short-term "'solutions du jour' which the popular media serve up in remarkable proportions" (Evernden 1999, xii) don't work.

We also need an "ecology of place" (Billick and Price 2011) that focuses on producing local as well as general knowledge. This perspective stresses an interdisciplinary approach and also makes one think of the world as an interconnected community whose very existence depends on our paying close attention to *all* of nature and her intricate webs, the exact opposite of what happens when we ignore and compartmentalize nature and exempt ourselves from being a member of a vast and interconnected community.

One of my favorite bumper stickers is "Nature bats last." We can try to outrun and outsmart nature but in the end she always wins. Rewilding our hearts will foster being nice, kind, and compassionate to all other beings, including their homes, and will get us out of the mindset of domination and exploitation. I trust this book has made a clear and strong case, indeed many cases, for the need for a social movement that leads to considerably more empathy and compassionate conservation as we move deeper into the twenty-first century and well beyond. There really seems to be no other alternative viable mindset. One can hardly be against adding compassion and empathy to the world. There is a clear and compelling case for compassionate conservation to be the rule, rather than the exception, so let's get on with it now.

References

Barringer, Almut. 2007. "The 'Spiritual Handshake': Toward a Metaphysical Sustainability Metrics." *Canadian Journal of Environmental Education* 12: 143–99.

Bekoff, Marc. 2010. *The Animal Manifesto: Six Reasons for Expanding Our Compassion Footprint.* Novato, CA: New World Library.

———. 2011. "Who Lives, Who Dies, and Why: Fishy Speciesism in the Real World and Its Wide-Ranging Moral Implications." Paper read at the Arcus Foundation Roundtable "Rethinking the Species Interface." The Desmond Tutu Center, New York.

———. In press. *Rewilding our Hearts.* Novato, CA: New World Library.

Bekoff, Marc, and Jessica Pierce. 2009. *Wild Justice: The Moral Lives of Animals.* Chicago: University of Chicago Press.

Berry, Thomas. 1999. *The Great Work: Our Way into the Future.* New York: Bell Tower.

Billick, Ian, and Mary V. Price, eds. 2011. *The Ecology of Place.* Chicago: University of Chicago Press.

Cohen, Joel. 2009. "Human Population Grows Up." In *A Pivotal Moment: Population, Justice, and the Environmental Challenge*, edited by L. Mazur, 27–37. Washington, DC: Island Press.

Compassionate Conservation. 2010. http://compassionateconservation.org/.

Ehrlich, P. 1997. *A World of Wounds: Ecologists and the Human Dilemma.* Oldendorf/Luhe, Germany: Ecology Institute.

Ellis, Erle. 2011a. "Forget Mother Nature: This Is a World of Our Making." *New Scientist* (June 11): 26–27. http://www.newscientist.com/article/mg21028165.700-forget-mother-nature-this-is-a-world-of-our-making.html.

———. 2011b. "The Planet of No Return." http://breakthroughjournal.org/content/authors/erle-ellis/the-planet-of-no-return.shtml.

Evernden, Neil. 1999. *The Natural Alien: Humankind and Environment.* 2nd ed. Toronto: University of Toronto Press.

Goodall, Jane. 1999. *Reason for Hope: A Spiritual Journey.* New York: Warner Books.

Goodall, Jane, with Thane Maynard and Gail Hudson. 2009. *Hope for Animals and Their World: How Endangered Species Are Being Rescued from the Brink.* New York: Grand Central Publishing.

Keltner, Dacher. 2009. *Born to Be Good: The Science of a Meaningful Life.* New York: W. W. Norton.

Kelsey, Elin. 2012. "Do Not Despair." *New Scientist* (January 7): 24–25.

Rifkin, Jeremy. 2009. *The Empathic Generation: The Race to Global Consciousness in a World in Crisis.* New York: Jeremy P. Tarcher/Penguin.

Sennett, Richard. 2011. *Together: The Rituals, Pleasures, and Politics of Cooperation.* New Haven, CT: Yale University Press.

Sharot, Tali. 2011. *The Optimism Bias.* New York: Pantheon Books. See also http://www.time.com/time/health/article/0,8599,2074067,00.html.

Shipman, Pat. 2011. *The Animal Connection: A New Perspective on What Makes Us Human.* New York: W. W. Norton.

Siddique, Abu Bakar. 2011. "Radio Collar Can Save Bengal Tiger." http://www.daily-sun.com/?view=details&type=daily_sun_news&pub_no=183&cat_id=1&menu_id=2&news_type_id=1&index=3&archiev=yes&arch_date=10-04-2011.

Soulé, Michael. 2002. "History's Lesson: Build Another Noah's Ark." *High Country News* 24, no. 9, May 13, 2002. Available at http://www.hcn.org/servlets/hcn.Article?article_id=11219.

Sundarbans Tiger Project. 2011. http://sundarbantigerproject.info/news.php?readmore=77.

Tobias, Michael. 2011. "Pro-Planet Is Pro-Choice?" http://blogs.forbes.com/michaeltobias/2011/05/09/pro-planet-is-pro-choice/.

Vidal, John. 2011. "Bolivia Enshrines Natural World's Rights with Equal Status for Mother Earth." http://www.guardian.co.uk/environment/2011/apr/10/bolivia-enshrines-natural-worlds-rights.

de Waal, Frans. 2010. *The Age of Empathy: Nature's Lessons for a Kinder Society.* New York: Three Rivers Press.

Williams, Terry Tempest. 2008. *Finding Beauty in a Broken World.* New York: Pantheon Books.

About the Contributors

LIV BAKER is a PhD candidate in the Animal Welfare Program at the University of British Columbia. Her work is in collaboration with the San Diego Zoo Institute for Conservation Research. Specifically, she has been involved with a translocation program for the endangered kangaroo rat, *Dipodomys stephensi*. Her interest is in applying animal welfare science to improve the treatment and survival of animals involved in conservation management programs.

MARC BEKOFF is Professor Emeritus of Ecology and Evolutionary Biology at the University of Colorado, a Fellow of the Animal Behavior Society, and a former Guggenheim Fellow. In 2000 he received the Exemplar Award from the Animal Behavior Society for major, long-term contributions to the field of animal behavior, and he and Jane Goodall cofounded Ethologists for the Ethical Treatment of Animals. In 2005 Bekoff was presented with The Bank One Faculty Community Service Award for the work he has done with children, senior citizens, and prisoners as part of Jane Goodall's Roots & Shoots program, and in 2009 he was presented with the Saint Francis of Assisi Award by the Auckland (New Zealand) SPCA. Bekoff has written numerous scientific and popular essays and books, including *The Emotional Lives of Animals, Animals At Play: Rules of the Game* (an award-winning children's book), *Animals Matter, Wild Justice: The Moral Lives of Animals* (with Jessica Pierce), *The Animal Manifesto: Six Reasons for Expanding Our Compassion Footprint, Rewilding Our Hearts* (in press), and the *Encyclopedia of Animal Behavior*, the *Encyclopedia of Human-Animal Relationships*, and the *Encyclopedia of Animal Rights and Animal Welfare* (two editions). His websites are marcbekoff.com and, with Jane Goodall, ethologicalethics.org.

DROR BEN-AMI is a Research Associate at the Institute of Sustainable Futures at the University of Technology, Sydney (UTS). His Honors and PhD dissertations at the University of New South Wales (UNSW) were on the eastern gray kangaroo and the swamp wallaby. He is an ecologist with interests in macropods, applied ecology, environmental management, and conservation biology. His current focus is on integrating science, policy, and nonlethal wildlife management. Ben-Ami is a cofounder of the Think Tank for Kangaroos (THINKK) and a director of Voiceless, which aims to reform policy, develop the animal protection movement, and inform about legalized animal cruelty.

JOEL BERGER is the John J. Craighead Chair of Wildlife Conservation at University of Montana. He loves working with animals in the wild, getting boots muddied, and moving from science to conservation actions. He has written scholarly or trade books on rhinos, moose, bison, and wild horses as well as coediting one on large carnivores and biodiversity. His field projects have taken him from Greenland and Svalbard to Mongolia and Tibet, and from Africa to Alaska. Much of his work however concentrates in western North America and areas where humans have imprudently ignored past lessons from nature.

SARAH BEXELL is the Director of Conservation Education at the Chengdu Research Base of Giant Panda Breeding, Sichuan Province, China. She is also a research scholar in the Institute for Human-Animal Connection at the University of Denver. She has worked in endangered species conservation, humane education, and conservation education for twenty years. She combines her background in animal behavior, child development, and science education to facilitate the human-animal bond to promote animal welfare and conservation.

DANIEL T. BLUMSTEIN is Professor and Chair of the Department of Ecology and Evolutionary Biology and Professor in UCLA's Institute of the Environment and Sustainability. He studies the evolution and maintenance of social behavior, communication, and antipredator behavior in a variety of taxa and integrates behavioral principles into conservation biology. His two most recent books are *A Primer of Conservation Behavior* (with Esteban Fernandez-Juricic) and *The Failure of Environmental Education (And How We Can Fix It)* (with Charles Saylan).

KEELY BOOM is a Research Fellow with the Think Tank for Kangaroos (THINKK) at UTS. Boom is an environmental and animal law expert whose research focus is critically assessing the law and policy that governs wildlife management. Her research findings have sparked academic and public discourse about the management of kangaroos, particularly on the need for compassionate conservation through the integration of environmental and animal protection.

SUSAN CLAYTON, Whitmore-Williams Professor of Psychology and Chair of Environmental Studies at the College of Wooster, has a PhD in social psychology from Yale. With Gene Myers, she is author of *Conservation Psychology: Understanding and Promoting Human Care for Nature* (2009, Wiley-Blackwell). A fellow of the American Psychological Association, Clayton has served as president of the Society for Population and Environmental Psychology and is currently the editor of *Human Ecology Review*.

EILEEN CRIST's research focuses on animal behavior science and the ways scientists understand, or avoid, the question of animal mind. She has been teaching at Virginia Tech in the Department of Science and Technology in Society since 1997, where she is the advisor for the undergraduate program Humanities, Science, and Environment. She is author of *Images of Animals: Anthropomorphism and Animal Mind* and coeditor of *Gaia in Turmoil: Climate Change, Biodepletion, and Earth Ethics in an Age of Crisis*. She is author of numerous academic papers and contributor to the late journal *Wild Earth*.

DAVID CROFT is a Visiting Fellow in the School of Biological, Earth, and Environmental Sciences at UNSW. After completing a PhD at the University of Cambridge, Croft has had a long career at UNSW teaching subjects in vertebrate biology, animal behavior and ecology, and natural resource management in the arid lands. He has published research on invertebrates, various marsupials, sheep, marine mammals, and primates. His specialty is the behavioral ecology of the kangaroos and their kind with a recent focus on interactions with people in livestock enterprises, on roads, and in wildlife tourism.

BRIAN CZECH is the founding president of the Center for the Advancement of the Steady State Economy, a conservation biologist with the US Fish and Wildlife Service, and a Visiting Professor in the National Capitol Region of Virginia Tech. Czech is the author of *Shoveling Fuel for a Runaway Train*, which calls for an end to reckless economic growth. (Note: this chapter does not represent the policy of the US Fish and Wildlife Service.)

MARCO FESTA-BIANCHET is a professor of ecology at the Université de Sherbrooke, Canada. He completed postdoctoral studies at the Large Animal Research Group, University of Cambridge, and a PhD at the University of Calgary. He was for four years the chair of the Committee on the Status of Endangered Wildlife in Canada and chairs the IUCN Mountain Ungulates Specialist Group. He is interested in the evolutionary ecology of large herbivores and has worked on bighorn sheep, mountain goats, Alpine ibex, chamois, caribou, and eastern gray kangaroos, mostly through long-term monitoring of marked individuals.

CAMILLA H. FOX is the founding executive director of Project Coyote (www .ProjectCoyote.org), a national coalition of scientists and educators promoting compassionate conservation and coexistence between people and wildlife through education, science, and advocacy. Fox is also a wildlife consultant for the Animal Welfare Institute and has been involved in environmental and wildlife advocacy for twenty years. She is the coauthor of two books, *Coyotes in Our Midst* and *Cull of the Wild*, and the coproducer of the award-winning companion film *Cull of the Wild: The Truth behind Trapping*.

CARRIE PACKWOOD FREEMAN is Assistant Professor of Communication at Georgia State University and researches media ethics, communication strategies for social justice movements, and media coverage of nonhuman animal and environmental issues, in particular, animal agribusiness and veganism. Active in the animal rights movement for two decades, she served as a volunteer director for local grassroots groups in Florida, Georgia, and Oregon and currently cohosts weekly Atlanta-based community radio programs on animal and environmental protection.

JASON LEIGH JARVIS is a PhD candidate in the Public Communication program at Georgia State University. Jarvis's research interests include environmental rhetoric and the rhetoric of international anti-American movements. He holds an MA in Communication from Wake Forest University where his thesis examined the intersection of religion, communication, and the environment.

DAVID JOHNS is a cofounder of the Wildlands Network and Yellowstone to Yukon Conservation Initiative and served as president of both organizations. He is author of *A New Conservation Politics* (2009). He has published in *Conservation Biology*, *Environmental Ethics*, *Wild Earth*, and has authored op-ed pieces and news analysis. Johns teaches politics and law at Portland State University. He has degrees in Political Science, Anthropology, and Law from Portland State University and Columbia University, where he was also an International Fellow. He is recipient of the Denver Zoological Foundation's Conservation Award, 2007.

BARBARA J. KING is Chancellor Professor of Anthropology at the College of William and Mary, where she has taught biological anthropology for twenty-five years. Early on, she studied baboons in the wild and African apes held in captivity. Now, her interests include animal cognition and emotion, defined more broadly than just primates, and the ethics of humans' treatment of other animals. King's books include *How Animals Grieve* (2013), *Being with Animals* (2010), *Evolving God* (2007), and *The Dynamic Dance* (2004). She posts about anthropology and animal behavior at NPR.org's 13.7 blog and reviews science books for The TLS.

SARAH KING has always been interested in the behavior and ecology of animals, particularly in relation to conservation. After earning a PhD, her interest in endangered species and how the behavior of animals relates to their habitat led her to Arizona where she worked on the Mt. Graham Red Squirrel Monitoring Project. Following this, Sarah returned to Mongolia to lead a UK DEFRA Darwin Initiative funded ecology and conservation awareness project for the Zoological Society of London. She then worked on a Przewalski's horse research and conservation project, one of the first scientific studies of this animal in the wild, in western Mongolia for Association Takh. Sarah is currently working at the University of Colorado at Boulder studying small mammal species distributions in the southern Rocky Mountains to examine patterns of biodiversity and the effect of climate change.

PETER J. LI is Associate Professor of East Asian Politics at the University of Houston–Downtown. His primary research interests are China's animal protection policy, wildlife trade, and social change in the course of a rapid economic transformation. Dr. Li also consults with international animal welfare organizations. He is a China policy specialist for Humane Society International. He is currently completing his book *In the Name of Development: Animal Welfare Crisis and China's Politics of Economic Modernization.*

VIVEK MENON is a wildlife conservationist, environmental commentator, author, and photographer with a passion for elephants. He has been part of the founding of five environmental and conservation organizations in India. The winner of the 2001 Rufford Award for International Conservation for his work to save the Asian elephant, Menon is the Executive Director and CEO of the Wildlife Trust of India as well as Regional Director and Advisor to the International Fund for Animal Welfare. He is the author or editor of eight wildlife books including the best-selling *Field Guide to Mammals of India*, scores of technical reports, and more than a hundred articles in various scientific and popular publications.

BEN A. MINTEER is Associate Professor of Environmental Ethics and Policy in the Center for Biology and Society and the School of Life Sciences at Arizona State University, and Senior Sustainability Scholar in ASU's Global Institute of Sustainability. He is author or editor of several books, including *Refounding Environmental Ethics* (Temple University Press, 2011) and *Nature in Common? Environmental Ethics and the Contested Foundations of Environmental Policy* (Temple University Press, 2009).

OLIN E. "GENE" MYERS JR. is Professor of Environmental Studies at Huxley College, Western Washington University. He is past president of the Society for Human Ecology. With Susan Clayton he recently coauthored *Conservation Psychology: Understanding and Promoting Human Care for Nature* (Wiley-Blackwell, 2009). One major branch of his research concerns child development and human-animal relations and their implications for conservation (*The Significance of Children and Animals: Social Development and Our Connections to Other Species*, Purdue University Press, 2007).

MICHAEL P. NELSON is the Ruth H. Spaniol Chair of Natural Resources in the Department of Forest Ecosystems and Society in the College of Forestry, and the lead-principle investigator for the H. J. Andrews Long-Term Ecological Research site, at Oregon State University in Corvallis. His most recent book is titled *Moral Ground: Ethical Action for a Planet in Peril* (Trinity University Press), edited with Kathleen Dean Moore.

JOLIE NESMITH received her Master of Social Work degree at the University of Denver in May 2011. NeSmith has established a student organization called ECO Conscious at the university in efforts to promote Conservation Social Work. NeSmith's background includes wilderness therapy, experiential education, and a great passion for the environment and all its inhabitants.

JOSPHAT NGONYO, a social science and communications graduate, is the founder of Youth for Conservation, and founder and Executive Director of the Africa Network for Animal Welfare. He also serves as Honorary Warden of the Kenya Wildlife Service, member of the Global Task Force for Farm Animal Welfare and Trade, member of the United Nation's Food and Agriculture Organisation (FAO) Gateway to Farm Animal Welfare Editorial Board, and an International Representative with Compassion in World Farming. Josphat won the prestigious Eastern Africa Environmental Leadership Award in 2003 and the Middle East Animal Welfare Award in 2007, in recognition of his dedication to animal welfare work.

DALE PETERSON teaches English at Tufts University and writes about animals, conservation, natural history, and animal science or scientists. His recent biography, *Jane Goodall: The Woman Who Redefined Man* (2006) was a *New York Times* Notable Book of the Year. His book about the Central African bushmeat trade, *Eating Apes* (2003), was named by *The Economist*, *Discover*, and the *Globe and Mail* as Best of the Year. His latest is *The Moral Lives of Animals* (2011).

DANIEL RAMP is a conservation biologist. As a Senior Lecturer in the School of the Environment at the University of Technology, Sydney (UTS), his research focuses on science that informs compassionate conservation initiatives aimed at understanding, mitigating, and adapting to environmental change. With a long interest in marsupials from the family Macropodidae, including a PhD from the University of Melbourne on eastern gray kangaroos, Daniel is a cofounder of THINKK—the Think Tank for Kangaroos, a multidisciplinary academic forum promoting greater understanding of kangaroos. He is also a Director of Voiceless, the Animal Protection Institute.

ANTHONY ROSE, social psychologist, conservationist, and writer, has been promoting the synergy of humanity and nature for three decades. His explorations of biosynergy in tropical rain forests and his investigations of interspecies empathy have been published in scores of books and professional journals. Rose's work as wildlife protector and conservation educator has introduced psychosocial and spiritual dimensions into a field in dire need of new visionary approaches. His search for biosynergy and compassion begins in Palos Verdes and Warner Springs, California, and reaches around the world.

GABRIELA ROSE is a developmental psychologist and horsewoman devoted to facilitating the learning of people challenged by special needs. She lives in Santa Cruz, California, teaches at a school for autistic children, and is training in the practice of equine assisted therapy. Gabriela's interest in compassionate bonds between people and other animals began in childhood with horses in the United States and was reinforced by personal interactions with gorillas, chimpanzees, elephants, and hippos in Africa and with dolphins and monkeys in Mexico and Central America.

PHILIP SEDDON is Associate Professor in Zoology and director of the Postgraduate Wildlife Management Program at the University of Otago in New Zealand. Seddon has contributed to species restoration and protected area management projects in New Zealand, Australia, the sub-Antarctic, South Africa, Namibia, Saudi Arabia, and the United Arab Emirates. He is chair of the Bird Section of the IUCN Reintroduction Specialist Group. He is also a member of the IUCN Species Survival Commission and the World Commission on Protected Areas.

ERIC J. SHELTON is on the staff of the Department of Tourism, University of Otago, Dunedin, New Zealand, and teaches a postgraduate course: Tourism and the Natural World. Shelton's research interests focus on the interaction of ideas of tourism, habitat, flora, and fauna within various philosophies of environment. Also, Shelton is actively involved in a whole-of-ecosystem project to produce habitat that encourages sea bird recolonization.

BRON TAYLOR is Professor of Religion and Environmental Ethics at the University of Florida, and a Fellow of the Rachel Carson Center in Munich, Germany. His research focuses on the emotional and spiritual dimensions of environmental movements. His books include *Dark Green Religion: Nature Spirituality and the Planetary Future* (2010) and *Ecological Resistance Movements* (1995). He is the founder of the International Society for the Study of Religion, Nature and Culture, and editor of its affiliated journal, as well as of the Encyclopedia of Religion and Nature (2005) and Ecological Resistance Movements (1995). For more information see www.brontaylor.com.

PHILIP TEDESCHI has studied and teaches about the intricate relationship between people, domestic and wild animals, and the natural world. Philip Tedeschi is an Animal-Assisted Social Work and Experiential Therapy Specialist and cofounder of the Institute for Human-Animal Connection programs at the University of Denver Graduate School of Social Work (GSSW). He teaches Master of Social Work courses in forensic social work and experiential therapy approaches, with emphasis on conservation and environmental social work and the inclusion of animals in therapeutic settings. He has many years of experience in nontraditional therapeutic approaches with children, adults, and families, as well as program development and intervention in interpersonal violence, including assessment and intervention with animal abuse, attachment, trauma disorder, and sexually abusive youth and adults.

YOLANDA VAN HEEZIK is Senior Lecturer in Zoology at the University of Otago. She has worked in Europe, the Middle East, Antarctica, Southern Africa, and New Zealand on a wide variety of conservation management projects on taxa ranging from penguins and wading birds to reptiles, small mammals, and invertebrates. Van Heezik's current focus is on urban ecology, where she is working with colleagues across the biological and social sciences to quantify the determinants of urban and peri-urban biodiversity and to understand and influence public attitudes toward urban-based wildlife.

JOHN A. VUCETICH is Associate Professor and population ecologist at Michigan Technological University. He co-leads research on the wolves and moose of Isle Royale, which represents the longest continuous study of any predator-prey system in the world. His scholarship also includes work on conservation ethics and the relationship between environmental science and ethics.

PAUL WALDAU is an educator-scholar-activist working at the intersection of animal studies, ethics, religion, law, and cultural studies. He is Associate Professor at Canisius College in Buffalo, New York, and principal faculty member for a two-year online Master of Science program in Anthrozoology. President of the Religion and Animals Institute, former director of the Center for Animals and Public Policy at Tufts University Cummings School of Veterinary Medicine, and author or editor of five books, Waldau has also served repeatedly since 2002 as the Barker Lecturer in Animal Law at Harvard Law School.

MARIAM WANJALA is an intern at the Africa Network for Animal Welfare (ANAW), a Pan African NGO based in Nairobi, Kenya, that is dedicated to working with communities to promote humane treatment of animals and conservation. She has a bachelor's degree in Conservation Biology from Makerere University, Kampala, Uganda. Wanjala has participated in ANAW projects of de-snaring, community conservation, and humane education, conference organization, and biodiversity management. She has also undertaken a project planning and management course and has also been an intern at Ecotourism Kenya that emphasizes sustainable tourism. Wanjala is a member of Nature Uganda, AISEC Uganda and ANAW.

Contributors' Contact Information

Liv Baker
University of British Columbia
Animal Welfare Program
Faculty of Land and Food Systems
Vancouver, BC Canada V6T 1Z4

Benjamin B. Beck
Scientist Emeritus, Smithsonian Conservation Biology Institute
9346 Cropper Island Road
Newark, MD 21841 USA

Marc Bekoff
Ecology and Evolutionary Biology
University of Colorado
Boulder, Colorado 80309-0334

Dror Ben-Ami
THINKK, Institute for Sustainable Futures
University of Technology, Sydney, Australia

Joel Berger
John J. Craighead Professor of Wildlife Conservation
Division of Biological Sciences and Senior Scientist Wildlife Conservation Society
University of Montana
Missoula, Montana 59812

Sarah M. Bexell
Director of Conservation Education
Chengdu Giant Panda Base
1375 Panda Road, Northern Suburb
Chengdu, Sichuan, PR China 610081

Daniel T. Blumstein
Professor and Chair
Department of Ecology and Evolutionary Biology

University of California, Los Angeles
Los Angeles, California 90095

Keely Boom
THINKK, Institute for Sustainable Futures
University of Technology, Sydney, Australia

Susan Clayton
Whitmore-Williams Professor of Psychology
The College of Wooster
930 College Mall
Wooster, Ohio 44691

Eileen Crist
Department of Science and Technology in Society
Virginia Polytechnic Institute and State University
Blacksburg, Virginia 24061

David B. Croft
School of Biological, Earth, and Environmental Sciences
University of New South Wales
Sydney 2052 NSW, Australia

Brian Czech
President, Center for the Advancement of the Steady State Economy
5101 S. 11th Street
Arlington, Virginia 22204

Marco Festa-Bianchet
Département de biologie
Université de Sherbrooke, Sherbooke
Québec, J1K 2R1, Canada

Camilla H. Fox
Project Coyote and the Animal Welfare Institute
PO Box 5007
Larkspur, California 94977

Carrie Packwood Freeman
Department of Communication
Georgia State University
662 One Park Place
PO Box 4000
Atlanta, Georgia 30302

Jason Leigh Jarvis
Department of Communication
Georgia State University
662 One Park Place
PO Box 4000
Atlanta, Georgia 30302

David Johns
Portland State University
Portland, Oregon 97201

Barbara J. King
Chancellor Professor of Anthropology
College of William & Mary
Williamsburg, Virginia 23187

Sarah King
University of Colorado Museum
Boulder, Colorado 80309
sarah.king@colorado.edu

Peter J. Li
Suite N1009F
Social Sciences Department
University of Houston–Downtown
One Main Street
Houston, Texas 77002

Vivek Menon
Executive Director, Wildlife Trust of India and Regional Director, International Fund for Animal Welfare (IFAW) Wildlife Trust of India
PO Box 3150
New Delhi, India 110003

Ben A. Minteer
School of Life Sciences, Assistant Professor
Arizona State University
Tempe, Arizona 85287-4501

Gene Myers
Huxley College of the Environment, Associate Professor
Western Washington University
Bellingham, Washington 98225-9805

Michael P. Nelson
Ruth H. Spaniol Chair of Natural Resources
Department of Forest Ecosystems and Society
College of Forestry
Oregon State University
Corvallis, Oregon 97331

Jolie NeSmith
University of Denver
855 W. Dillon Road
Louisville, Colorado 80027

Josphat Ngonyo
President, Africa Network for Animal Welfare
PO Box 3731
00506 Nairobi, Kenya

Dale Peterson
English Department
Tufts University
33 Richdale Avenues
Cambridge, Massachusetts 02140

Daniel Ramp
University of Technology, Sydney
PO Box 123, Broadway NSW 2007, Australia

Anthony L. Rose, president and CEO
The Biosynergy Institute
PO Box 3430
Palos Verdes Peninsula, California 90274

A. Gabriela Rose
The Biosynergy Institute
PO Box 3430
Palos Verdes Peninsula, California 90274

Philip J. Seddon
Department of Zoology
University of Otago
Dunedin, New Zealand

Eric J. Shelton
Department of Tourism
University of Otago
PO Box 56
Dunedin, New Zealand

Bron Taylor
Department of Religion
University of Florida
PO Box 117410
Gainesville Florida 32611-7410

Philip Tedeschi
Graduate School of Social Work
University of Denver
2148 S. High Street
Denver, Colorado 80208

Yolanda van Heezik
Department of Zoology
University of Otago
Dunedin, New Zealand

John A. Vucetich
Michigan Technological University
School of Forest Resources and Environmental Science
Houghton, Michigan 49931

Paul Waldau
Institute for the Study of Human-Animal Relations (ISHAR)
Canisius College
Buffalo, New York 14208

Mariam Wanjala
Africa Network for Animal Welfare
PO Box 3731
00506 Nairobi, Kenya

Index

Abram, David, xxii
Aceh, 365
activism, advice regarding, 383–85
Adorno, Theodor, 56
advocacy, advice regarding, 383–85
African Network for Animal Welfare, 233
agriculture: animal threats to, 119–20, 123; capital requirements of, 85–86; China's Great Leap Forward and, 318–19; climate change and, 228; industrial, 29, 30–31, 56; landscape transformation and, 302
Albrecht, Glenn, 225–26
Alligator River National Wildlife Refuge, 155
altruism, 35, 214, 217
Amazon rain forest, 206–7
American Academy of Religion, 40
American Fisheries Society, 177
American Indian Movement (US), 242
American Indians. *See* Native Americans
American Society of Mammalogists, 119, 177
Ammann, Karl, 63–65
animal behavior: affective states in, 159; evolution and, 159–60; horse social systems and, 157; hybridization and, 156; individual variation in, 159–60; reintroduction of species and, 99–100, 155–56; research

practices and, 153–54, 157; translocations and, 160–63
Animal Damage Control Act, 119–20
animal law, 31, 36–37
animal minds. *See* mental life of animals
animal movement: ancient roots of, 29; animal welfare versus animal rights in, 78–80; compatibilist ethics and, 77–78; conservation and, 27–28, 30–37, 39–40, 42–43, 77–83; diversity of, 28–29, 34; ecosystems and, 34–35; impact of, 32; individual animals and, 35; meaning of term, 27–28; reasons for opposition to, 30; violence and, 42; wild animals ignored by, 169; world we want and, 42–43. *See also* animal welfare and protection
Animal Planet, 262, 265–66
animal protection. *See* animal welfare and protection
animals: agency of, 273–74; bridging gap between humans and nature, 200; in captivity, 110, 155, 162; charismatic traits of, 278; children and, 202–3, 217–18, 272; coevolution of with humans, 199, 205–6; coherence displayed by, 273–74; as commodities, 73; as common patrimony, 271; as companions, 37–39, 367–68; cruelty to, 232–33; human

animals (*continued*)
attunement to, 231; human concern and
lack of concern for, 212–13; humans as,
213; humans' connection with, 207–8,
217; human spokespeople for, 266–67;
individual differences among, 100,
159–63; othering of, 218; privacy rights
of, 263, 264–65; as resources, 55–57;
rights of, 31; services of, to humans, 368;
taking perspective of, 214, 219; terms for
human outgroups and, 213; as threats or
pests, 259–60; word choice about, 309;
as world-poor, 59, 60n3. *See also* animal
behavior; animal law; animal movement;
animal therapy; animal welfare and protection; mental life of animals
Animal Studies Group, 56
animal therapy, 214, 274, 373–74
animal welfare and protection: academic
study of, 37; versus animal rights, 78–80;
animals' similarity to humans and, 214;
anticruelty and, 29; attacks on humans
and, 380–81; children's character and,
40; China's role in, 317; conservation
and, xi–xii, 4, 6–7, 15, 88, 308–9; economic growth and, 178–80; harmful
forms of, 37–39; human health and,
231–33; at individual-creature level,
3; landscape management and, 304;
livestock-raising practices and, 306; versus social conservation, 85; social realities created by, 39; steady state economy
and, 178; type of Nature and, xi; utilitarianism and, 80; for wild animals, 169;
wildlife management and, 295–96. *See
also* animal movement
animism: *Avatar* (film) and, 367–68; great
environmental thinkers and, 356; viable
equivalent to, 353, 357
Annan, Kofi A., 227
Antarctica, xviii
anthrarchy, 307–8, 309
Anthropocene period, xvi–xvii, xx, 207–9,
384
anthropocentrism: of documentaries, 338–
39; reasons for ignoring nature and, xv–
xvi; of religion, 354–55, 356; of wildlife
management, 295
anthropomorphism, 23n8

anticolonialism, 249
Aotearoa. *See* New Zealand
apartheid and anti-apartheid movement,
238–39, 242–44, 247
apartheid, human-animal, 48, 59
aquariums, 212
archaeology, 205–6, 207, 369
Aristotle, 16
Armstrong, Karen, 29
Asibey, Emmanuel, 67–68
Asoka (Indian leader), 333, 334, 336, 339n7
Australasia: anthrarchy in, 307–8; bushmeat
trade in, 305; drivers of wildlife management in, 300–302; geological origins of,
296–97; hunting in, 304–5; landscape
management in, 302–4; pest control in,
297–99; useful species turned threats in,
299–300; wildlife management in, 296,
305–10. *See also* New Zealand
Australia. *See* Australasia
Avatar (film): biosynergy in, 292–93, 364–
67; blindness to nature and, 361–62,
367; fight or flight in, 363–64; human-
horse connection in, 368–74; interspecies bonding in, 367–68, 374; mutual
seeing in, 374–76
awareness raising, media and, 257–58, 267

Babbit, Bruce, 250
BACI design, 105, 107
Baker, Liv, 100
Bal, R., 107
Bangladesh, 379–83
Bardige, B., 282
Barringer, Almut, 386
bears, 290, 317, 320–21, 325
beavers, 142, 143
behaviorism, 54
Bekoff, Marc, 122
Bennett, E. L., 83, 86
Bennett, Elizabeth, 66
Bentham, Jeremy, 17, 35
Benz-Schwarzburg, Judith, 5
Berenguer, J., 214
Berger, Joel, 97–98
Berry, Thomas, 379
Besthorn, F., 232
Bexell, Sarah, 219, 233, 277–78
Bijker, W. E., 107

191; connections with, 200, 213–15, 219; conservation economy and, 186–89; economy of, 175–76; estrangement from, 137; expectations of, 137–38; immanence of, 186; interconnectedness and, 189, 355–56; many Natures and, xi–xii; meanings of, 188–89, 190, 191–92, 200, 213; natural capital and, 237; objectification of, 5; as pathway to spiritual truth, 354; purpose of, 186; redecorating, xix, xx; reified, 189; as resources, 55–56; rules for interacting with, xxi; sensory exploration of, 355–56; wilderness as stand-in for, 169, 187; wilderness debate and, 186. *See also* ignoring of nature
Nelson, M. P., 186
Nelson, Michael, on empathy, 3–5
Neoplatonism, 52, 54
net primary product (NPP), 237
New Zealand, 110–11, 169–70, 183–84, 186–87, 190–94. *See also* Australasia
Ngonyo, Josphat, 291
Ngorongoro Conservation Area (Tanzania), 348
NGOs (nongovernmental organizations), 338–39, 340n15, 348–49
NHAs (nonhuman animals). *See* animals
Nichols, J. D., 110
Nietzsche, Friedrich, 16–17
nitrogen cycle, 227
nongovernmental organizations (NGOs), 338–39, 340n15, 348–49
nonsentient beings, empathy with, 19
nonviolence, 242
Northern Rangeland Trust (Kenya), 350
Norway, 127
Nouabale-Ndoki National Park, 87
NPP (net primary product), 237

O'Barry, Ric, 267
Oates, J. F., 331–32
Obama, Barack, 246
organicism, 356–58
Orr, David, 32
outdoor recreation, 174
overconsumption, xiv

Pacelle, Wayne, xix
pandas, 111

Paolinetti, Jamie, 369
Papua New Guinea. *See* Australasia
Pavlov, Ivan, 159
PEC (principle of ethical consistency). *See* ethics
penguins, xviii
People for the Ethical Treatment of Animals (PETA), 32–33
Perry, Dan, 78
Perry, Gad, 78
pest control, 295–96, 297–99
PETA (People for the Ethical Treatment of Animals), 32–33
Peterson, Dale, 5–6, 83, 289
pets, 37–39, 212–13, 231–32, 367–68
Pierce, Jessica, xxi
Pierson, David, 262
PIEs (profound interspecies events), 370, 373–74
pigs, 299–300
Pinter-Wollman, N., 161–62
Planet Earth (television series), 264
Planet Green (television channel), 265–66
plants: as common patrimony, 271; overgrazing and, 114
poisons, 120–21, 123
polar bears, 317
politics: changing what's possible in, 238–39; contentious nature of, 240; emotion and, 212; good and bad governance and, 347; insider and outsider strategies and, 240–43; timing of critique of nature and, 189–90
pollution, xviii, 174
Pony Boy, Gawani, 371
population size, xix–xx, 231, 331–32, 382–83
population viability, 11–12
poverty, 233, 289, 331, 334–36, 382–83
power lines, 173
pragmatism, 18, 83–90. *See also* consequentialism
precautionary principal, 108–9
predators: control methods for, 120, 123; hyperpredation and, 109–10; management of, 98, 115–16, 119–23; sustainable harvest rates and, 126
preservationism, 183, 191
primates, 67, 74–75, 82
Prodohl, P. A., 162

Lightning Source UK Ltd.
Milton Keynes UK
UKOW04f1913111017
310814UK00001B/91/P

9 780226 925356